SECRETS
OF THE SOIL

Other books by Peter Tompkins

To a Young Actress
Shaw and Molly Tompkins
The Eunuch and the Virgin
A Spy in Rome
The Murder of Admiral Darlan
Italy Betrayed
Secrets of the Great Pyramid
The Secret Life of Plants (with Christopher Bird)
Mysteries of the Mexican Pyramids
The Magic of Obelisks

Also by Christopher Bird

The Divining Hand: The 500-Year-Old Mystery of Dowsing

SECRETS
OF THE SOIL

Peter Tompkins and Christopher Bird

HARPER & ROW, PUBLISHERS, New York
Grand Rapids, Philadelphia, St. Louis, San Francisco
London, Singapore, Sydney, Tokyo

FIRST EDITION

Designed by: Sidney Feinberg

Library of Congress Cataloging-in-Publication Data

Tompkins, Peter.
  Secrets of the soil.

  Bibliography: p.
  Includes index.
  1. Soils.   2. Soil management.   3. Organic farming.
4. Agricultural ecology.   I. Bird, Christopher, 1928–   II. Title.
S591.T64   1989   631.4   87-54080
ISBN 0-06-015817-4

89 90 91 92 93 DT/MPC 10 9 8 7 6 5 4 3 2 1

So long as one feeds on food from unhealthy soil,
the spirit will lack the stamina to free itself
from the prison of the body.

—RUDOLF STEINER

Contents

Acknowledgments

Our grateful thanks to all those who have helped and contributed to this effort, especially for reading the manuscript and making helpful suggestions; Eddie Albert for his encouragement while the going was rough; Dr. Bargyla Rateaver for sharing her encyclopedic knowledge of the soil and how to care for it organically; Christian and Joanna Campe for their invaluable assistance over the years, particularly with the remineralization of European soils; Sara Sorelle for her cheerful and indefatigable help in researching the book; Jerree Tompkins for the endless processing of material through her IBM PC; and, as always, the staff of the Library of Congress for their invaluable facilities so courteously offered.

We are also much beholden, for their hospitality and/or logistic support, to: bioscientist and author Dr. Alexander P. Dubrov, and to historian Dr. Pavel Pozner, in Moscow, U.S.S.R.; engineer and dowser Wlodimierz Szwarc; author Lech Stefański and his wife Helena, in Warsaw, Poland; and to Dr. Zdeněk Rejdák, President, International Association for Psychotronic Research, and his able interpreter Eva Roubalova in Prague, Czechoslovakia; Ljerka Radović, leading translator of contemporary American and British fiction, in Belgrade, and Mr. and Mrs. Mato Modrić in Rovinj, Yugoslavia; Dr. Maria Felsenreich, fighter for eco-agriculture, in Gänserndorf, Austria; Lile and Hans Schulyok in Lucerne, and Pierre Lehmann, nuclear physicist turned environmental activist, in Vevey, Switzerland; Detlev Moos, publisher, in Gräfelfing,

G.F.R.; Jerome Dumoulin, senior editor, *L'Express* and Dr. P. B. Laffout, senior member of the Ecole Française D'Extrême Orient in Paris, France; John W. Mattingly and his wife Frieda, in Loveland, Colorado; Hannah Campbell and David Bird, Bison Associates, in Cambridge, Massachusetts; and Rita A. McBrayer, astrologer and "eco-warrior," in Maryland. Our thanks also to Mark Medish of Harvard University and to his father Vadim Medish, professor of Russian language and literature at American University in Washington, D.C., for his assistance in deciphering highly technical documents and tapes from the Soviet Union.

Introduction

No CREATURE, not even swine, befouls its nest with such abandon as does *homo sapiens*, poisoning his habitat with fiendishly concocted chemicals and their deadly toxic waste. A morass of rotting human flesh awaits us all unless the antidotes are rapidly applied. Providentially, they exist, they work, and as detailed in these pages, can bring us back to health.

That the earth is ailing—almost beyond repair—was clear enough as early as 1912 to Nobel Prize winner Dr. Alexis Carrel. In *Man, the Unknown* this eminent French scientist warned that since *soil is the basis for all human life* our only hope for a healthy world rests on reestablishing the harmony in the soil we have disrupted by our modern methods of agronomy. All of life will be either healthy or unhealthy, said Carrel, according to the fertility of the soil. Directly, or indirectly, all food comes from soil.

Today soils are tired, overworked, depleted, sick, poisoned by synthetic chemicals. Hence the quality of food has suffered, and so has health. Malnutrition begins with the soil. Buoyant human health depends on wholesome food, and this can only come from fertile and productive soils. Minerals in the soil, said Carrel, control the metabolism of cells in plant, animal, and man. Diseases are created chiefly by destroying the harmony reigning among mineral substances present in infinitesimal amounts in air, water, food, but most importantly in soil. If soil is deficient in trace elements, food and water will be equally deficient.

Carrel then came to the point: chemical fertilizers cannot restore soil fertility. They do not work on the soil but are enforcedly imbibed by plants, poisoning both plant and soil. Only organic humus makes for life.

Plants, said Carrel, are the great intermediaries by which the elements in rocks, converted by microorganisms into humus, can be made available to animal and man, to be built into flesh, bone, and blood. Chemical fertilizers, on the contrary, can neither add to the humus content of soil nor replace it. They destroy its physical properties, and therefore its life. When chemical fertilizers are put into the soil they dissolve and seek natural combination with minerals already present. New combinations glut or overload the plant, causing it to become unbalanced. Others remain in the soil, many in the form of poisons.

Plants that are chemically fertilized may look lush, but lush growth produces watery tissues, which become more susceptible to disease; and the protein quality suffers. Chemical fertilizers, said Carrel, by increasing the abundance of crops without replacing *all* the elements exhausted from the soil, have contributed to changing the nutritive value of our cereals: "The more civilization progresses, the further its gets from a natural diet." Our present diet consists of adulterated and denatured foods, from which the most precious essential factors have been removed by coloring, bleaching, heating, and preserving. Pasteurizing milk kills the enzymes vital to nutrition, leaving only the rotted corpses of bacteria. Milk is also the second leading source—after meat—of pesticide residues in the U.S. diet. White bread, spuriously "enriched," has its germ, which contains the vital nutrients, ritually removed, a deliberate castration.

Anyone alive before World War II, especially in Europe, knows that bread, fruit, vegetables, and meat bear no relation to what they were before the war. Our crop yields may have doubled or even tripled, but their nutritive quality has diminished progressively. Visual impression of foods has become the most important factor, though anyone with a glimmer of second sight will pass up, as no more alive than the products of Madame Toussaud's wax museum, the cosmetic and congealed displays of the grocery store today.

Abundance does not mean the food contains a sufficient amount of needed elements and vitamins. There is no doubt, says Dr. Melchior Dikkers, Professor of Biochemistry and Organic Chemistry at Loyola University, that malnutrition is the most important problem confronting mankind at the present time. Every two seconds a child dies of starvation; a staggering sixty million adults die each year. The United States, despite its boasted food production, is grossly undernourished. And, though the per capita expenditure on health care in the United States is the highest in the world, so is the incidence of cancer, obesity, heart, and circulatory diseases.

Amazingly, Dr. Joseph D. Weissman, associate professor at the

UCLA College of Medicine, a specialist in preventive medicine and immunology, has discovered, after years of research, that nearly all the noninfectious diseases that presently plague mankind are of recent origin, developed during the nineteenth and twentieth centuries, and that the billions of dollars spent on research, newer diagnostic techniques, organ transplants, coronary bypass procedures, chemotherapy, radiation, and all the various drugs, have not appreciably altered the advance of these killer diseases, merely enriched the chemist and the medical practitioner.

Dr. Weissman argues that most of today's killer diseases are caused by environmental toxins produced by our industrial society. Many doctors agree, aware that the great increase in diseases of degeneration, such as cancer and heart disease, undeterred by the advances of modern medicine, are primarily due to extensive use of synthetic chemicals in our daily diet, food preservatives, insecticides, fungicides, pesticides, and so on.

Most people, says Weissman, assume their ailments arise from causes beyond their control, unaware that they can choose a life of excellent health, remaining active, trim, and alert into their second century. He believes that choices of diet and lifestyle in our industrial societies play a large part, perhaps the largest, in whether or not we remain vibrant past our prime.

But doctors in general know very little about food. Average training during four years of medical school is 2.5 hours per physician. Yet Dr. Robert S. Mendelsohn, Associate Professor of Preventive Medicine at the University of Illinois School of Medicine—described as a member of a small fraternity dedicated to freeing the healing art from the domination of drug companies—lays the blame on the plethora of *mis*information on nutrition put out in medical schools, suggesting they might do better not to teach the subject at all.

Even more amazing, Dr. Weissman's research reveals that many of the killer diseases have developed only within the last hundred years, demonstrably through toxic chemicals introduced into the environment and food supply as by-products of the Industrial Revolution—chemicals such as chlorine and its compounds, coal-tar derivatives, pharmaceuticals, petrochemicals, and so on.

The emergence of industrialization, with its massive toxic wastes, coincided with the appearance of many of the new diseases. Our ancestors may have had a shorter average life span, largely owing to infant mortality, says Weissman, but, like present-day primitive peoples, they were virtually free of "degenerative" diseases.

A hundred years ago coronary heart disease was virtually unknown in Europe and America. The first case described in medical literature surfaced in 1910. Today it is the leading cause of death. Cancer, which today is responsible for 3.4 percent of all deaths in Europe and America, was responsible for only 1 percent a hundred years ago. Today even newborn and very young children are victims of cancer and leukemia. Diabetes, the third most common cause of death, once struck only one in fifty thousand Americans; now it strikes one in twenty.

Water in primitive lands—as was the case in developed countries before the late nineteenth century—needed no disinfection. Where there are no industries or factories pouring waste pollutants into the environment, plants, marine life, and land animals are not tainted by dangerous chemicals.

Now, not only water but soil and air are everywhere polluted, a pollution that is transmitted via plant and animal to man. In the developed world, says Weissman, there is virtually no clean soil or water left: toxins are in all the food we eat, the water we drink, the air we breathe. Fruits, vegetables, grain, fish, poultry, meats, eggs, dairy products are all affected. And some foods are concentrators and magnifiers of the pollution, the greatest concentrations of toxins occurring in animal fat and cholesterol. Mother's milk could not be legally sold in the supermarket; it would not pass the government's safety test. The percentage of U.S. mothers' milk containing dangerous levels of DDT is a stupefying 99 percent.

Protection against disease, says Weissman, is more important, and more effective, than later therapy. And protective medicine starts in the soil.

Poisoning of the soil with artificial agricultural additives began in the middle of the last century when a German chemist, Justus von Liebig, known as the "father of chemical agriculture," mistakenly deduced from the ashes of a plant he had burnt that what nourished plants was nitrogen, phosphorus, and potash (or potassium carbonate)—the NPK of today's chemical agriculture.

Liebig's dicta—and he wrote profusely—led to a vast and profitable commercial development of synthetic chemicals. Lulled by propaganda, world farmers became dependent on German mines for supplies of potassium salts, known as "muriate of potash," without which they were told that nothing on their farms would grow. When World War I interrupted exports from Germany, prospectors located deposits in the United States, launching American companies into rapid exploitation of this bonanza of unnecessary chemicals.

From the amount of phosphoric acid also found in the ash of his burnt plant, Liebig further concluded that phosphorus must be a prime requirement for the growth of plants. Since Roman times, farmers had been using ground-up bones to obtain their phosphorus. By treating bones with sulfuric acid Liebig created what he called a "superphosphate." When vast quantities of sea-derived calcium phosphate were discovered—believed to be the skeletons of sea animals collected over millions of years—a whole new industry of artificial "mineral manures" was launched.

Up until Liebig's time, it was believed that because virgin soils were highly fertile, and contained much humus, the various stages of this brown decaying organic matter must be the principal source of nourishment for plants. Liebig attacked the notion with vehemence. Of humus and of the humic acid derived from it, he wrote: "There is not the shadow of a proof that either of them exerts any influence on the growth of plants either in the way of nourishment or otherwise."

As William Shestone put it in his 1875 biography of Liebig: "These were the facts and arguments by which, once and for all, Liebig rendered the humus theory untenable by any reasonable human being."

That the secret to fertilizing soil lay precisely in this organic excreta, not chemicals, Liebig only concluded ten years later. Too late. By that time the chemical companies were off to such a profitable start there was no stopping them in their headlong race to destroy the soil and all that it supports.

The first chemical produced on a commercial scale in the incipient "age of chemicals" was the sulfuric acid used by Liebig to produce his "superphosphate," a clear, corrosive, oily liquid still the most widely sold chemical today, basic to the manufacture of a host of other chemical substances, along with the production of dyes, drugs, paper, pigments, and explosives.

Next most important among the chemicals concocted in the lab for commercial use was alkali, a soluble mineral salt, named by the Arabs from the sea-beach saltwort plant from whose ashes they first derived the substance. While it was at first primarily used in the manufacture of soap and glass, by mid-nineteenth century all the major chemical agents in use were connected in one way or another to alkali. Britain's United Alkali Corporation, set up in 1891, became the world's largest chemical enterprise, with forty-three firms employing fifty chemists and twelve thousand plant workers, eventually to be swallowed up by the giant government-sponsored amalgam of Imperial Chemical Industries.

Accidentally, a whole new branch of chemistry was developed in the mid-nineteenth century by a young English chemistry student working in a makeshift lab in his father's house during the Easter vacation of

1856. Experimenting with coal tar, William Henry Perkin produced a mauve dye from its constituent benzene, the first of the so-called aniline dyes, remarkable for the way it held fast and would not wash out as did natural colors. Patented, his mauve became fashionable at the courts of both Victoria and Napoleon III, obtaining for Perkin a fortune and a knighthood. Soon aniline red, yellow, and black followed mauve; and millions remained to be made from synthesized indigo, the color of jeans.

When a disciple of Liebig, Friedrich von Kekule, realized—in what has been called "the most brilliant piece of prediction to be found in the whole range of organic chemistry" and one that would elevate him to the nobility—that six atoms of carbon in the benzene molecule could be linked together in a circle, with a hydrogen atom attached to each, German chemists saw their way to the construction of endless new compounds by artificially uniting carbon in their test tubes with nitrogen, hydrogen, sulfur, chlorine, etc., in what amounted to a heyday for sorcerers' apprentices.

Drugs were soon added to the inventory of chemical-company products, as German and Swiss dye companies found endless new ways of turning coal tar and other waste products into a health-debilitating but highly profitable pharmacopoeia. In the United States alone, $8 billion are spent yearly on so-called medicines. And coal tar had further lethal uses, chemically essential to the vast expansion of explosives.

It remained for a German chemist, Fritz Haber, to discover in 1905 a laboratory process for turning the endless tons of free nitrogen in the air into liquid ammonia, 82 percent of which is nitrogen. By 1915 Karl Bosch, a German engineer, joined Haber in designing the first synthetic ammonia plant in the Reich, enabling the German High Command to indulge in the Kaiser's war. German dye firms, banding together for patriotism and for profit, produced explosives, chemical fertilizers, drugs, and, as a bonus, the poison gases responsible for some 800,000 casualties in World War I.

With the end of hostilities, the huge amounts of gas left over were redirected to the insect—but on a wider scale, thanks to the improved methods of dusting and spraying developed for use on humans by the military. Increased doses of nitrogen, no longer needed for explosives, were indiscriminately dumped on crops, weakening their resistance to insects, creating a vicious circle that snowballed as it endured, progressively more profitable for the few as it poisoned soil and aquifer for the many.

German chemical companies, with money from their opposite numbers in the United States—who had made equally enormous profits from the war—amalgamated in 1925 to form the I. G. Farben conglomerate, soon the largest chemical enterprise in Europe, closely bonded with its U.S. partners. Together these conglomerates funded Hitler, rearming his *Wehrmacht* as a "bulwark against the Soviets." And with petroleum, courtesy of Standard Oil of New Jersey, Hitler was enabled to roll his tanks into Poland and into World War II.

While loyal GIs desperately struggled with their lives to undo this handiwork, at Auschwitz I. G. Farben, with slave labor guaranteed by Himmler, produced a special gas to exterminate millions of unwary victims, mostly Jewish.

From World War II, American chemical companies, which had boomed between the wars, derived an even greater bonanza from the free ammonia Bosch had prestidigitated from the air. A million tons of bombs were dropped on Germany alone, causing millions of dollars to be funneled by U.S. taxpayers into chemical-company coffers, as America paid with blood and money for the greed of these treasonous companies.

At war's end, eighteen new ammonia factories, developed in the U.S. at taxpayers' expense to manufacture explosives, were obliged to find a market for their surplus. Du Pont, Dow, Monsanto, American Cyanamid, with their vast wartime profits, produced ever more fertilizer to dump on the unwary farmer, who dumped it onto his fields to kill the goose that laid the golden egg.

As a by-product of the war, to keep fleas, lice, and other insects from contaminating GI troops, one of the most toxic pollutants ever invented was produced by a Swiss chemist, Paul Mueller, who chose to give the secret of its manufacture to the Allies: DDT. Derived entirely from the test tube, it was the most potent insecticide yet seen, capable of killing all sorts of bugs in a broad spectrum with astonishing speed and efficiency. On the home front, with manpower critically short, farmers used it against insects to increase crop yields and save on labor.

Following the Allied victory in 1945, DDT began to be used like water, until the toxin seeped into every animal and human body in America. Everywhere, chemical firms reinvested their wartime gains to launch into unparalleled growth in a massive quest for new synthetic broad-spectrum pesticides. The farmer, fearing disaster—his plants, weakened by a surfeit of chemicals, were attracting more and more bugs —turned to even more chemicals. Complacently, the companies brought out new products by the score, mostly chlorinated hydrocarbons similar to DDT, such as chlordane, heptachlor, dieldrin, aldrin, endrin; and "organic phosphates" such as parathion and malathion.

In an attempt to beat the game by ever greater production, trusting farmers in America, prodded by bankers, chemical companies, and the manufacturers of agricultural machinery, changed from a subsistence way of life to commercial enterprises, investing large cash payments in new land and equipment, going heavily into debt on fertilizers, pesticides, and herbicides—and, in so doing, sealed their own doom.

That chemicals were pointlessly poisoning the soil, killing microorganisms, stunting plants, and proliferating degenerative disease in man and beast was perfectly clear to a whole group of sensitive minds in Europe and America as early as World War I. Distinguished, distressed, and well-informed, several authors on both sides of the Atlantic were speaking up and propagandizing for a viable alternative method of agriculture requiring no chemicals.

Their main premise was that in soil properly nourished with adequate supplies of humus crops do not suffer from disease, and do not require poisonous sprays to keep off parasites; that animals fed on these plants develop a high degree of disease resistance, and that man, nurtured with such plants and animals, can reach an extraordinary (and in fact quite natural) standard of health, able to resist disease and infection from whatever cause it may derive.

One of the first to sense that the use of chemical fertilizers was doing more harm than good, that it was destroying the life and vitality of topsoil, momentarily stimulating plant growth but actually inviting disease, was Sir Albert Howard. As a British colonial officer in India, with the high-sounding title of Imperial Chemical Botanist to the Government of the Raj at Pusa, Sir Albert had the rare opportunity of being free to carry out experiments without restraints, enabling him to grow whatever crops he liked in any way he liked with land, money, and facilities provided by the government. He was thus able to observe, dispassionately, and with no axe to grind, the reaction of suitable and properly grown varieties of plants when subjected to insects and other potential pests.

He found that the factor that mattered most in soil management was a regular supply of freshly made humus, prepared from animal and vegetable wastes, and that the maintenance of soil fertility was the fundamental basis of health. He claimed that his crops, grown on land so treated, resisted all the pests that were rife in the district and that this resistance was passed on to the livestock when they were fed on crops so grown.

He noticed that the natives never used artificial fertilizers or poison sprays, but were extremely careful in returning all animal and plant residues to the soil. Every blade of grass that could be salvaged, all leaves that fell, all weeds that were cut down found their way back into the soil, there to decompose into humus and reenter the cycle of life.

Sir Albert proved that livestock fed on organically grown fodder were disease resistant, as were his oxen, which even during an epidemic of hoof-and-mouth disease rubbed noses with infected neighboring stock with no ill effects. "The healthy, well-fed animals reacted towards the disease exactly as improved and properly cultivated crops did to insect and fungi—no infection occurred."

As a result of his experiments, Sir Albert reached the conclusion that crops have a natural power of resistance to infection, and that proper nutrition is all that is required to make this power operative. "But the moment we introduce a substitute phase in the nitrogen cycle by means of artificial manures, like sulphate of ammonia, trouble begins which invariably ends with some outbreak of disease, and by the running out of the variety."

Crops and livestock raised on land made fertile by his methods of humus treatment attained a high measure of immunity from infective and parasitic, as well as from degenerative diseases. Further, his treatment appeared to be curative as well as preventive.

By 1916 Sir Albert was lecturing that chemical fertilizers were a waste of money, maintaining that organic matter, along with the good aeration it promoted, was alone enough to allow microbes to provide sufficient amounts of nutrients to feed the world.

Returning to England in 1931 after thirty years in India Sir Albert became known as the founder of the "organic" movement and set about popularizing his ideas. By the beginning of the Second World War he had brought out his *Agricultural Testament*, followed, when the shooting was over, by *The Soil and Health*, a book in which he warned that the use of synthetic chemical fertilizers leads to imperfectly synthesized protein in leaves, and thus results in many of the diseases found in plants, animals, and human beings. As a healthy alternative he pleaded for a simple system in which these proteins are produced from freshly prepared humus and its derivatives, in which case he averred that "all goes well; the plant resists disease and the variety is, to all intents and purposes, eternal."

In vain did such stalwart supporters of Sir Albert as Lady Eve Balfour do battle for his cause in Britain, organizing the Soil Association, and producing a thoroughly convincing work entitled *The Living Soil*. It validated Howard's basic premise that humus confers on plants a power of disease resistance amounting almost to immunity, something that cannot be obtained with artificial fertilizers.

In lucid terms Lady Eve pointed out that the action of compost is not due to the plant nutrients it contains, but to its biological reaction, which has the effect of fundamentally modifying the soil microflora. "All these substances are merely some of the raw materials from which humus can be made. They cannot become humus until they have been metabolized by soil organisms."

But the odds were too heavily stacked against her. Imperial Chemicals forged ahead unmolested. In the United States, J. I. Rodale picked up the banner and launched a movement with his *Organic Gardening and Farming* magazine, its tenets supported by *Pay Dirt*, published in 1945. At Emmaus, Pennsylvania, Rodale created an experimental organic farm and was active in organizing organic garden clubs throughout the United States. He pointed out that in China organic agriculture was able to feed a population of nine hundred million, nearly as many livestock, and, on about the same amount of arable land as is available in the United States, three times the number of hogs.

He quoted reports from travelers to China to the effect that there was no starvation, poverty, or the like, all without huge doses of chemicals, insecticides and heavy, petroleum-gobbling machines, but only by careful composting of all organic stuff and a labor-intensive method.

Scientific support for the argument for organic farming came in lapidary language from one of the most brilliant soil scientists produced in America, Dr. William A. Albrecht, Chairman of the Department of Soils at the University of Missouri, with four degrees from the University of Illinois. Widely traveled, he had studied the soils of Great Britain, the European continent, and Australia, drawing conclusions seasoned by a farm boy's upbringing. His extensive experiments with growing plants and animals substantiated his observation that a declining soil fertility, due to a lack of organic material, major elements, and trace minerals, was responsible for poor crops and in turn for pathological conditions in animals fed deficient foods from such soils, and that mankind was no exception. Degenerative diseases, as causes of death in the United States, had risen from 39 percent in the decade 1920–29 to 60 percent in the year 1948.

Organic matter, said Albrecht, may be called the *constitution* of the soil. And a good constitution, he added wryly, is the capacity, according to its meaning as used in the medical profession, of an individual to survive despite the doctors rather than because of them. Insects and disease, he pointed out, are the *symptoms* of a failing crop, not the cause. "The use of poisonous sprays is an act of desperation in a dying agriculture. Fertilizer placement is the art of putting salt in the ground so that plant roots can somehow manage to avoid it!"

In sum he preached that weeds are an index to the character of the soil.

It is therefore a mistake to rely on herbicides to eradicate them, since the chemicals deal with effect, not cause. Insects and nature's predators are disposal crews, summoned when they are needed, repelled when they are not. Crop losses in dry weather, or during mild cold snaps, are not so much the result of drought and cold as of nutrient deficiency. NPK [nitrogen, phosphorus, potassium] formulas, as legislated and enforced by State Departments of Agriculture, mean malnutrition, attack by insects, bacteria, and fungi, weed takeover, crop loss in dry weather, and general loss of mental acuity in the population, leading to degenerative metabolic disease and early death.

The vast bibliography of Albrecht's scientific and popular papers reveals a lifetime of meticulous scientific investigation into the chemistry and biology of the planet, highlighting the fundamental necessity for feeding plants, animals, and humans through ministrations to the soil itself, correcting deficiencies of diet at their point of origin: the soil.

In 1939 Louis Bromfield, author of *The Rains Came*, etc., returned from the India of Sir Albert Howard to his Malabar Farm in Pleasant Valley, Ohio, to put Howard's agricultural philosophy into practice. Working with Albrecht, he bought up several worn-out farms and produced abundant crops with organic techniques. In a practical way he proved that insect damage and disease could be controlled with humus, good plant nutrition, and sound soil management.

Were Thomas E. Dewey to have defeated Harry S. Truman in 1948, Bromfield was slated to become U.S. Secretary of Agriculture, with every intention of "derailing the fossil-fuel technology that had taken command of the education machine, USDA, Extension, and the farm press."

But Truman's triumph brought in the policy of deliberately banishing small farmers to industrial centers and of unleashing the petrochemicals. Through Truman's creation of the CIA and of a National Security Council trained for "dirty tricks," the multinationals were able, often through the guise of AID, to impose their deadly chemicals not only on America, North and South, but on all the Third World markets. Sir Albert's Indians were brainwashed and corrupted into dousing their healthy plants with all kinds of poisons. Chemical-fertilizer consumption in India rose from 1.1 million tons in 1966–67 to 50 million tons in 1978–79.*

* During the late 1960s the United States and the World Bank applied pressure on India to allow Western chemical companies such as Standard Oil of California and International Minerals & Chemicals to build fertilizer plants on the subcontinent. Collusion is indicated by the fact that farmers received subsidies from the Indian government of 10 to 20 percent on fertilizers and 25 percent on pesticides, plus government-backed loans to pay for them. As a result, fertilizer consumption in one area of India rose between 1969 and 1979 from 3.5 to 50 kilograms per hectare (a hectare is about an acre and a half).

While Albrecht was the leading scientific supporter of organic farming in America, no modern voice has spoken out against social injustice, environmental deception, and commercial hypocrisy as applied to agriculture more candidly, clearly, and trenchantly than Charles Walters, Jr. A Kansan of Volga Germanic stock, Walters since 1971 has edited and published a straight-punching and hard-hitting monthly, *Acres U.S.A.: A Voice for Eco-Agriculture*, the *Eco* standing both for economic and ecological.

Schooled in economics by his Jesuit professors, Walters has almost single-handedly fought the Truman heritage of diminishing the farmer, supporting instead the principle of agricultural *parity*, a concept so easy to understand that most economists and financial writers eschew it as "simplistic." † Walters's slogan, "Cheap food means sick or hungry people," dramatically emphasizes his belief that a Kansas farmer can no more collect a fair price for his production than the Zulu tribesman could pay for it so long as the price of food is arbitrarily kept below its fair market price.

With the publication in 1962 of Rachel Carson's startling exposé *Silent Spring*, the public was awakened with a shock to the danger of the situation, and organics took on new meaning in America. Great pressure had been put on *The New Yorker* magazine by the chemical companies to prevent her articles from being published, and legal action was threatened to prevent Houghton Mifflin from bringing out the book, accusing her of being a Communist. ‡

Yet, in 1963, Dr. Jerome Weisner, science counselor to President John F. Kennedy, reporting to a commission assembled to examine the

† Sitting in his Raytown, Missouri, offices, we heard Walters take a call from one of hundreds of farmers seeking to elucidate the mystery of parity, and get to the nub of the matter in a few words: "When there is par exchange," said Walters, "that is to say when the farmer gets a full and not an arbitrarily discounted price for his produce—on a par with the price he has to pay for all that he imports onto his farm—then he prospers. He can pay off his debts. He can enjoy the earnings of the just. On the other side of that equation—and here's the rub—bankers and money-lenders must go hungry. The farmer doesn't need their loans. The international manipulators lose their base, the munitions makers see their profits eroded by peace. When basic raw commodities move across borders at less than equity of exchange, armies follow." Such talk, eminently reasonable today, in the McCarthy era could get one into trouble.

‡ Jack Doyle, Director of Agricultural Projects for the Environmental Policy Institute, in his brilliant *Altered Harvest*, reports that shortly after Carson's book called attention to the acute toxicity of some pesticides and their indiscriminate use throughout the country, several members of the National Academy of Sciences engaged in pesticide problems and related fields attacked the book. I. L. Baldwin, chairman of the pest-control committee, wrote a long, critical review for *Science* magazine. Another member of the pest-control committee, economic entomologist George C. Decker—who had also been a frequent consultant to the chemical industry—called the book "science fiction," comparable in its message to the TV program *Twilight Zone*. At congressional hearings, Mitchel R. Zavon, a consultant for the Shell Chemical Company, also a member of the academy's pest-control

premises of *Silent Spring*, declared: "Use of pesticides is more dangerous than atomic fallout."

Carson had written: "We are rightly appalled by the genetic effects of radiation. . . . How then, could we be indifferent to the same effect from farm chemicals used freely in the environment?"

The meaning of this strange language, as Charles Walters was quick to point out in *Acres U.S.A.*, proved elusive, until an Italian scientist, Amerigo Mosca, winner of the chemistry prize at the Brussels World's Fair, presented certain startling findings. Mosca stressed the point that toxic farm chemicals are radiomimetic in that they ape the character of radiation.

The damage resulting from nuclear radiation is the same as the damage resulting from the use of toxic genetic chemicals, said Mosca. And the use of fungicides of organic syntheses (Zineb, Captan, Phaltan, etc.) annually causes the same damage to present and future generations as atomic fallout from 29 H-bombs of 14 megatons—damage equal to the fallout of 14,500 atomic bombs of the Hiroshima type.

Mosca computed that in the United States in the 1970s, yearly use of toxic genetic chemicals was about 453,000 tons, which caused damage equal to atomic fallout from 145 H-bombs of 14 megatons, or 72,000 atomic bombs of the Hiroshima type. And in charts, graphs, and statistics —all of which appeared as part of his running story—the Italian scientist revealed that mentally retarded babies had reached the stunning statistic of 15 percent of live births. He concluded that damage to plants, crops, and soil fertility, coupled with water pollution, was practically incalculable. The sperm count of the average American male is down 30 percent from thirty years ago, attributable to chlorinated hydrocarbon pesticides (PCBs, DDT, etc.), and 25 percent of male college students are now sterile. Continuation of the scenario would see the destruction of the American people within a matter of a generation.

Mosca's full report was classified by the Italian government, not to be revealed for fifty years—by which time, perhaps, it was hoped that sinister allegations about Montedison—producer of megatons of fertilizers, pesticides, and herbicides—would be glossed over and forgotten.*

---

committee, characterized Carson as one of the "peddlers of fear" whose campaign against pesticides would "cut off food for people around the world." Two other academy scientists suggested that Carson's work suffered from ignorance and bias and that she had ignored the sound appraisals of pesticides conducted by responsible bodies such as the academy!

* In 1975, Cesare Merzagora asserted that when he took over as president of Montedison his predecessor handed him four paper bags filled with records of the company's slush fund operations.

Driving over hundreds of miles of country roads, Walters could not help noticing increasing numbers of funerals due to death by cancer among his farmer friends and a host of "scrambled children tetratogenically birthed, bodily deformed or mentally retarded." Grieved by the untimely lingering cancer death of his sister, exposed to agricultural chemicals in the factory where she worked, he bluntly entitled one front-page article: "Is Modern Agriculture Worth Having?"

And Walters was among the first to expose the dangers behind the now highly propagandized irradiation of foodstuffs to kill pathogens and extend shelf life.

When I saw this process proposed behind the scenes [said Walters], I cited dozens of scientists who warned about some of the consequences of eating irradiated food: embryonal damage, reduced digestibility, malignant lymphomas in mice, changes in organs, and more. Since the after-effects of the consumption of irradiated foods on living tissue are similar to those of direct radiation, the relevant problems, which include an eventual reduction of the resistance against infectious diseases, AIDS included, deserved attention, but the Svengalis of science defend irradiation as cheap.

That all this horror is unnecessary, redundant, and avoidable has now been demonstrated by a band of happy warriors in their battle for organic farming. Healthy and economic alternatives do exist, though some of them appear extraordinary. To discover what they might be, we crisscrossed the planet up and down. To describe them we have produced this book, along with an appendix on where and how to apply the knowledge. With a little effort the planet *can* be saved from destruction by corruption, poison, and pollution. The Garden of Eden is not forever lost. The secret to its revival lies buried no deeper than the first few inches of your soil.

SECRETS
OF THE SOIL

CHAPTER 1

Cornucopia

ONE WARM December morning, solstitial sun sparkling on the wooded hills of southern Virginia, six of us sat in a circle looking, no doubt, like a coven of warlocks and witches, stuffing freshly gathered cow manure into desiccated cow horns.

We were on the hundred-acre farm of a former U.S. naval officer, Hugh J. Courtney, gray-bearded and easy-going in his blue denim coveralls. For almost ten years Hugh has been devoting his retirement to producing the various biodynamic preparations recommended over half a century ago by the clairvoyant Austrian scientist Rudolf Steiner as a prime remedy for our planet's sickening soil.

By three o'clock, as the sun slanted deeper through the pine groves, we had stuffed 850 horns bought by our host, over a period of years, from a cooperative slaughterhouse at fifty cents apiece. No longer foul smelling as when first collected, the horns had been processed by curing and drying, then slapped hard together until the bone core popped. The manure—fresh from a small herd of Angus-Guernseys leisurely browsing the biodynamically fertilized meadows that ran down to a meandering creek—was surprisingly sweet to the nostrils. Some fifty gallons of it filled various crocks and pails, awaiting processing.

Scooping up a spoonful of mushy cow manure and packing it into a firmly held cow horn, our host explained how he'd first gotten into biodynamic farming in College Park, Maryland, when he chanced to find on the shelves of the Beautiful Day Trading Company a hard-to-find volume on agriculture written by Rudolf Steiner. It was just one phrase, he explained, one short phrase in this thin but explosive little book which

1

had spurred him to action! Steiner's injunction that the earth could only be healed and the nutritive quality of its produce be made healthy again if the benefits of his extraordinary agricultural preparations were made available to the largest possible areas over the entire earth.

"Many spiritual and occult disciplines," said Hugh, his smile angelically benign, "speak peripherally of agriculture. But this is the only one I've found that puts it all together. When I fully realized the implications I decided to do what I could to promote it."

Steiner's booklet is indeed quite startling, requiring more than a single reading. Conceived in June of 1924—just a year before he died at the age of sixty-four—it came in answer to the plea of a group of German and Austrian farmers worried about the plight of European agriculture. Seed stock had dangerously degenerated and a crippling increase in animal and plant disease was ravaging the countryside. Steiner replied with eight lectures delivered in the Silesian town of Koberwitz, now a part of Poland. Bound together with the simple title of *Agriculture*, the lectures now constitute the basic and extraordinary primer for what has come to be known as biodynamic gardening and farming, the essential remedy, according to its practitioners, for the planet's dying soil.

Mosaically pointing out the danger of chemical fertilizers and the importance of good compost and humus for a healthy agriculture, Steiner anticipated such pioneers of "organic" agriculture as Howard, Balfour, and Rodale. But Steiner went further, much further, by attributing the effectiveness of his biodynamic method to cosmic, telluric, and spiritual influences on soil and plant. And so weird were Steiner's recipes and explanations, later embraced and brought to America by his Austrian protégé Ehrenfried Pfeiffer, that early practitioners of biodynamic farming in the United States behaved virtually as a secret society for fear of being accused by their more orthodox chemical neighbors of practicing witchcraft.

"You can give people," said Courtney, "only what they are ready to receive." His kindly eyes, magnified by heavy horn-rimmed spectacles, darted from side to side as if testing the environs for the approval of unseen listeners. "More intrepid souls," he went on, "saw in biodynamics a means of working with the energies which create and maintain life. To them, Steiner's spiritual science is a desperately needed human service offered to a dying earth, to aid nature where she is weak after so many centuries of abuse. And that's how it stands today."

Steiner's declared aim, as was Alexis Carrel's, was to work with the soil as the true foundation of human health. This meant restoring to the soil the organic matter it needs to hold its fertility, and restoring to the soil a balanced system of functions by treating it not merely as a mixture or aggregation of chemicals, whether mineral or organic, but as a truly living system.

Like his fellow organic enthusiasts, Steiner insisted on avoiding chemicals, concentrating instead on natural composts inoculated with the products of certain processed and revivifying herbs. These he selected to help microorganisms quickly decompose the raw organic matter of the compost heap into simple compounds, reassembling them into the ingredients of a long-lasting, earth-smelling, dark-brown, light-textured, friable humus, a substance which, because of its colloidal state, holds its structure, resists leaching, helps fix nitrogen directly from the air, and increases the availability of minerals to plants—the staff of life.

Stuffing fresh cow manure into desiccated cow horns. When dug up the following year, the manure will have turned to sweet-smelling friable humus.

As we sat in the noonday sun dutifully scooping our spoonfuls of khaki manure into conical cow horns from an apparently endless succession of burlap bags, an associate of our host, a biodynamic herbalist, Lee McWhorter, elaborated on the essential role of microbes in the soil. "Traditional agriculture," he explained, "depends entirely upon the recycling by bacteria and other microbes of various chemical elements— principally the nitrogen, sulfur, carbon, and oxygen on which plants are nourished. Nitrogen is of paramount importance to life on earth. It's an essential constituent of nucleic acids and amino acids, the building blocks of both proteins and enzymes, the source of sap and blood. But, although it is abundant in the air above every acre of land, it cannot be tapped by most plants without the aid of microbes. Hence the symbiotic relation, beneficial to both plant and microbe, which must have developed many millions or billions of years in the past."

"Did you know," asked one of Courtney's neighbors, Will Chapin, who had joined us to help with the horns, "that more microorganisms germinate in half a cup of fertile earth than there are humans on the planet, and that a hundred thousand or more of them flourish on every square inch of human skin?" He paused to let the weight of his figures make their impression, then added: "The combined weight of all the microbial cells on earth is twenty-five times that of its animal life; every acre of well-cultivated land contains up to half a ton of thriving microorganisms, not to mention up to a ton of earthworms, which can daily excrete a ton of humic castings."

With his gloved hand he pared the excess manure from a freshly filled horn.

"But producing humus," interjected our host, raising his spoon for emphasis, "is only part of the solution. As basic as is the presence in the soil of teeming microorganisms for the creation of good friable humus, it is merely an indication that more powerful forces are creatively at work, both cosmic and telluric. That, in essence, is what Steiner's book on agriculture is really all about."

As we rested from our Augean efforts, Courtney's wife, Liz, a cheerful and attractive teacher of dramatic arts, announced a break for lunch. During the course of it, Courtney described the true purpose of what we were doing: preparing the first and perhaps most important of Steiner's remedies for a dying earth, arbitrarily called "preparation 500," an alchemical rather than chemical potion. Our cow horns, like so many ice-cream cones filled at a Tastee-Freez emporium, were to be buried in the ground for the winter, during which time our host assured us that cosmic and telluric influences, called by Steiner *formative forces*, would transform or even transmute the manure to a dark, earthy, and odorless substance, a fourth of a cup of which, stirred into three gallons of rain water, in extremely dilute, literally homeopathic, amounts, would be capable—in conjunction with such regular biodynamic practices as composting, crop rotation, and deep rooting—of revivifying an entire acre of dying land.

The rest of the preparations, as Courtney described them, BD 501 to 508, sounded arcane enough to have been added to the "eye of newt, and toe of frog . . . finger of birth-strangled babe" stirred into a potentizing cauldron by the witches in *Macbeth* to ensnare the Thane of Cawdor.

BD 501, perhaps the least exotic, is simply quartz crystal ground to a fine powder. It too is buried in a cow horn, but throughout the summer, *not* the winter. A quarter teaspoon of it, stirred into three gallons of water, is sprayed in the spring or early summer on one acre of growing plants. Its function, according to Steinerians, is to "enhance light metabolism in the plant, stimulating photosynthesis and the formation of chlorophyll." It is intended to influence the color, aroma, flavor, and keeping qualities of crops.

The next five preparations—502 to 506—were explained as being designed to be inserted into a compost pile to help microorganisms transform it quickly into fertile humus, somehow drawing on what Steiner calls *etheric formative forces*. Considerably more exotic, 502 and 506 are usually treated as a pair: the first consists of blossoms of yarrow stuffed into the bladder of a buck deer. The bladder, obtained from a hunter, is blown up like a balloon and allowed to dry before being stuffed. BD 506 is the flower of dandelion, to be inserted into a cow's mesentery—that tenuous membrane which surrounds the animal's internal organs. It is essential, or so our host insisted, that the dandelion be placed in the *inner* side of the mesentery. Reversed, it is inclined to putrefy. Bladder and mesentery, suitably stuffed, spend the winter buried beneath the soil, there to be worked on by the mysterious forces of the cosmos, which Steiner describes as pullulating with life beneath the snowy, frozen soil of winter.

BD 503, the flower of camomile, is stuffed into a bovine intestine, as with sausage meat—"a charming operation" according to Steiner—and must be buried all winter in a sunny spot where the snow will remain over it for long stretches at a time.

The stinging nettle, annoying in the field, turns out to be a boon to its neighbors, according to Steiner, as a great enlivener of the soil, stimulating its health and helping to provide plants with the individual components of nutrition they most need. As preparation 504, it is buried without benefit of sheath other than peat moss, or layers of netting— iron netting, warned our host, not copper. Iron, he explained, is related to the planet Mars, which goes well with the nettle, whereas copper is less good because of its association with the planet Venus—strange astrological notions later to be validated scientifically by a variety of supporting sources.

BD 505 is the bark of an oak tree, preferably a white oak, ground up and placed in the skull of a domestic animal—cow, sheep, goat, or pig. Put into the earth under a layer of peat moss, it is to be irrigated with plenty of water to acquire a coating of slime.

Last of the compost inoculants, 507, is also the simplest, being merely the juice of valerian blossoms, a wild plant that grows abundantly in the northeast United States, especially in Maine and New Hampshire.

Finally there is Steiner's preparation 508, which does not go into the compost. It is common or garden horsetail, *Equisetum arvense*, brewed into a strong tea to be sprayed onto plants and trees in the spring and summer to work prophylactically against fungus molds.

Our host, aware that his explanations were straining our credulity, relented and informed us it was time to bury the horns, a performance he wished to carry out while the sun was still up, before the ground began to freeze.

Stacked in wooden crates, our handiwork was loaded into a Ford pickup and driven down to a spot in the valley by a stream where a circular hole twelve feet wide and two feet deep had been dug into the soft alluvial soil.

Starting from the center, one by one, the well-stuffed horns were placed tip down in a growing circle until all 850 stood neatly stacked. The lot was then covered with about eight inches of dirt, and we were invited to return in the spring to see for ourselves how the tellurian forces

Lee McWhorter and Will Chapin burying some 850 cow horns filled with cow manure. They will stay below the ground from shortly after the Winter solstice until St. John's Day or Midsummer Day (June 24) of the following year.

of winter had miraculously transformed this cow shit into manna.

As Courtney leveled the fresh soil on the pile, he fortuitously struck a small horn left over from the previous year. Shaking out a small amount of dark friable matter into the cup of his hand he assured us it was sufficient to bring new life to a whole acre of land. But first it would have to be stirred homeopathically into three gallons of water for an hour, twenty seconds one way, twenty seconds the other, in order to be "potentized with the required forces of the cosmos." Arcane as it sounded, this too would be explained in terms to satisfy the orthodox.*

Back on the hill by the house site, in a root cellar dug into the earth, walled with stone and roofed with cement, our host opened deep bins to show us by the light of a lantern the various Steinerian preparations lying in earthenware crocks surrounded by damp peat moss to keep them moist and protect them from such noxious effects as gasoline fumes or electric current. The 500, in a forty-gallon crock, seemed to radiate energy, waiting to be potentized by homeopathic stirring. The 501, or sun-imbibing quartz, stood immobile in a sun-lit window.

Such a wild alchemical approach to gardening and farming made it easier to understand why early biodynamic farmers had chosen to do their stuff on the q.t. But it was a challenge to us to find out if and how such homeopathic wizardry might actually work in practice to effect, as claimed, a revolutionary approach to modern agriculture.

To show us his own quiet method of performing this apparent magic, Lee McWhorter invited us to his herb farm in the Shenandoah Valley, strangely named La Dama Maya, in honor, as he explained it, of a Californian biodynamic flower girl he had met and married in Mexico, motivated, he believes, by some metempsychotic Mexique past.

* The essence of homeopathy—generally described as a system of medical practice that treats a disease by administering minute doses of a remedy that would in a healthy person produce the symptoms being treated—is that the smaller the amount of matter the more powerful the force exerted, as if power were a prisoner of matter, that can be loosened and freed by vigorous succussion, or shaking.

CHAPTER 2

Pulse of Life

ONE GRAY DAY between Christmas and New Year's we set off across the Blue Ridge Mountains to the McWhorter farm in the Shenandoah Valley. Our objective was to learn to make regular biodynamic compost and its homeopathic substitute known as "barrel compost."

Lee and his wife, Maureen, were waiting in their two-story Victorian farmhouse with a jug of herbal tea. Our various astrological signs determined to the satisfaction of Lee, a Virgo, and Maureen, a triple Scorpio, we were led to a greenhouse that stretched the length of the house. Trays of nursling herbs lay in phalanxes, redolent with the scent of rosemary, wormwood, myrtle, thyme, Greek oregano, golden sage, salvia, santolina, and a delicate winter savory, all vibrantly alive and flourishing despite the season.

And so began the lesson in how to brew the prime and most important ingredient in biodynamic farming: Steiner's preparation 500. Into a five-gallon bucket containing three gallons of rain water, Lee poured half a handful of the black 500—enough, he assured us, to spray a whole acre. Rhythmically he began to stir it with a long stick, first clockwise to create what looked like a deep vortex, then counterclockwise to create a seething, turbulent "chaos," followed by another swirling vortex.

"I use this special stick," said Lee, "to get the vortex to the bottom of the bucket. It's just a plain old stick. But it has a curve that helps. You have to get the outside moving. Steiner claimed the vortex was the rhythm of life, that a lot of seeds have the shape of a vortex. This particular motion seems to energize the water."

8

He was sitting on a milk stool, silhouetted against a backdrop of snowflakes falling gently beyond the greenhouse door. "I stir it less than a minute one way before reversing. When I get the rhythm I just sit and contemplate what I am doing. The important thing is to put your thoughts and determination into what you are doing. Steiner says you should put your very life into it, so that what comes back out of the earth is a reflection of your own effort and spirit. The foundation of Steiner's thought is that all aspects of the physical world are permeated and guided by the spiritual. He believed that the soil, in addition to being supplied with nutrients, microorganisms, and humus, needed to be affected by the will and spirit of the farmer or gardener, as well as by intangible forces stemming from the moon, the other planets, the sun, and the stars. I'm putting energy into this stirring. I know I'm turning the stick; but what's turning me? Perhaps the bluefish I ate for dinner last night. It goes around and around, the same basic rhythm in the universe, a pulse that I can only call life. If Steiner is right, that vortex I'm looking at in this bucket is drawing the forces in from the air, from the cosmos. Those are life-giving forces, not death-dealing. Somehow they're there: kinetic, potential."

He paused, reflectively, then went on stirring. "But an hour is a long

Lee McWhorter creating the vortex to suck cosmic and planetary forces into three gallons of water to create BD 500.

time to keep this up," he said with a sigh. "It's not so bad if you've got three or four people to help. Steiner envisioned getting his guests to do it after Sunday lunch, as a form of entertainment. If you'd like to participate I've got some other sticks, and you can see for yourselves which you prefer. It's easier with a longer stick. You can make a smaller circle and get a bigger vortex."

When the hour was up, Lee paused to admire what looked like a bucket of plain water with a few dregs of dirt in the bottom. "This is it!" he said. "The magical potion. But it's not all that powerful all by itself. We've learned that it's more effective on soil that has been treated with compost made with the other preps from 502 to 507, or sprayed with an infusion of the barrel compost which already contains all the preps. So I'm going to mix into this 500, homeopathically, an ounce of barrel compost. It will only take another twenty minutes."

We had seen the compost he was referring to, dark and earthy, at Hugh Courtney's, lying in a barrel dug into the soil down by the stream: but we had not understood its function. Lee explained that it had been developed by a German follower of Steiner, Maria Thun, who had observed and experimented with plants for some ten years on a German government research farm near Kassel. The barrel compost was a simpler method for getting *all* the Steinerian preparations into the soil homeopathically. Not that one didn't want to spread normal biodynamic compost, he added quickly. It was a question of acreage: if one had too much land, and couldn't afford to make that much regular compost, the barrel mixture had a great effect, especially good for the changeover from orthodox chemical fertilizing to the biodynamic method.

"Steiner," said Lee, "explicitly stated that as a result of the concentration and subsequent dilution of the preps it was what he called the radiant effect that was doing the work. No longer the substance itself."

Maria Thun's barrel compost is made by inoculating a gram or so of each of the preparations 502–506 into a mixture of cow manure, fresh eggshells, and ground-up basalt, a volcanic rock that contains all the elements that become clay after dissolution.

Its advantages, Lee explained, are that it can be prepared at any time. It needs less incubation than regular biodynamic compost to be complete, perhaps three months instead of six, and is not only powerful as a fertilizer but is said to put the earth in a state of defense against the pernicious intrusion of radioactivity, especially against the fatal fixation of strontium-90/2 in the bones. Eggshells, he said, add to the soil the element of calcium, and plants grown in high-calcium soils have less radioactivity, especially in Europe. Suitably stirred, for a good hour, the mixture of cow dung and additives is placed in a barrel—open top and bottom—which has been buried in the earth to its waist, then banked with earth to within an inch of its top, there to spend the winter.

"A barrel of this compost, when fermented," said Lee, "can contain anywhere from fifteen hundred to two thousand ounces of the finished product, each ounce of which will care for an acre of land. That means that one barrel can take care of a pretty big spread, some thousands of acres. Just the essence of the preparations, even in the tiniest doses, appears to have a regenerative effect on the soil and make the 500 more productive. In a pasture, everywhere the cows have dropped their dung there will be dead things. Those dead things will biodegrade into healthy humus much faster after you've sprayed with the essence of the barrel compost. If you have a lot of acreage you can't possibly make enough actual compost, let alone spread it: it might take tons and tons. But an ounce—or a third of a cup—of this barrel compost, stirred into water and stimulated by the forces freed through the compost, can generate a billion microbes in each teaspoon of soil. You get an idea of what's happening if you realize that each and every one of those microbes has a mouth and is eating debris on the ground, then laying down its corpse as organic residue, often in a matter of minutes. Pretty soon your fields are rich in humus. But Steiner, mind you, was perfectly clear that the presence of abundant microorganisms in the soil is merely an indication that cosmic forces are at work. It's like flies: they only come if there's dirt. Ditto with the microorganisms: they only proliferate if the forces are there. And the preps mediate the forces."

As soon as he'd finished stirring the potion of barrel compost into the bucket of 500, Lee took a piece of cheesecloth and strained the contents into a backpack sprayer, leaving about a quart of liquid in the bucket. "I don't want any residue to clog my nozzle," he explained. "I'm going to spray one acre of my outdoor herbs, as an autumn spray. In Steiner's day, before they invented the backpack sprayer, they used a pail and a big brush to swish the stuff onto the ground. Maureen still likes to do it that way, and it works just as well."

Pail and brush in hand, Maureen looked like an Andrew Wyeth painting as she set off into the field.

"We spray in the autumn after everything is clear," said Lee, "before the ground freezes. I want to get the farm sprayed so that when the earth breathes in again in the summer it will breathe all these forces back into the earth. Then things will come back to life. Normally we wouldn't spray in the middle of the day, like this. I'd do it early in the morning when there's rime or dew on the ground. But with so little sun the stuff's not about to evaporate."

A light snow was still falling as we followed Lee into the garden to watch him spray his dormant herbs.

"We haven't put anything on this soil but biodynamic preps for ten years," he said. "The full effect of the method appears in the course of the first three or four years. It consists of a continuous increase in fertility, and an improvement in the quality and flavor of the produce. And we harvest on those calendar days when the moon is influencing certain portions of plants and giving them a better keeping quality. Our customers all remark about the beauty and power of our herbs, but if I try to explain to them that it comes from cosmic forces, my wife makes me hush. She says the customers aren't ready for such talk. At least not yet. It's too far out."

Lee moved about the aisles between his rows of herbs, pumping a fine mist into the air. "After many years of careful research," he went on, "Sherry Wildfeuer of the Camp Hill Village in Kimberton Hills, Pennsylvania, produced a calendar which shows specific days (and hours) judged best for working on the leaf, root, fruit, or flower portions of crops: one works with spinach on a leaf day, potatoes on a root day, peaches on a fruit day, blossoms on a flower day. The days succeed each other in cycles of nine, or just over a third of a moon period. We do our transplanting only on a root day, so the roots can dig right in. The place I've noticed it most is with broccoli seedlings. It takes most people a couple of weeks to get them up. We do it by the moon and they come up in three days. Everyone knows that apples are pruned and cow horns sheared when the moon is waning, lest the moon's powerful tide-raising forces cause both tree and cow to bleed away without healing. If you want your hair to grow luxuriant, you cut it only at the very moment of full moon. So we figured: why not pay attention to the rest of the calendar? Once you tune into the cycles of the moon you quickly see the effects they are having, even on the weather."

Lee smiled at us broadly: "The mechanics may appear complex, but the premise is simple. This planet, and everything on it, is an integral part of both the solar system and the cosmos: every last blade of grass is affected by the whole. With his clairvoyant vision, Steiner could describe the formative forces as they operate in cycles."

"What about the 501?" we asked.

"Ah!" said Lee. "That's for spring and early summer. When the leaves are beginning to sprout, when the plants have three or four sets of leaves and are starting to grow, then we spray with the silica 501. A teaspoon of ground-up quartz crystal mixed into three gallons of water, stirred alternately clockwise and counter-clockwise for an hour, just as with the 500, makes a potentized liquid. Scrutinized under the microscope, the 501 would most likely show nothing but three gallons of plain old rain water.

But its effect is immediate and noticeable. Steiner says it enhances light metabolism in the plants, and stimulates photosynthesis and the formation of chlorophyll. I know that it influences color, aroma, and the keeping qualities of all our crops. And you can spray 501 throughout the summer to keep bringing in the forces of light."

Lee put down his backpack while there was still some fluid in the tank. "Now I'll show you how to make a biodynamic compost pile." His enthusiasm was in no way flagging. "I don't make them very big. I don't need to for just nine acres, especially when I spray with the barrel compost. But I like to put a handful of the regular biodynamic compost in the hole when I transplant an herb. If you didn't use the barrrel compost by the ounce you might have to use as much as five hundred to a thousand pounds of the regular biodynamic compost on each acre. On a thousand acres it would mean you would have to be making tons and tons of the stuff. So you see the advantage of barrel compost? Luckily, once you start spraying and composting, either with one or the other, each year you need less and less. If you don't mind standing out here in the snow, I'll build you a small heap, right here and now, and we'll inoculate it."

Picking what he considered a suitable spot for the pile, Lee cleaned off an area three by four feet to reveal the bare earth. "The pile has to be in direct contact with the soil," he explained, "so that the earth forces can work their way up to affect the pile, and the earthworms and microorganisms can have access to decompose the material."

With a wheelbarrow he set about collecting various ingredients to build up in layers, first one of dried weeds, which he pulled from the edge of a field. These he covered with a couple of inches of earth, followed by a layer of straw, followed by a layer of cow manure. "It's biodynamic manure," he said with a happy grin. "I brought it here from Hugh Courtney's in the back of my van. I should be putting in a layer of quicklime, but I don't have any, and with this soil it'll do just as well without."

Lee dug up some thistles, followed by some green weeds that were alive and doing well. "This is white yarrow," he said, tearing it up by the roots and waving it in the air. "It grows wild like a weed. But I've got it planted around here; we harvest it in the spring, especially for Hugh. It's one of the preps that goes into a stag bladder. And we use it in our herbs. It's a good expectorant to get rid of phlegm, and it's also a diathoretic to make you sweat. Growing medicinal herbs is a matter of intent: you have to know just what effect you're looking for; and the herbs ought to be biodynamically grown."

Lee wheeled his barrow to where he had three compost piles already working, separated by three pallets. The first was completed, as could be seen from the shovelful of dark-brown, friable material he produced from it, sweet-smelling and earthy, crawling with earthworms. Next to it was a second pile in the process of biodegrading; he figured it would be ready in a couple of months. The third pile, exuding fragments of orange peel and eggshell, was clearly being built up from kitchen residue.

"I've got other small piles of compost like this all over the property," said Lee. "When they're ready, I can spread them right on the spot." He shoveled some of the biodegrading material onto his barrow and started to wheel toward the fresh pile he was building. "A little of this older stuff will act as a starter in the newer pile. But the real starters are the Steinerian preps, the one Pfeiffer kept talking about but whose actual content he tried to keep restricted. I'm going to insert them as soon as the pile is about four feet high and has a skin."

By skin, Lee meant a layer of straw and earth to insulate the pile from the fiercer elements, but not the quickening rain. This he arranged cosmetically with his shovel before using the handle to dig five holes a few inches apart, about eighteen inches deep into the pile.

"Into each of these holes I'm going to put one of the preps, 502 to

A shovelful of compost inoculated with six biodynamic preparations. It has turned into humus alive with earthworms, the basis for biodynamic farming and gardening.

506," he explained, producing from his pocket five little plastic containers, not much bigger than teabags, the contents of each of which, about a gram or so, he shook into each hole before sealing it.

The last preparation, 507, *Valeriana officinalis*, or the juice of the valerian flower, he poured from a small quarter-ounce vial into his backpack sprayer, which still contained the remnants of the mixture of barrel compost and 500. This he sprayed over the whole pile in a fine drizzle.

"The compost has to be damp," he explained, waving the nozzle, "or it won't heat up. But it mustn't be too wet, or the aerobic microbes won't have enough air to breathe; fermentation won't take place, and the compost will rot instead of biodegrading."

He looked up with a satisfied grin. "Come back in the spring and you'll see a lovely pile of degraded compost, ready to give life to the soil. This pile will have turned to soft, dark, friable humus, the secret of agricultural health on this planet—courtesy Dr. Steiner, his cosmic forces, and billions or trillions of cooperative microorganisms."

"Who else," we asked, "is farming this way in America?"

"Ahaa!" came the answer. "For that you will have to go to Kimberton, to the Bio-Dynamic Association, and talk to the man who runs it now, Roderick Shouldice."

CHAPTER 3

Moonshine

THIRTY MILES WEST of Philadelphia, in the low-lying hills of Kimberton, one of the first biodynamic training farms in America came into being in the late 1930s, dedicated to an ecological, nonsynthetic, sustainable agriculture. On a thousand acres of what might be the rolling downs of Surrey, Alarik Myrin, of the Sun Oil Company, first gave sanctuary to Steiner's early follower, Ehrenfried Pfeiffer, an anti-Nazi biochemist and pragmatic farmer, who fled the Nazi millennium to bring his master's agricultural message to America.

The old mansion house, overlooking a sprawling four-hundred-acre farm, now houses a large group of cheerful handicapped persons of various ages—frail human beings, exterminatable under Hitler's genetic laws, but cherished by Steiner, who devoted some of the best years of his life to the art and science of successfully rehabilitating the handicapped. Camp Hill Village, now one of many such communities stretching from Ireland to Botswana, is a nonprofit venture partly funded by government. We found its members happily attending to their chores about the farm, which economically includes a bakery, a cheese factory, and even a store, where only biodynamic products are sold, including delicious home-baked bread and remarkably tasty vegetables.

Of the Bio-Dynamic Association not a sign. A wild goose chase, we wondered, forewarned that the practitioners were wary of publicity? Then we met in the wood-paneled, picture-windowed coffeeshop a quiet, outgoing young man in his thirties, Roderick Shouldice, recently appointed administrator of the association. He explained that the original acreage had been divided by Myrin at his death into two farms, side by side, both biodynamically run. One was for the handicapped group, where visiting students could also learn the biodynamic method, and from which extra income could be derived by trucking fresh milk, yogurt, and cheeses to Philadelphia—the state of Pennsylvania being one of three in the Union that still allows raw milk for sale. The other, Seven Stars Farm, of five hundred adjacent acres, is run as a dairy farm to nourish the students of the Steinerian Waldorf School—from kindergarten through twelfth grade.

Aiming at self-sufficiency as well as health, the farms grow their own wheat, rye, barley, oats, soybeans, and corn, along with most of their fruits and vegetables. Many of the fruit trees, deliberately planted by Pfeiffer in the early 1940s on the strip-cultivated slopes to help avoid erosion, are still bearing fruit: apple trees and pear trees, their bark painted with a special Steinerian prep to prevent mildew: in this case a mixture of fine clay and fresh cow manure dissolved with 1 percent equisetum tea and one portion of stirred 500 reduced to a consistency that can be used as paint.

In a well-designed root cellar built by the Myrins near one of the tenant houses—handy in those days as a potential wartime bomb shelter—Steiner's various preparations, 500 to 507, lay awake in earthenware pots surrounded by peat moss, their lids daubed with yellow paint to differentiate yarrow from oak bark or nettle, waiting to be inoculated into compost, or to be sprayed homeopathically onto an avid soil.

Seven Stars Farm, as Rod Shouldice explained, does not use Maria Thun's barrel compost, but has enough cows of its own to spread the real thing regularly on all its acres from a series of fifty-foot-long compost heaps inoculated with the preps. Rod and the farmers do this, injecting piles that have been front-loaded into windrows by two professional biodynamic farmers who oversee the land and care for the livestock. Brown, Swiss, Guernsey, and Hereford cows, pigs, sheep, goats, chickens, all produce biodynamic manure for biodynamic compost, it being a part of Steiner's notion that every farm should be fertilized by the animals that live on it, creating a cycle of ever-increasing richness in soil untainted by foreign or chemical elements.

To stir the 500, which is made right there on the farm, Rod uses a system with a platform six feet above ground, surmounted by a trellis from which a ten-foot pole dangles into a 135-gallon barrel. The system allows Rod to stir ninety-five gallons at a time, with little exertion, being relieved of the suspended weight of the pole. The resulting 500 is easily poured by gravity into a sprayer hooked to a tractor, which can spray five gallons onto each of nineteen acres. The entire farm is fertilized with biodynamic compost from the windrows, plowed into the topsoil in the

Underground vault created by the Myrins in 1942 to house the various Steiner preps in earthenware pots surrounded by peat moss.

spring, and regularly stimulated by the cosmic forces released through the spraying of 500 and 501.

To these biodynamicists, the whole earth is but a reflection of what is taking place in the cosmos, an idea that goes back at least to Ancient Rome, when no one denied that the moon, sun, and the seven known planets greatly affected life on earth.* But at Camp Hill the effects of the cosmos on growing plants have been confirmed by intensive laboratory and field experiments carried out during the past half century. As an indication that man once understood the power of the sun's intercourse with soil and plant, they point to the Egyptian glyph of the Sungod Ra, depicted with rays that end in manipulative hands.

* In his *History of Nature*, the Roman scholar Pliny the Elder, a contemporary of Jesus Christ, took up the influence of the moon's phases on vegetal and animal life. If one wanted juicy and good-looking fruits and vegetables for sale or consumption, the optimal time to pick them was at full moon, just when ants were busiest in their hills, even at night, and marine animals such as oysters were in their period of burgeoning growth. At new moon, the same ants were seen to be listless and the growth of sea creatures slack. Fruit, said Pliny, is much less susceptible to rotting at new moon and can be easily and efficiently dried.

Pliny's compatriot, the Mantuan poet Virgil, born just before the historian died, made plain, in a discourse on agriculture, how husbandmen took cues from heavenly spheres

Every twenty-four hours, sun, moon, planets, and stars, as observed
by these stargazers, have a strong and pronounced effect upon the
growth of plants. Planting and harvesting at Camp Hill Village Farm are
done as strictly as possible in conjunction with the Kimberton Hills Calendar, whose editor, Sherry Wildfeuer, and several contributors, are in
residence nearby. Steiner and his followers maintain that as the sun, the
moon, and the planets move through the twelve zodiacal segments of
the sky they have differing effects on the earth and on the plants that
grow in it.

Because each day the sun rises four minutes later than the stars
against which it moved the previous day, in a year it gradually traverses
the entire circle of the zodiac, spending a month in each constellation.
The moon, in its monthly cycle, spends only about three days in each.
Apart from its obvious waxing and waning—the influence of which on
tides and growing plants is no longer in dispute—the moon has two other
motions that are said by Steinerians to affect both plants and planet. The
moon's elliptical orbit brings it at times nearer and at times farther from
the earth, creating a harmonic effect comparable to the earth's drawing
nearer to and farther from the sun in summer and winter. When the
moon reaches its nearest point, or perigee, and a "winter" moon is
evoked, the plants' relation to the sun, according to biodynamic farmers,
is hindered in that seeds put down at the moon's perigee produce plants
that tend to be vulnerable to fungus diseases and pests.

European biodynamic moon watchers such as compost developer
Maria Thun point out that with the ascending moon plant forces and
saps flow upward more strongly to fill the plant with vitality. But, when
the moon has reached its highest point and begins to go down again, the
plant, they say, orients itself toward the root, a time more favorable for
transplanting because it enables the plant quickly to form rootlets with
which to anchor itself. As the sap flow weakens, it is also a suitable time
for pruning trees or clipping hedges. When the moon runs low on the
horizon, echoing the sun's influence in autumn and winter, the vitality
of plants is said to concentrate in their lower parts: time to concentrate
on manuring, rooting, cutting, composting, and harvesting root crops.

Biodynamicists point out that the forces streaming continually from
the direction of the constellations are "focused" by the moon and are

---

and constellations to tell them when to sow their crops, certain seeds being best put into
the ground when "glittering Taurus opens the year with his golden horns."

With the dawn of the so-called age of reason and the enthronement of "science" in the
sixteenth and seventeenth centuries, such astrological notions began to be rejected as
superstition, until, in 1983, an affirmation was signed by 121 scientists, many of them Nobel
laureates, deriding astrology as so much humbug.

THE KIMBERTON HILLS AGRICULTURAL CALENDAR

A BEGINNER'S GUIDE
FOR UNDERSTANDING THE INFLUENCE
OF COSMIC RHYTHMS
IN FARMING AND GARDENING

1988

Published by Kimberton Hills Publications.

able through its power to become directly effective in plant life. They therefore choose to work with plants on those days when the moon has entered a zone of the sky that especially enhances the growth of that part of the plant they wish to encourage.

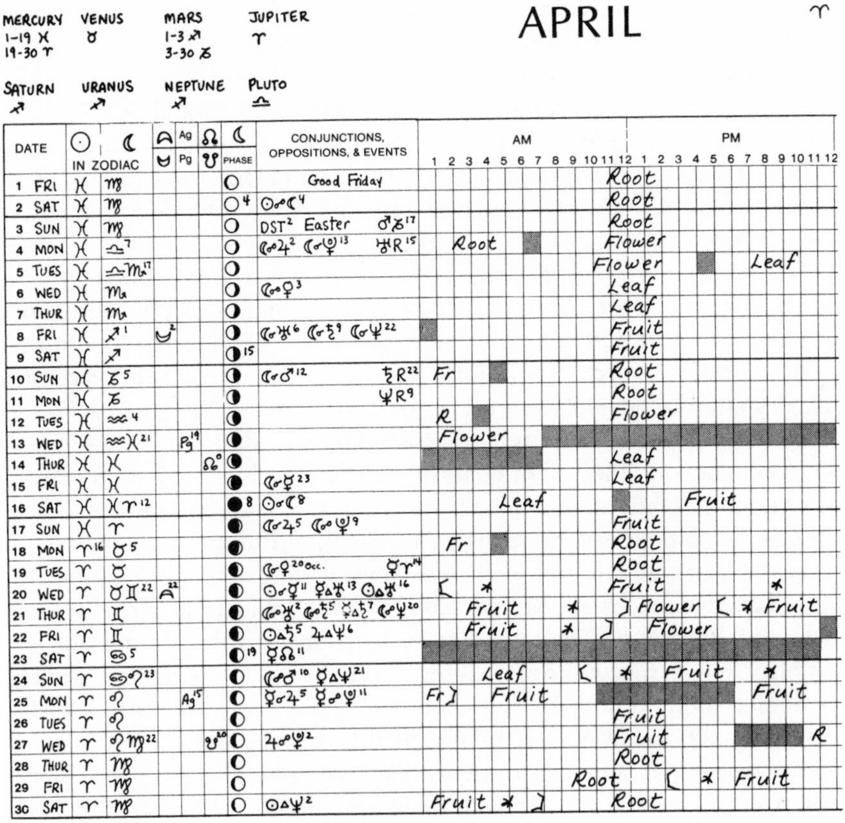

MERCURY 1-19 ♓ 19-30 ♈	VENUS ♉	MARS 1-3 ♐ 3-30 ♑
SATURN ♐	URANUS ♐	NEPTUNE ♐

JUPITER ♈   **APRIL**   ♈

PLUTO ♎

DATE	☉	☾ IN ZODIAC			☾ PHASE	CONJUNCTIONS, OPPOSITIONS, & EVENTS	AM / PM (planting)
1 FRI	♓	♍			O	Good Friday	Root
2 SAT	♓	♍			O⁴	☉⚹☾⁴	Root
3 SUN	♓	♍			O	DST² Easter ♂⚹♅¹⁷	Root
4 MON	♓	♎⁷			O	☾⚹♃² ☾⚹♀¹³ ♅R¹⁵	Root / Flower
5 TUES	♓	♎-♏¹⁷			O		Flower ... Leaf
6 WED	♓	♏			O	☾⚹♀³	Leaf
7 THUR	♓	♏			O		Leaf
8 FRI	♓	♐¹	☽²		O	☾⚹♅⁶ ☾⚹♄⁹ ☾⚹♆²²	Fruit
9 SAT	♓	♐			O¹⁵		Fruit
10 SUN	♓	♑⁵			O	☾⚹♂¹² ♄R²²	Fr / Root
11 MON	♓	♑			O	♆R⁹	Root
12 TUES	♓	♒⁴			O		Flower
13 WED	♓	♒¹	Pg¹⁹		O		Flower
14 THUR	♓	♓		☊°	O		Leaf
15 FRI	♓	♓			●	☾⚹♀²³	Leaf
16 SAT	♓	♓♈¹²			●⁸	☉⚹☾⁸	Leaf / Fruit
17 SUN	♓	♈			●	☾⚹♃⁵ ☾⚹♀⁹	Fruit
18 MON	♈¹⁶	♉⁵			●		Fr / Root
19 TUES	♈	♉			●	☾⚹♀²⁰ occ. ☿♈⁴	Root
20 WED	♈	♉♊²²	A		●	☉⚹♀¹¹ ♀⚹♅¹³ ☉△♅¹⁶	Fruit
21 THUR	♈	♊			●	☾⚹♃² ☾⚹♄⁵ ♀△♄⁷ ☾⚹♀²⁰	Fruit / Flower / Fruit
22 FRI	♈	♊			●	☉△♄⁵ ♃△♆⁶	Fruit / Flower
23 SAT	♈	♊♋⁵			●¹⁹	☿☊¹¹	
24 SUN	♈	♋♌²³			O	☾⚹♂¹⁰ ♀△♆²¹	Leaf / Fruit
25 MON	♈	♌	Ag¹⁵		O	☿⚹♃⁵ ♀⚹♅¹¹	Fr / Fruit / Fruit
26 TUES	♈	♌			O		Fruit
27 WED	♈	♌♍²²	♉°		O	♃⚹♆²	Fruit / R
28 THUR	♈	♍			O		Root
29 FRI	♈	♍			O		Root / Fruit
30 SAT	♈	♍			O	☉△♆²	Fruit / Root

One month in the Kimberton Calendar, originated by Maria Thun in Germany, showing planetary days and exactly which part of what day to plant or harvest root, leaf, fruit, and flowers.

Knowing which days are especially favorable for leaf or fruit vegetables, Rod explained that at Kimberton Hills they sow cabbage seeds on a leaf day and wait for a fruit day for tomatoes. The four aspects of a plant —root, leaf-stem, flower, and fruit-seed—have distinct qualities and functions. In cultivating plants, biodynamicists try to encourage the *root* growth of carrots, beets, turnips, and potatoes; the *leaf* growth of lettuce, spinach, and grass; the *blossoms* of their favorite flowering plants; and the *fruit* and *seed* formation of such crops as wheat, corn, tomatoes, beans, and squash. Steinerians point out that the moon itself, acting not only as a reflecting mechanism, but of its own power, works on the earth through the element of water, available to plants either through rain or irrigation, and is a carrier of those forces that tend to influence reproduction and vegetative growth.

Another basic rhythm is brought about by the daily rotation of the earth upon its axis. Within each single day, says Maria Thun, a plant is affected by the earth's complete revolution—an effect comparable to what biodynamic practitioners describe as the yearly in-and-out "breathing" of the earth. They say that from about three in the morning until midday this daily rhythm causes sap to rise, and that from three in the afternoon until the middle of the night, the rhythm tends to influence the lower parts of the plants.

When Steiner noted that the icy crystals that form in winter on the panes of windows are different on a flower shop from those appearing on the windows of a butcher shop, he suggested to Pfeiffer and to another of his early followers, Lily Kolisko, that they experiment in the lab with the formation of crystals as a means of demonstrating what he called his "formative forces" in nature. To satisfy Steiner's desire to demonstrate what lies "beyond the veil of matter," Pfeiffer developed a system known as chromatography in which various solutions of mineral salts, imbibed by rounds of paper, were made to reveal the differing patterns left by forces at work in soil and plant life. With a similar method, known as "capillary dynamolysis," Kolisko went further, validating not only Steiner's but Paracelsus's dictum that each planet is linked to a metal on which it has a special effect, varying with the motions of that planet. With her method Kolisko was able to explain such mysteries as why highly refined metal objects such as ball bearings can turn out badly if manufactured at an inappropriate planetary phase. The same subtle forces she found to be at work in the living cells of plants, affecting both their growth from seed and the quality of fruit and vegetable, and even such odd facts as that lumber, to be lasting, *must* be cut in the appropriate moon phase.†

Rod told us how Kimberton Hills student farmers are trained in the observation of color, movement, and the gradual transformations of nature. By these means they are able to recognize what they call the "plant being" that lives and unfolds through all the forms from seed to root to leaf, and back to seed again. Taking advantage of cosmic cycles, they

---

† Many painstaking experiments to validate the actual effect of stars, sun, planets, and moon on the germination of seeds and the growth of plants were carried out over a period of thirty years by Lily Kolisko, as described in detail in *Agriculture of Tomorrow* written with her husband, Eugen.

like to harvest lettuce and other upper-plant crops in the morning when they are most full of vitality, and root crops in the late afternoon when they are most full of life. Transplanting they consider best done in late evening when the downward flow of sap can be used for establishing roots in their new environment. By performing normal agricultural practices such as seed sowing, transplanting, cultivating, and harvesting at times when the appropriate element is working strongly from the cosmos, they have shown that one can enhance the size, form, flavor, and storage quality of crops.

It has been shown that plants react strongly to changes in the weather by rushing sap to their roots in anticipation of a coming storm. This enables them to send sugar back up to leaves and boughs when things have quieted, and so repair any damage by the storm. The mystery remains as to how the plants can tell that a storm is coming, but an empirical test is easily made from the sugar content of a plant just before, during, and after any storm.

Taking a leaf from Steiner and such occult masters as Paracelsus, Kimberton farmers are aware, as Rod explained, that the four parts of the plant correspond to the four classic elements of nature: earth, water, air, and fire. Clearly, they say, the root of a plant is concentrated in the element of earth, the green vegetative parts are linked to the flow of moisture, the flower opens into the airy element, and the fruit is slowly ripened by the warmth of the sun, which seals the seed.

Constellationally—and enigmatically—they associate the four elements with four segments of the zodiac: earth with Taurus-Virgo-Capricorn; water with Cancer-Scorpio-Pisces; air with Gemini-Libra-Aquarius; fire with Aries-Leo-Sagittarius. And they conceive that the moon, passing before these constellations, enables those particular elemental forces to work more strongly into plant life. Maria Thun, observing and experimenting for ten years on a West German government research farm, has accumulated data on the effect on the growth of plants of planetary oppositions, trines, and conjunctions. Whereas oppositions and trines are said to be positive, nodes (where orbits intersect), occultations (when one celestial body passes in front of another), and outright eclipses (when one body obscures another), are generally considered unfavorable for plant work, usually causing adverse effects, especially in sowing.

Just as the rise in traffic accidents occurring on node days has been confirmed statistically, so have the effects of the moon on cattle breeding been established by Steinerian farmers over the years. If a bull is taken to the cow on a node day, either the cow remains barren or, worse, the calf is born with undesirable characteristics. Experience has shown that

Zodiac Sign Influences

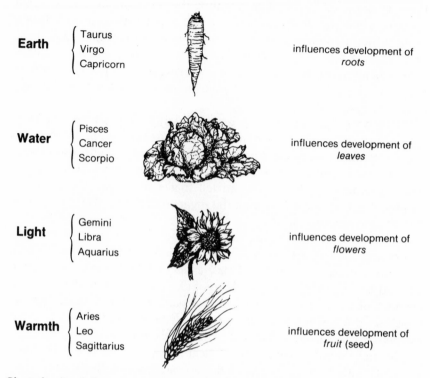

Earth	Taurus Virgo Capricorn	influences development of *roots*
Water	Pisces Cancer Scorpio	influences development of *leaves*
Light	Gemini Libra Aquarius	influences development of *flowers*
Warmth	Aries Leo Sagittarius	influences development of *fruit* (seed)

Chart showing influences on the growth of plants attributed by biodynamic farmers and gardeners to zodiacal signs and their respective elements.

when planets enter into opposition at 180 degrees, be it with the sun or with one another, the life forces of the plant are increasingly intensified, beginning several days before the actual event. Maria Thun claims that the forces of both opposing planets, influenced by the impulses from their respective zodiacal constellations, interpenetrate, fructifying and augmenting one another in their effect upon the earth. The moon's effect, she says, can sometimes be enhanced through oppositions, but on other occasions be diminished.

All of this lore, which echoes ancient astrological knowledge—as Rod admitted—is actually the result of many, many years of careful experiments, with strictly scientific methodology carried out by devoted followers of the Steinerian vision. When Maria Thun first began her studies of the rhythmic cycles of the moon, she was unaware that her Kassel vegetable garden was near the center of a ring of hills, each connected since ancient Celtic and Druidic times with a different zodiacal sign. But she soon found that the ancient history of these "mystery centers" enabled her to interface the effects of moon, sun, and planets, enclosed as they are within the twelve zodiacal constellations, various sections of which are visible at night at different times of year. The result was the creation of a calendar, the Biodynamic Sowing Chart, incorporating all the data of the three moon rhythms, the planetary positions, and the zodiacal signs, by which the members of such communities as Kimberton can plant and harvest their superior products.

By creating her sowing chart on this basis, Thun was careful to be correct, not only geometrically and theoretically, but actually. To sow seeds by the wrong *moonshine*, without paying attention to the exact scope of each particular celestial sign, would be to escape its influence and, like Procrustes, not make beds to fit people of different heights, but cruelly stretch or amputate them to fit a single arbitrary bed.

CHAPTER 4

Golden Garbage

IN THE PEACEFUL woodlands of southern New York, just west of the Hudson River, a sprightly lady in her eighties, Margrit Selke, presides over an A-frame hut that resembles more the home of the Good Witch in a Grimm fairytale than an establishment for the pursuit of science. It is Pfeiffer's laboratory at Threefold Farm. All in diminutive storybook form it contains the equipment for the manufacture of the Pfeiffer "starter": alembics, retorts, microscopes, heating and cooling containers, great mixing vats, specially designed, and a line of refrigerators stocked with vials of multiple strains of incubating or dormant microorganisms designed to chew up each and every component of any kind of biodegradable garbage set before them.

Also an émigré from Hitler's Germany, Margrit Selke continues to produce Pfeiffer's starter, still in demand by those determined enough to request it. On the day we visited, a thousand units lay in plastic bags in a box half the size of a footlocker, ready to be shipped to Saudi Arabia, and another thousand were ready for Dubai, each parcel designed to revivify a thousand acres of barren land. For the more volatile market in this country, Ms. Selke continues to keep the strains alive, along with cherished memories of their developer.

Pfeiffer was barely twenty when he first worked with Steiner at Dornach just outside Basel, Switzerland, in 1919. The master, then in his fifties, quick to see the young man's talent, turned his interests from electronics and physical chemistry toward what he considered the more vitally important study of biochemistry, biology, and their application to agriculture.

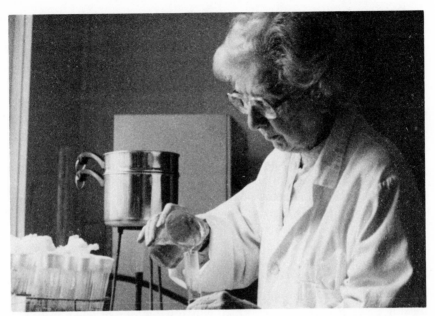

Margrit Selke in her lab at Threefold Farm, N.Y.

By 1925 Pfeiffer had set up his own Bio-Chemical Research Laboratory in a shed of the Weleda Drug Company, manufacturers of non-chemical medicines, at Arlesheim, near Dornach, intent upon following Steiner's lead not only into anthroposophy, but into his newly expressed theories about the vital state of agriculture. Just before his death, Steiner had made the same request of Pfeiffer he had made of Kolisko, that he search for some chemical agent that would reveal the formative forces in biological substances. Spurred by this need to show, by strictly orthodox scientific methods, the existence of whatever it might be in living matter that differentiates it from the inorganic, Pfeiffer investigated a great number of chemicals, finally settling on copper chloride as the most suitable agent. With it Pfeiffer was able to develop a method similar to Kolisko's to analyze biological substances, which came to be known as "sensitive crystallization."

To test the effectiveness of Steiner's preps, Pfeiffer carried out a series of experiments by placing chopped potatoes in thin dilutions of 500 to 507 and studied the development of roots, discovering that 500 had a particularly stimulative effect on root growth, causing numerous fibrous roots, whereas 501 increased the assimilative activity of plants, and 504 had a special influence on flavor; all the others strengthened growth compared with control samples.

Pfeiffer claimed that in the last half century, since the advent of hybrid fertilization and of insecticides, humanity all over the world had been receiving increasingly deficient protein in food, and that there was nothing normal anymore in what was being grown. By comparable testing of chemically-grown wheat seeds and biodynamic wheat seeds, Pfeiffer showed that, on the seventh day after sprouting, biodynamic seeds contained 42 percent protein compared with 23 percent for chemically-grown seeds. Harvested in the summer, the biodynamic wheat had an almost gloss-like grain, and 12 to 18 percent protein compared to 10 to 11 percent for chemically-grown wheat, which led Pfeiffer to remark that with such nutritious wheat one could almost avoid eating meat.

Subjected to 100 degrees of heat for half an hour, the biodynamic kernels still sprouted, whereas artificially fertilized kernels were dead.

Other experiments were made in fields where parcels of land side by side were repeatedly cultivated in the same way, planted with the same crops, but fertilized differently to demonstrate the difference.

From 1926 to 1938 Pfeiffer also ran several farms in Holland, totaling eight hundred acres, complete with their own flour mill and bakery for making especially wholesome bread, which fed a total of seven hundred families. Biodynamically-produced vegetables, distributed under the Demeter brand—named for the Greek goddess of agriculture—at first avoided by outsiders as too expensive, were soon forced onto their tables by children who refused to eat any other kind of vegetable put before them.

For fear of startling Americans, Pfeiffer's first book published in the United States, Bio-Dynamic Farming and Gardening, so bowdlerized Steiner it avoided offending anyone squeamish about metaphysics, and could thus persist, without causing the slightest reaction, as the guiding light for a few hearty souls. There was no mention of buried cow horns, stag bladders, or the slimy skulls of animals. The preps were politely described as generic concoctions of herbs.

While World War II raged on in Europe, Pfeiffer was establishing his bona fides in America by obtaining an honorary degree in medicine from the Hahneman Medical College in Philadelphia—at one time almost wholly devoted to homeopathy—in recognition of his remarkable biological research with the use of crystallizations. Using a single drop of a sick patient's blood, he made it possible to diagnose many forms of disease, including cancer.

By 1944, as an American citizen, after a falling-out with his patron at Kimberton Hills, Pfeiffer moved to a 285-acre farm of his own in Chester, New York, which he operated biodynamically. But the land was stony and sterile, the cattle riddled with the dread Bang's disease, and only

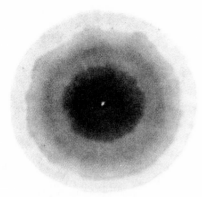
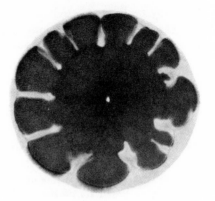

The left chromatogram shows margarine, lacking in vitamin and enzyme formation. On the right, fresh homemade butter from unpasteurized milk shows what Pfeiffer calls a healthy abundance of vitamin influences.

Pfeiffer's cheerful determination, along with his applied scientifically-made composts, enabled him within two years to restore the land, feed the grain grown from it to his cattle, and cure them without medication. "They cured themselves," was Pfeiffer's smiling explanation. "We simply provided nutrition and care."

Greatly preoccupied with the general condition of soils in America, Pfeiffer found a third of the United States on the way to becoming useless from erosion and bad agricultural practices. In the Middle West, dust storms had been moving eastward at the rate of forty miles a year. Because of unbalanced overgrazing, and a greedy tillage of the soil, with no care given to protect it, the sod was gradually being loosened and the protective soil covering wafted into the atmosphere by drought and wind. A biologically beneficial form of the balanced, diversified farm, with heavy legume plantings, meadows, and green manuring, was being given up in favor of a one-sided monoculture. Between 1935 and 1938 some 90 percent of the settlers of the Great Plains abandoned their farms, tragically described by Steinbeck in *The Grapes of Wrath*.

To make things grow on a piece of soil, Pfeiffer pointed out, was not necessarily farming: it could be destroying the earth's fertility. A cultivated field, he insisted, was a living organism, a living entity in the totality of its processes. "A well-tended field does not have an unlimited capacity for increased productivity, especially not one directly proportional to the amount of fertilizer applied. An increase in production may be obtained, but the soil, as an organic structure, can go to pieces."

Pfeiffer preached repeatedly that soils intensely treated with chemical fertilizer or orchards sprayed for a long time with chemicals no longer have any biological activity. They begin to die. Also, vineyards treated for years with copper and lime solutions become devoid of earthworms, losing their capacity for creating humus.

Strong doses of chemical fertilizers, said Pfeiffer, especially those containing soluble salts like potassium or ammonium sulfate, or highly corrosive substances such as nitro-phosphates, or poisonous sprays, like arsenic and lead preparations, injure and destroy the microorganic world. And he was quick to point out that every measure that disturbs life in the soil and drives away the earthworms and bacteria renders the soil more lifeless and less capable of supporting plant life. The only healthy alternative, he advocated, was Steiner's biodynamic compost.

But opposition to methods that smacked of witchcraft was still too virulent in North America; and the strain of a constant sixteen-hour day divided between science and farming was too great for Pfeiffer's weak constitution. From congenital diabetes and advanced tuberculosis he collapsed and was confined to a sanitarium for a year. An indefatigable researcher, Pfeiffer took the blow as an omen to change his approach. Out of bed, he headed for the sanitarium's bacteriological laboratory to study under the microscope the very tubercular bacteria that had been gnawing at his lungs, particularly fascinated by the way specimens from digestive tracts break down and digest waste material—all of which gave him food for thought. Already at Dornach, Steiner had suggested he search for a method of transforming vaster quantities of organic matter into fertilizing compost. To Pfeiffer this meant identifying and isolating the individual strains of bacteria that could digest and transform the various ingredients in an entire city dump.

Sufficiently recovered, Pfeiffer founded his Biochemical Research Laboratory at Threefold Farm in Chester, New York, in a Steinerian center organized in the 1920s by a disciple of anthroposophy who was happy to offer Pfeiffer the opportunity for developing Steinerian notions.

With the help of Margrit Selke, Pfeiffer patiently sought out and isolated the various strains of bacteria which, rendered dormant and added to Steiner's traditional biodynamic preparations, he predicted would produce a bacterial "starter" so vital it would be called "bio-dyna-mite," capable of transforming vast amounts of city garbage and slaughterhouse refuse into valuable organic fertilizer at competitive prices that would lead to the demise of chemicals.

It is said that Pfeiffer was also led to this development of Steiner's original preps because he found it difficult to convince American farmers of the validity of having to potentize the various ingredients by stirring for an hour, let alone bury cow horns, oak bark, or skulls. In his starter he maintained the preps were already correctly potentized by stirring.

Dr. Pfeiffer at Threefold Farm, Chester, New York in the 1950s.

Only a select group of dedicated biodynamic enthusiasts continued, unostentatiously, to use the original preps, stuffing and stirring as Steiner had recommended, while Pfeiffer launched into the large-scale whole-sale reduction of city refuse.

By 1950 he had approached Tony Dalcino, president of an Oakland, California, garbage-collecting outfit, and offered to translate into fertil-izing humus as much as possible of the four hundred tons a day of garbage collected by Dalcino's trucks from the city of Oakland. To drive home to Dalcino the folly of the present system of handling refuse and the advantages of his offer, Pfeiffer explained how it costs Americans, as taxpayers, billions of dollars a year to cart away as garbage precious minerals and organic material taken out of the soil in the form of food, while it was costing farmers some $7 billion a year to put chemical fertil-izers back into the ground.

For funds to construct a factory to process this material, Pfeiffer convinced the owner of a wastepaper business in Buffalo, New York, Richard Stovroff, by offering him as a carrot the separated wastepaper to recycle. Seven stockholders, mainly paper processors who saw in the garbage the promise of reusable pulp, were persuaded to spend a total of $150,000 to start the Comco Company.

Shortly thereafter a small slate-gray building rose on a peninsula on the edge of San Francisco Bay. In it a hundred tons a day of wilted refuse was fed onto conveyor belts, to come out as compost inoculated with the Pfeiffer starter.

Dr. Pfeiffer's first municipal composting plant, Oakland, California, 1950–52.

As described by a contemporary journalist, lines of heavily laden garbage trucks would rumble down to the Comco factory, jouncing across the refuse-strewn yard to dump their aromatic loads. Tractor plows nosed the muck into piles to be pushed into a long trough leading to the plant. Inside, a pair of giant suction fans hanging over the conveyor belt, like outsize vacuum cleaners, sucked up most of the wastepaper. Huge magnets scanned for metal objects, and ten workers rummaged through the refuse with gloved hands to pull out glass or wooden objects. Properly picked over, what was left got dropped through a chute onto waiting trucks to be carted to a hopper. There enormous steel blades, rotating against stationary blades, chewed it like hamburger, while nozzles showered it with water spiked with bacteria, a tablespoon of microbes to every ton of gook.

Action of the bacteria was immediate. Within two to four days they could multiply 300 million times, with metabolic action so intense the mixture heated up to more than 150 degrees as various strains of furiously procreating bacteria decomposed and digested the garbage, producing enzymes to speed up the digestive process and make possible chemical changes—a weird spectacle against the San Francisco landscape as the mountainous piles actually cooked, throwing off dense clouds of steam.

In less than a week, with decomposition completed, the piles would shrink and cool. As Pfeiffer explained the process: "During the digestive period, food-building bacteria have begun to grow. Their function, as in the life process itself, is to use the decomposed matter to build living organic matter, store up nutrients in the mass to be used by growing plants, changing basic elements so they can be absorbed into plant roots."

Such bacterial life, Pfeiffer explained, is present in virgin soil; but in the garbage compost the concentration is several hundred times greater. "After the first week of violent decomposition, the garbage has ceased to be rotting material and has become stabilized plant food. It has no odor. Actually it repels vermin; and carrion birds will hover around the piles but not venture to alight on them."

The result of all this was that some three weeks after an Oakland housewife had scraped from her dinner plates into the garbage can the remains of a meal, this waste, reconditioned, could be shipped as a sweet-smelling black earth to farms and nurseries anywhere in the country. The Ferry-Morse Seed Company, distributor of the product, also used it to cultivate its own prize grass and flower seeds in Salinas, California. It even sprayed Pfeiffer's bacteria directly from a helicopter over several thousand acres of its own farmland.

Vegetables grown with the converted garbage were found to weigh 25 percent more than those grown with conventional fertilizers, and had as much as three times as much vitamin A. Grain showed a consistently higher protein content. Laboratory experiments showed that Pfeiffer's mixture could restore even sterile sand to vigorous fertility, eventually transforming desert into rich farmland so long as adequate water was available. Organic matter, mineral balance, and essential structure were restored, permitting the absorption and retention of moisture.

The hope was to provide the nation with a cheap supply of natural organic matter to anchor the topsoil and reverse the trend toward a continental dust bowl. "If all the U.S. garbage were processed each year," said Pfeiffer, "we would have about thirty million tons of compost, enough to fertilize ten million acres of land. Garbage dumps would just about disappear."

The head of the Oakland Sanitation Department, Walter F. Gibson, called the plant "a boon to any municipality. . . . Economically sound, it can be operated in any area." The Oakland *Tribune* chided San Francisco for not getting a compost plant, commenting: "California does have its backward cities!"

Lady Eve Balfour, organizing secretary of the Soil Association Ltd. of Britain, climbing gingerly around the hillocks of garbage, singled out the Comco plant as the high spot in her U.S. tour.

But Pfeiffer had bitten off more than either he or the phalanxes of his microbes could chew. Concerted opposition by the producers of chemical fertilizers, worried about losing business, was too tough to digest. Within two years the Oakland Comco company had closed its facilities for good.

Nearly forty years later, no really effective effort has been made to take up where Pfeiffer left off. According to a recent article in the *New York Times*, the next decade's principal agenda for New York City will be the disposal of its garbage, twenty-one billion pounds a year; by 1990, all available landfill space will have run out.

"Like time," said the head of the city's Sanitation Commission as he sat in his seventh-floor office surrounded by an Art Deco collection of miniature trash cans, "like time, the flow of garbage never stops."

Back in his lab at Threefold Farm, gravely disappointed but unflinching, Pfeiffer continued to research, battling for his cause from podium to podium, facing odds strongly stacked against biodynamic farming. Along with the efforts of other organic farmers, he was derided by the powerful chemical companies with their vast assets and overt and covert financial control of agricultural colleges, newspapers, magazines, and publishing houses.

By 1961 Pfeiffer had succumbed, ostensibly to tuberculosis, leaving Margrit Selke to carry on alone. In her good witch A-frame lab, she reminisced about the highlights of her hero's life. Doc, as his co-workers liked to call him, had always been ready, she told us, to help with his knowledge of biodynamics, and would devote as much care to a small backyard garden as to a several-hundred-acre farm, giving instant advice from his cornucopia of experience. Shown a garden with a very high toxic iron content, he would prescribe the Steinerian remedy of planting stinging nettles around the entire perimeter, with the result that the next soil reading would show a 40 percent decline in iron, and stay that way. For an infestation of pea aphids he would suggest, equally successfully, a tincture of green soap, diluted five ounces to a gallon of tap water.

Among his more arcane suggestions was the caveat never to take a rock smaller than a man's fist from the soil, as rocks are a valuable source of minerals, and their loss impoverishes the soil. As an excellent means of controlling pests in a garden, Pfeiffer recommended young turkeys "because they clean up insects without disturbing growing plants, and talk to each other all the time in the sweetest voices, a joy to hear!"

Biodynamicist Helen Philbrick, author of a charming book on companion plants, recounts how Pfeiffer cured for her a peach tree that had suffered an injury to its trunk and developed a large decayed area, which, no matter how carefully pruned and excavated, continued to languish. Pfeiffer, she says, looked closely at the tree, up and down, all around, studied the soil, the hillside, the sky, and the adjoining woods. Presently he stepped out onto the road and examined the drainage ditch and fences.

Beginning to suspect he had forgotten all about her tree, Helen Philbrick was about to remind him when he finally returned and said: "The trouble with your peach tree is that row of fence posts over there."

He must, says Helen, have read disbelief in our faces, followed by our immediate question: "What could fence posts up there have to do with a tree down here?"

With his accustomed patience, and with no suggestion that anyone else's knowledge of biology might be disgracefully limited, Pfeiffer quietly explained: "The fence posts are hosts to white shelf fungus, which you can plainly see. The function of shelf fungus is to reduce decaying wood to topsoil so that it can be returned to earth to start its cycle again. The shelf fungus from the dead fence posts spreads into the decayed wood in the damaged peach tree and the wood thus continues to decay. If you will either remove the fence posts or treat them with crankcase oil to stop the fungus, there, you will save your peach tree, here." His advice being followed, says Helen Philbrick, many good peaches were gathered from that tree for many years to follow.

But Pfeiffer's advice was not always so promptly followed. On the whole he failed to convince his fellow Americans of the tenets of biodynamic farming as taught to him by Rudolf Steiner. So few of his listeners dared follow his prescriptions he began to water them down and failed to talk openly of what BD preps were actually made. Farmers were too fearful of being caught by their neighbors burying cow horns, and too lazy—or overworked—to spend an hour potentizing the 500 by stirring it into water. Also, they had no way to assure themselves of a ready supply of stag bladders, bovine mesenteries, or cows' intestines. So BD was barely kept alive by a few devoted farmers who practiced on the sly without publicizing what they were up to.

For many years those who—for whatever reason—could not handle the preparations for themselves relied on the efforts of a devoted spirit that shone among the hills of Pennsylvania just south of the Poconos, tucked into a petite five-foot-one-inch body with radiant, peach-complexioned features, Josephine Porter. The most industrious and indefatigable supporter of Steinerian biodynamics, she kept the preparations alive for all of a generation, almost single-handedly.

From the middle of the 1940s when she was first apprenticed to Pfeiffer at his Chester, New York, farm until she died in 1985—of emphysema brought on by regular milking of eighty goats to which she was allergic —Josephine took care of producing all the preps. With her tiny gloved hands she collected cow manure from a herd of her own Angus, scrounged cow horns from local slaughterhouses, wheedled stag bladders from friendly neighboring hunters, raised stands of yarrow, camomile, nettles, equisetum, and valerian, drying, pressing, stuffing, sewing with her finely shaped fingers, day in and day out, year in and year out, for a lifetime, until she put a notice on the BD bulletin board at Kimberton Hills and obtained the devoted discipleship of our naval host. Hugh Courtney came regularly to Pennsylvania to learn from her the secrets of Steinerian magic, as convinced as she that in the proper manufacture of the preps lay the healing of America's soil and the future of the country's agriculture.

But, despite their efforts, for years BD in America remained virtually as unknown as the Rosicrucian Ordo Templis Orientis to which Steiner belonged at the turn of the century, a spinoff of the Golden Dawn, with whose warlock leaders, such as McGreggor Mathers and the fabulous Aleister Crowley, few dared to have themselves associated.

CHAPTER 5

Microcosmos

PFEIFFER'S PREOCCUPATION with microbes may have been less morbid than prophetic; for the little creatures— largely responsible for everything that lives on earth—may yet save the planet from destruction.

Not only is the soil their natural habitat: they invented it, as a base for all that lives. Toothless and mouthless, they ingest through their membranes and chew up with chemical action the seasoned elements from the hard, bare rock of a planet they inherited before it could be advertised as earth, laying down their carcasses to produce a living soil of humus in which the earliest plants, the lichens, progenitors of the giant forests and the majestic redwoods of the West, first established a tenuous toehold.

As Charlie Walters, editor of *Acres U.S.A.*, puts it in his Kansas drawl: there are more kinds and numbers of minute livestock hidden in the shallows and depths of an acre of soil than ever walk the surface of that field. The weight of microorganisms busy under grassland is far greater than that of all the large mammals, cows, horses, rabbits, mice, gophers, toads, snakes, birds, grasshoppers, spiders, and other types of animal life that run above it or take shelter in it. A single microbe reaching maturity and dividing within less than half an hour, can, in the course of a single day, grow into 300 million more, and in another day to more than the number of human beings who have ever lived. As computed by Lynn Margulis and her son Dorion Sagan in their brilliant *Microcosmos*, bacteria, in four days of unlimited growth, could outnumber all the protons and even all the quarks estimated by physicists to exist within the entire universe.

37

That we should be so cavalier about these microorganisms is ironic, if not tragic; for the human body may be a direct descendant of the first single-celled bacterium to inhabit planet earth three and a half billion years ago, inventor of life's miniaturized chemical systems, a feat unrivaled in the universe as we know it. Man's body (and woman's too), along with every organ in it, is composed of quadrillions of animal cells and a hundred quadrillion of bacterial cells. The intestinal tract is lined with vital microbes that alone enable food to digest. Microbes proliferate throughout the length of the intestine. And every inch of skin teems with friendly but unseen creatures by the billion. Though we kill them religiously with soap, left alone, they clean the skin as well as any cream produced by Arden or Weleda.

As master chemists, bacteria have transformed the planet from a cratered moonlike terrain of volcanic glassy rocks into the fertile globe we know, with its entrancing landscapes upon which we walked so freely until the advent of the petrochemists.

Life, in its basic form of microbes, has been a companion of the earth from shortly after the planet's inception. And so close is the vital bond between the environment of earth and the microscopic organisms thriving upon it, it is virtually impossible for biologists to give a concise definition of the difference between that which is living and that which is not, or to tell whether microscopic algae—an ancient, ubiquitous, perennial source of life—are truly animal or plant.

Within the soil, procreating in high concentrations, bacteria ensure fertility, recycling the elements through the chemical laboratory that constitutes their bodies, making them available to plants. Nitrogen and carbon aren't alone in requiring the help of microbes before they can be rendered fit for plants. As nitrogen is converted into nitrate, phosphorus is turned into phosphate, sulfur to sulfate, chlorine to chloride, boron to borate, molybdenum to molybdate, and so on through the elements, thanks to microbes.

While a minority can subsist on mineral inorganic detritus, or rock dust, the majority feed on organic compounds, degrading organic molecules in the soil deposited from plant and animal tissues, recycling dead cells into mineral substances in a solution that can readily be reassimilated, first by plants, and so on, up the ladder of life.

Microbes first attack the substances that decompose most readily, such as sugars and cellulose. When these are used up, most of the microbes die, making up with their bodies half the total of the soil's organic matter; and a staggering number of dead microbes are decomposed and consumed by other microbes, in never-ending cycles. The oxidation of plant tissues being incomplete, lignin, tannins, fluvic acid, kerogen, and waxes—which resist the action of the microorganisms—are formed into a humus that undergoes a slower degradation, conferring to the soil its hydrophilic, water-loving capacity, its colloidal structure, and its resistance to erosion.

Animal excretory products, such as urea and uric acid, are transformed by brigades of bacteria working in relays into ammonia or ammonium salts, which are then converted by other bacteria into nitrates.

Without the bacteria's ability to capture nitrogen directly from the air, the earth's life forms would long since have died from nitrogen starvation.

Beer, wine, bread, and cheese would not exist without the intervention of the microbe, which makes of man as great a scatophage as is the microbe. For alcohol and carbon dioxide—as in a cooling *Spritze*—are but the excrement and exhaled breath of living microbes.

So varied is microbes' alchemy they can convert corn-steep liquor— a waste product—into penicillin, or dead cuttlefish into perfume. And they reveal, as shown in Brian J. Ford's highly readable and informative *Microbe Power*, an astonishing level of over-production: the B-2 that microbes produce is *ten thousand times* as much vitamin as they require for their own metabolism.

Coal, limestone, and iron ore, three basic ingredients of an industrial society, have all been bequeathed to man by microbes. It is possible that all the commercial oil fields are derived from the remains of ancient, extinct microbic diatoms. Fossil beds contain billions of cubic feet of microbes. And sulfur is another element we owe largely to bacteria. Texas deposits, which yield some 90 percent of the world's sulfur, were converted by bacteria from calcium sulfate, the substance known as gypsum or plaster of Paris.

Eventually the total amount of work to be done in the soil by what Walters calls the "farmer's unpaid workers," is divided up among the microbes. Some collect nitrogen from the air. Some serve as scavengers. Some release ammonia from protein substances. Some change the ammonia to nitrite and to nitrate. In the struggle for existence, some find it simpler to feed on the dead bodies of their fellows than continue to synthesize new supplies of food from purely inorganic materials. Others manage to attach themselves as parasites to other living microbes, which serve them as willing or unwilling hosts.

Margulis and Sagan are certain that all visible organisms have evolved through symbiosis from their invisible predecessors, their coming together leading to mutual benefit through the permanent sharing of cells. They are convinced that all existing photosynthesis—which they call undoubtedly the most important single metabolic innovation in the history of life on the planet—first occurred not in plants but in bacteria. This harnessing of discrete particles of light to reduce carbon to its energy-rich form, a photo-chemistry that is still not thoroughly understood, evolved, they say, first in bacteria, and only later in plants and algae. Even the minute green chloroplasts within the leaf cells, which actually trap the radiating solar energy, may originally, according to Margulis and Sagan, have been independent microbes of a sort, enslaved and put to work within the plant for its own benefit.

Photosynthetic bacteria appear to be vestiges of ancestors that played an important role before cyano-bacteria appeared and oxidized water to produce the free oxygen they learned, and then taught us, to breathe. "The production of food and oxygen from light were to make microbes the basis of a global food cycle that extends to us today; animals could never have evolved without the food of photosynthesis and the oxygen in the air."

Now plants carry out the greater part of photosynthesis on earth, while most bacteria, and to a lesser degree some fungi, ensure the conversion of wastes from all living organisms into mineral substances available to plants. Thus the biosphere is all of a piece, an immense, integrated, living system, an organism in itself, comprising man, beast, plant, worm, and microorganism, each animated by a different facet of the one.

And, if microorganisms built the soil, the real tiller and fructifier of the soil is not man, but worm. "It may be doubted," wrote Charles Darwin, "whether there are many other animals which have played so important a part in the history of the world as have these lowly organized creatures." All the vegetable mold of England, Darwin concluded, passed through, "and will again pass many times through, the intestinal canals of worms."

Impaling this frantically wriggling body on a fishhook or slicing its spread-eagled body on the dissecting table of some retarded, superannuated high school biology class may be the closest most people ever get to an earthworm: unjust, unfair, and unwarranted retribution against what may be man's most useful ally in his struggle to survive, considered by Darwin the greatest plowman, an animal of greater value than the horse, relatively more powerful than the African elephant, and more important to man than even the cow.

Despite his classic work on the earthworm, *The Formation of Vegetable Mould Through the Action of Worms with Observations on Their Habits*, written shortly before his death in 1881, and despite a comprehensive bibliography of research papers on earthworms prepared for the Darwin Centenary Symposium on Earthworm Ecology, containing two thousand references covering the period 1930 to 1980, prepared by Satchell and Martin in 1981, the earthworm has been grossly neglected and cruelly mistreated in the course of modern agriculture practice.

Even Darwin missed the prime asset of the worm: that within its digestive tract it incubates enormous quantities of the microorganisms which, in its castings, become the base for fertile humus.

It took a French scientist and ecologist, André Voisin, author of the insightful *Soil, Grass and Cancer*, to point out that the earthworm, and in particular the slippery lumbricid, most common in the United States and Europe, is not only essential to good agriculture but is the very foundation of all civilization. In *Better Grassland Sward*, Voisin traces man's civilizations in relation to the distribution of active earthworms, of which he lists some three thousand species.

Among the most ancient of terrestrial animal groups, several hundred million years old, they come in various colors and sizes: brown, purple, red, pink, blue, green, and light tan, the smallest barely an inch long, the largest a ten-foot giant in Australia, though South African newspapers reported a boa-constrictor-sized monster twenty feet long, a yard wide through the middle. The most common European and American earthworm, *Lumbricus terrestris*, grows barely longer than six inches.

Ten thousand years ago, immediately after the last ice age, the lumbricid earthworms were to be found only in certain restricted areas of the planet, such as in the valleys of three great civilizations—the Indus, the Euphrates, and the Nile—where crops grew almost without cultivation in a soil of immensely fruitful richness.

As Jerry Minnich points out in *The Earthworm Book*, other areas of the earth offered ideal climates and rich soils, but produced, with the exception of China, no such civilizations. The Egyptian experience alone, says Minnich, is strong indication that a complex civilization cannot develop until the basic agricultural needs of its people are met, and that requires the earthworm.

Not that the point was entirely overlooked by the USDA. An agricultural report on investigations carried out in the valley of the Nile in 1949, before the folly of the Aswan Dam, indicated that the great fertility of the soil was due in large part to the work of earthworms. It was estimated that during the six months of active growing season each year the castings of earthworms on these soils amounted to a stunning 120 tons per acre, and in each handful of that soil are more microorganisms than there are humans on the planet.

Thirty years before the birth of Darwin, as the American colonists were breaking away from the mother country, an English naturalist, Gilbert White, was writing:

Worms seem to be the great promoters of vegetation, perforating and loosening the soil, rendering it pervious to rains and the fibers of plants by drawing straws and stalks of leaves and twigs into it; and, most of all, by throwing up such infinite numbers of lumps of earth called worm-casts, which being their excrement, is a fine manure for grain and grass. . . . The earth without worms would soon become cold, hard-bound, and void of fermentation, and consequently sterile.

That the phenomenon was understood before the time of Christ is clear from Cleopatra's decree that the earthworm be revered and protected by all her subjects as a sacred animal. Egyptians were forbidden to remove it from the land, and farmers were not to trouble the worms for fear of stunting the renowned fertility of the Nilotic valley's soil.

In the northern part of North America the last ice age so stripped the country bare of earthworms that in very few areas of what is now the United States were agricultural lands rich enough to support even moderately large populations of native American Indians. As Minnich says: "Before European contact, the only lumbricids native to the United States were some lacy species of *Bismatus* and *Eisenia*, essentially worthless as soil builders."

But wedged in the shoes of the colonists' horses were tiny lumbricid egg capsules, and in the root balls of European plants immigrant earthworms arrived to remedy the situation. In no time a rich but dormant soil was transformed into one of high fertility. The lush meadows of New England, the vast farmlands of the upper Midwest, the great wheat fields of Canada are all attributed to the introduction of the earthworm.

By the early part of the twentieth century, says Minnich, New Zealand soil scientists observed that European lumbricids were making vigorous inroads into the island's previously wormless soils. Hill pastures that could barely support a stand of grass were gradually becoming lush and green even though no fertilizer was applied. Counts of earthworms ran as high as over four million per acre, more than three times the maximum populations of the same species in their Old World habitats. The source of all this fertility was what the worms excreted in the form of castings, compost of the highest grade, containing mineral and organic matter in a soluble form, excellent as both a fertilizer and as a soil conditioner.

Earthworms can produce more compost, in a shorter time, with less effort, than any other method. As they burrow, they are constantly bathed in mucus, which helps them through the roughest ground. Continually rubbed off, this mucus helps cement the walls of their tunnels.

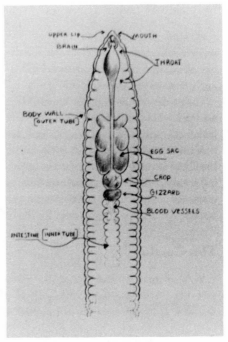

Internal organs of the earthworm
(*Lumbricus terrestris*). From *The
Earthworm Book* by Jerry Minnich.

And, while it helps a worm worm its way out of a predator's grasp, it also helps hold the soil firm, retaining moisture as it hardens.

In classical Greek times, Aristotle called the earthworm "the guts of the soil" because it produces particles that are smaller than when they enter, held together by the intestinal fluid that makes for·a finer-structured earth. An omnivorous and unfinicky eater, the eyeless earthworm ingests whatever appears before it in morsels fit for its toothless gums.

Muscularly pumping through the soil, it ingests not only organic matter but the raw earth itself, using sand and other mineral particles as grinding stones in its gizzard. Mixed in the crop with digestive chemicals and disintegrator bacteria, the elements come out in different combinations, more easily taken up by plants.

Worm castings, neutralized by constant additions of carbonate of lime from three pairs of calciferous glands near the worm's gizzard, and finely ground prior to digestion, are five times as rich in available *nitrogen*, seven times as rich in available *phosphates*, and eleven times as rich in available *potash* as anything else in the upper six inches of the soil, producing a nutrient in just the right condition for the plant to absorb. Real organic NPK! What's more, the castings are always more acidically neutral than the soil from which they were formed, naturally improving the local pH factor as armies of earthworms work to keep the soil in balance, neither too acid nor too alkaline for the growth of plants.

Could it be that these great sinusoid fertilizers actually transmute elements, as the French savant Louis Kervran would have it, or are they merely collecting, distilling, and rearranging them to fertilize the soil? The former would appear to be more likely.

Castings, usually deposited in old burrows, or by night crawlers on the surface when they come up to mate or draw leaves into their burrows, consist of about one-third of the contents of the worm's intestines, in pelletlike form, and have a third more bacteria than the surrounding soil.

Even when ample organic matter is available, earthworms consume large amounts of soil, and by mixing the two produce a rich humus, perfect in texture, with more plant nutrients than in the material from which it was derived. Castings contain a higher percentage of aggregates than is found in the surrounding soil—aggregates being the formations of individual particles of sand, clay, and silt, grouped into larger units, which help make a crumblike structure of the soil.

An earthworm is said to produce its own weight in castings each day it is on the prowl. Henry Hopp of the USDA estimates that one acre of good agricultural land can produce well over five tons of castings in a year, or more than 5 percent of the total soil volume to plow depth. In the process of producing its castings, on even an ordinary agricultural soil, earthworms are credited with turning more than fifty tons of soil per acre, and in the Nile Valley as many as two hundred tons, into a fructifying base.

Earthworms are prodigious diggers and earth movers, capable of burrowing down as deep as fifteen feet. They can squeeze between and push apart the soil crumbs, and one worm alone can move a stone fifty times its own weight. As they burrow, earthworms mix and sift the soils, breaking up clods and burying stones. Some carry down leaves and other organic matter; others bring nutrients and humus to the top. Tunnels held together by their mucus afford planted roots quicker avenues into the soil. And the mucus, forming humus, prevents erosion. Henry Hopp says these materials, once dried, do not dissolve again in water. Yet, while the soil thus treated holds the required moisture, the burrows drain superfluous water. Experiments have shown that soils with earthworms drain from four to ten times faster than those without. Conversely, in light sandy soils, where water tends to run straight through to the subsoil, the aggregates produced by earthworm castings act to improve the retention of water.

By digging into the subsoil, loosening it, and threading it with tunnels, earthworms gradually deepen the topsoil layer. By ripping up fine mineral particles and depositing them as castings on or near the surface of the soil, they are constantly adding nutrients to the zone in which plant roots feed, delivering mineral substances that would otherwise remain largely unavailable to most plants.

With their mixing, digging, burrowing, fertilizing, and humus-making activities, the worms have an immense impact on the soil, its texture, its fertility, and its ability to support everything that lives in or on it, especially plants that form the basis of our food supply. But the worms must be fed, proliferating in direct proportion to the amount of organic matter incorporated into the soil, a supply which must be kept up so long as one wishes to retain the earthworms. *Eisenia foetida*, a red manure worm that inhabits compost heaps, turning animal manure into sweet-smelling humus, grows to five inches, but cannot live without copious amounts of decaying organic matter.

Night crawlers, so named because they creep about at night on the surface of the earth, feed on leaves, which they drag down into their burrows, and even with their pinhead brains they have the wit to pull them by the narrow end—which shows more wit than the leaf-gathering suburbanite who regularly spends a fortune to deprive the earthworm of his autumnal fare.

In an orchard, during the three months of autumn, earthworms can dispose of 90 percent of the fallen leaves, dissolving even tough material such as stems and roots. Darwin, who reported seeing burrows plugged with twigs, bits of paper, feathers, tufts of wool, or horsehair, claims that worms, though congenital scatophages, showed a predilection for celery, carrot leaves, wild cherry leaves, and especially raw meat, including fat. Minnich reports that one Wisconsin commercial raiser of earthworms even chose to feed his charges ice cream as a treat on Saturday nights.

More surprising still is his report that a German researcher, C. Merker, writing in the 1940s, astounded fellow scientists by asserting that earthworms have voices, and can actually sing, their faint sound being "rarely in a solo number, but generally in series marked by a definity and changing rhythm." Dr. Merker claimed to be able to hear the sounds when within twelve feet of the worms, sounds produced not by chance but by the deliberate opening and closing of the earthworms' mouths.

How this could be, when earthworms have no lungs—breathing through the whole surface of their skin, moistened to dissolve oxygen, which is pumped through the bloodstream by five sets of double hearts in rings or segments close to the head—is all the more amazing.

A cleric contemporary of Darwin complained that earthworms are also "much addicted to venery." In suitable weather, night crawlers can spend a goodly portion of their nocturnal activities in the pursuit of sex, even an entire night coupled to a willing hermaphroditic mate, each possessing both male and female organs. With the undersides of their bodies held firmly together by tiny bristles, or setae, they lie with their heads pointing in opposite directions, touching in the region of the sper-mathecal openings, where the clitter—a white band a third of the way down their bodies—touches the surface of its mate.

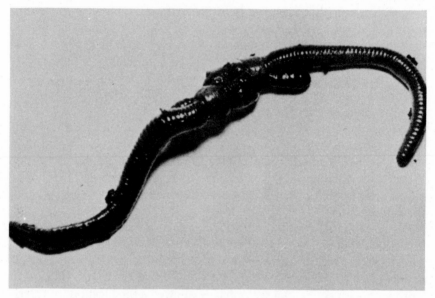

Mating earthworms. As hermaphrodites, they fertilize each other to produce more worms to fertilize the soil. From *The Amazing Earthworm* by Lilo Hess.

They copulate by exchanging sperm cells stored in cuplike hollows in the ninth and tenth segments, exuding a special mucus from the sexual region to protect the spermatozoa being mutually exchanged. More mucus secreted by the clitellum forms a jellylike ring, which picks up the worm eggs from ovaries and sperm cells from testes, slipping the ring off the body, to form a tiny yellow cocoon. Greatly enlarged, it looks like a lemon and contains scores of fertilized eggs, which can be found in the soil during the warmer months of winter. Under good conditions, an average red worm can produce from 150 to more than 200 young ones annually.

One of the principal functions of the earthworm is to consume avail-able mineral nutrients, and, by actions of enzymes in their digestive tract, render them water soluble, easily absorbable by the root hairs of plants, to be made available in turn to the cells of plants, animals, and man.

As Voisin points out, without earthworms there would be no civilization. But Minnich complains that, with the single exception of Dr. Henry Hopp, the attitude of USDA scientists, along with that of many of their associated colleagues in subsidized state universities, has traditionally been negative toward the earthworm.

> They have long begun with the assumption that earthworms are just one more facet of the "unscientific" cult of organic gardening and farming, and that this method of growing crops is antithetical to the "modern" methods of agriculture, including its principles of heavy chemical treatment, monocropping, and other facets of maximum-profit agribusiness. The earthworm, thus judged guilty by reason of association (with organic methods), the USDA has long discouraged serious investigation into the possible benefits of earthworms in agriculture, and has even gone so far as to denigrate or ignore the work of other researchers who have revealed such benefits. Since the USDA has either conducted or influenced the great bulk of agricultural research in this country during the present century, its position on any facet of agriculture or horticulture has broad, far-reaching, and determining effects on both scientific direction and public attitudes. . . . The USDA will sponsor no significant earthworm research, and its long tradition of ignorance is the chief reason why we know so little about earthworms, and why we have failed to utilize their power throughout the present century.

The seriousness of the situation was recently emphasized by Marcel B. Bouche, Secretary of the Soil Zoology Committee of the International Society of Soil Science, in his foreword to Dr. Kenneth E. Lee's last word on *Earthworms*, a book published in 1985 by Academic Press, which for the first time places the worm on a world-wide scale in the economy of nature.

> Humanity [writes Bouche] knows little about its most important commensals. We are unaware of the nocturnal, hidden, subterranean activity of the most important animal biomass that shares with us the earth's land surface. . . . Using increasingly powerful physical and chemical methods, we decide to remodel the landscape, to disturb the soils, to pulverize chemicals, to release fumes and waste water . . . ignoring the principal animal that inhabits the environments we alter. . . . If we compare, for example, the significance accorded to ornithology and the multitude of birdwatchers studying about one kilogram of birds per hectare, with the extremely limited number of research workers' interest in the hundreds of kilograms or tons per hectare of earthworms, we must conclude that our knowledge of ecosystems is funda-

mentally distorted by our above-ground, visual perception of nature and our ignorance of life below-ground.

Normally healthy and long-lived, earthworms are discouraged if not killed outright by many pesticides and most chemical fertilizers. Copper sulfate, in concentrations near the surface of the soil, even in only 260 parts per million, can drastically reduce the worm population, and any nitrogenous fertilizer will quickly wipe them out. Nearly all commercial brands contain high levels of nitrogen in the form of ammonia, which destroys earthworms by creating intolerably high acidic soil.

Yet the more organic material they receive the faster they proliferate. And, as they proliferate, so do their symbiotic progenitors, the micro-organisms, manufacturers of humus, the basis for a fertile soil. Steiner's premise was basic: that his biodynamic preps create the ambience for the infusion of the essential cosmic and telluric forces that generate this metabolic miracle.

CHAPTER 6

Miracle Down Under

WHILE ON THE vast continents of the Americas biodynamic agriculture seems hardly to have progressed beyond its introduction fifty years ago, recent rumor had it that on the world's smallest continent, a hemisphere away, it had successfully spread, in the same half century, to over a million and a quarter acres. To find out if this was so, and how it could have come about, we set off "down under" to investigate.

At the Melbourne airport we were met by a short, almost electrically wiry man with piercing brown eyes of a Slavic cast almost unadorned by eyebrows, Alex Podolinsky, founder of the Biodynamic Farming Association of Australia. Descendant of an ancient Russian-Lithuanian line who took refuge in Germany after the great October Revolution, Podolinsky was raised in Bavaria and educated in England and Europe, where he became exposed to the philosophical tenets of Steiner's peace-loving anthroposophy. Caught in Germany when the war began, as a stateless person, he was dragooned into the German *Wehrmacht*, a slave-soldier to the Nazis.

Following the Normandy landings, he was so badly injured by a bomb that his spine was permanently damaged. But somehow he survived —as if chosen to fulfill some higher function.

In Australia, in 1949, Alex realized that the essence of the semidesert island-continent on which he had landed lay not in its crowded cities but in the wide expanses of its endless acreage, and that his real calling was to work with nature.

49

Alex Podolinsky in 1986.

His start in agriculture came when a wealthy friend with a small rundown farm at Wonga Park, Victoria, a tiny lumbering center east of Melbourne, offered to loan him the farm for a year to see what could be done with its soil.

There Podolinsky began putting into practice the BD knowledge he had picked up in Germany, and to think about how it could be specifically adapted to Southern Hemispheric conditions. These included an intensity and quality of light unknown in northern Europe, short and long periods of severe drought, and soils of such fragility and shallowness that the dominating vegetation on the eastern part of the huge land mass is mainly represented by eighteen hundred varieties of tall "gums," or eucalyptus, which predominantly populate the forests of the island continent.

On the way from the airport Podolinsky spoke of the older farmers, who had noticed what was happening to their land, who could recognize the indicators. "In the 1920s, they started putting on superphosphate and had a tremendous result in clover which spurted ahead. Over the next decades, one bag of phosphate did less and less until finally five bags couldn't do the same. In irrigation country twelve bags couldn't do the trick. So they went into potash which after two years no longer had any effect. Next they tried nitrogenous fertilizer which gave a big boost in growth, seriously sickening their soils. Disheartened, and not knowing what to do, a lot of them turned to BD, not because of any anthroposophic or Steinerian philosophy, which in any case they couldn't understand, but as a matter of clear common sense."

With an absolute, almost dictatorial self-assurance, modified by a nurturing instinct, a desire to do right by the world, by plants, animals, humans, and above all by the soil of Mother Earth, Podolinsky made his point.

"With the pasture that grew on my first little farm you couldn't have fed two head of cattle." His clipped precise accent could pass for pukka English. "No compost could be made. Instantly I realized that what kept agriculture going in the 'old countries' of Europe was the cow; but in Australia there were no cows, only kangaroos, whose droppings gave no such fertility as exists in cow manure. So I sensed that what cow manure had done in Europe over thousands of years would here in Australia have to be replaced with the 'cow power' of Steiner's 500 to act as an igniting point, to achieve the same result quickly, giving the land a 'new impulse.' "

Over the past thirty-five years Podolinsky's biodynamic practice has expanded across the Australian continent, until now there are hundreds of hard-bitten, practical Aussies successfully farming his way, on all types of soil, coast and inland, elevated or at sea level, even though he spent not a single hour in publicly promoting, publicizing, or otherwise devoting to the method any form of salesmanship.

By word of mouth the biodynamic message spread from the first man who saw it working on the Podolinsky farm in the 1950s until today his association, to which only farmers are admitted, covers all of Australia. "It took resolution," says Alex. "In the early days people had very grave doubts."

For years Steinerians, or rather Podolinskians, stuck to their novel practices, doing their work unheralded, seeking no publicity and avoiding any salesmanship. Today they are speaking out. They're successful and confident. And they're growing in number.

What broke their seclusion was a thirty-eight-minute prime-time feature film on Australian television, A Winter's Tale, broadcast nation-wide in 1985, as one of a more than decade-long and highly popular series, Big Country. Opening with a haunting scene of Podolinsky on his tractor spraying 500 on a moonlit pasture with an owl hooting in the background, it went on to show Alex professionally stuffing and burying cow horns, communing with his herd of cattle in a lush and fertile landscape to which not a speck of chemical of any sort had been added in over twenty-five years. The film, unsolicited by Podolinsky but insisted upon by producer-director Paul Williams of the Australian Broadcasting Company, caused a sensation. No other single program in the history of the ABC, either on radio in the early days, or later on television, ever received as much response as the one finally made as a result of Williams's trip to Podolinsky's farm near Powelltown. Soon after the broadcast, the station got over six thousand letters, mostly handwritten, the largest number from farmers bursting to know more about what they'd seen and what many of them characterized as "just plain miraculous!"

Dusk was falling in mid-July, in the depth of Australia's winter, when we pulled into the drive of Alex Podolinsky's modest farmhouse in Powelltown, where he has lived twenty-four years and raised seven children. Nestling in a glen on a branch of the Little Yarra River, no more than a stream, the house is surrounded by small wooded mountains, reminiscent of the Welsh-English border or the foothills of the Poconos in eastern Pennsylvania.

A damp fog hung over the secluded valley, and a chilly though snowless atmosphere enveloped the house, permeating it except for the kitchen where Alex had prepared a meal of roast beef cut into slabs simmered in vegetable broth.

"BD carcasses," said Alex with gusto, "have much more meat on them and less fat than the average. And a far better taste. It cuts out better, and has none of that fatty 'marbleization' so prized in the States."

Up before dawn, like any dairyman, to milk his cows, year after year Alex had evidently established with his animals a relationship of a sort not normally granted to less sensitive herdsmen.

"If I'm not in a good mood in the morning," Alex shouted cheerfully, "I very soon am, once I get near my cows. They tell me exactly what I'm like, and they notice cosmic happenings. A cow relates to its total environment. To live in harmony with one's surroundings like that requires a purity, like the purity of a child." As he spoke, a roan Jersey looked over the fence at us, placidly chewing her cud, her limpid eyes full of peace.

Waving toward the cow's horns, Alex continued his monologue. "See that cow! Her horns are on her, so to say, to hold in what's taking place in her magnificent cosmos of digestion and metabolism. Deer have antlers reaching out, exactly the opposite from cow horns in expression. Unlike cows, the deer constantly communicate with nature in a state of alertness, with a certain amount of fear in the background, the exact opposite from a cow's placid demeanor. A mature cow has horns, and if they are later cut off, that polled cow will 'go dull,' and never be the same as when she was horned."

Alex turned to explain: "When one looks into such a cavity, washed clean, one gains the impression that if one fell into it one might never get out again."

Looking up, he pointed to the roan Jersey milk cow. "See that animal? In comparison with other ruminants, such as sheep or goats, which also have several stomachs, she typically just stands there, chewing her cud. Her legs are not meant for galloping about like a gazelle or a horse. She is steady, calm, placid, and harmonious. Her digestive system is a veritable cosmos in nature, the most refined on earth. The manure that comes out of her is up to 25 percent microbes."

Under a watery sun Alex walked to his acre-sized vegetable garden to demonstrate the difference between the land as he had found it and what it is now. Grabbing a spade, he drove its sharp blade a quarter of the way into a patch of resistant uncultivated ground bordering the garden. With a sharp crunching noise he revealed a sample of black sandy soil so loosely knit it could have been poured from the mouth of a narrow-necked bottle. Moving a few steps into the garden, Alex sent the spade full length, this time silently, into earth as spongily elastic as a mattress, to turn up a mass of soil the color of chocolate.

"It's the *humus* that gives it that richness," he said with satisfaction. "In Latin the word means 'earth' and gives rise to the adjective *humilis*, or humble. It's mostly a colloid, a substance like jelly, between a solution and a suspension, that can be carried in the hand without running through the fingers. The prime characteristic of humus is that its elements, which neither leach out nor evaporate, are solubly available to plants. Humus can hold water to 75 percent of its volume—water filled with essential minerals, available to plants in a balanced colloidal form."

Alex poked with his spade. "We've had lots of rain here recently, in ten-inch, not one-inch, amounts. Yet you see it hasn't water-logged the soil. In the garden, there is over 15 percent of organic matter, maybe up to 18 percent, whereas in the non-biodynamically-treated strip into which we first dug, there is less than 1 percent. All thanks to the application of 500, key to food production in the future. Its concentrated life force is so powerful a tiny speck will enliven dead soil."

He picked up a clod of earth from which a mass of thick plant roots protruded. "Note the white hair roots healthily growing even in the coldest part of winter. They're only a few days old, but very active. These feeder roots are different from their pipelike companions, which suck up water for transpiration and are the ones you'd mainly see in chemically-fertilized ground. These feeder roots are the key to biodynamic success because, symbiotically, it's not only the *soil that makes the plant, but the plant that makes the soil*. Roots and microbes live and die together, and together making humus they breed new life."

Strolling over to a shed he picked up a pint-sized glass jar. "Humus and roots are made for each other. A really perfect match. At the height of spring growth a jar full of humus is buried in a biologically active working pasture about four inches below the surface. Six weeks later, the brown spongy substance has completely disappeared and the jar is solidly packed with nothing but white hair roots that can be extracted from it only with great difficulty. Once pried out, they appear as spankingly clean as if they had been thoroughly laundered and dried. The roots have invaded the jar to ingest every smidgin of the humus, not just the soluble elements within it, clearly illustrating that humus is plant food of first priority and choice."

Pitching the jar, Alex walked around the shed, as if lecturing to a chemistry class: "In another jar, I sealed humus airtight for three days. Cut off from its oxygen supply, the stuff turned into water-holding globs of putrid, evil-smelling green particles. This kind of half-rotted matter, proliferating with harmful anaerobic pathogens—those thriving in an oxygen-deprived milieu—is what makes for dying soil. In living soils, the pathogens are neutralized by *aerobic* microbes. Sensing this, cows will readily drink from pails in which a quantity of humus has been liquefied, even though that humus was derived from their own manure, whereas they will not dare to taste water infected with manure in its still raw form."

Behind the shed, a plot of ground about sixty by seventy feet was covered with galvanized roofing and plastic sheeting: a weather-protected winter resting place for a staggering number of cow horns—160,000—all stuffed and buried by Alex with the help of a single hardworking assistant.

To accent the time and tedium involved in this arduous manual labor, Alex chuckled. "Not only does each horn have to be filled, it also has to be emptied. That's a lot of horns: 160,000 on my farm alone! The biggest lot interred in any one place in Australia, though many other people are now burying their own in other parts of the country. One of them, Max Chandler, with sixteen thousand acres near Monto in New South Wales, just delivered thirty bags of huge horns from an inland cattle breed. They're nearly twice as heavy as any I've ever seen. Beautiful! Way out west across the continent people are gathering them up. One of our members who goes all over the place collecting wild flowering plants has provided a lot of horns. And another, a sugar-cane farmer in northern Queensland, has put together a whole ton over recent months. We'll need every one of them and a whole lot more. Since the film was broadcast, many farmers, some of them growing up to twelve thousand acres of wheat, are asking to join with us, and we'd fail in our responsibility if we let them equip themselves to make the changeover, then be unable to supply them with sufficient 500 for their needs."

Was it, we asked, the shape of the cow horn that was somehow affecting the microbes through a process of mutation, or was it merely the burial process, which effected the alteration of manure into the product 500?

Alex hesitated. "It's the 'earth pull' to which the horns are subjected. At the right time of year, when vegetation is dormant, the earth has a strong pull. The process takes place, strangely enough, when the soil temperature is lower than 42 degrees Fahrenheit, so cold that no microbes could be active. This cold area of Victoria is the best I've seen anywhere in Australia." Then he added with finality: "As for the cow horn, it was most strange to those listening to Steiner when he first suggested its use; but all subsequent experiments have proved him out. We have performed tests with the same manure, on the one hand put into a cow horn, on the other into a wooden box, porcelain mug, or other receptacle, side by side in the ground, and only a few inches apart from one another. But the manure in the other containers never converts the way it does in the cow horn. Not that Steiner," he added with a smile, "didn't say to go out and experiment with other methods. He was only leading the way. He wanted the rest of us to look, and see, and try on our own."

On the way back to the shed, Alex told how he had once met a highly intelligent peasant, Ernst Jacobi, who had been one of the first to establish a successful biodynamic farm in southern Germany in the 1920s.

"Jacobi grew marvelous vegetables," said Alex. "Farouk of Egypt became one of his customers. The king invited Jacobi—a simple, gaunt fellow, and a real philosopher—to Cairo to set up BD gardens for the royal estates. And the king was very pleased."

Inside the shed, Alex displayed the vats in which the 500 is stored when it comes from the horns: cylindrical tanks put into cubic boxes, surrounded by peat moss and covered with lids some three inches thick that do not close tightly but allow air to circulate through a space no wider than a stove match.

In the tanks the 500, almost black, was moist to the touch. There were holes in the bottom of the bins to allow aeration from below, and a layer of empty cow horns had been placed at the bottom before the 500 was added, further to allow aeration.

"It's very important," said Alex, echoing the warnings of our Virginia expert, Hugh Courtney, "that the storage boxes with their tanks not be put close to any foul-smelling source of pollution, such as a jerry can or a pump for gas or diesel fuel, or near any source of electricity."

On a 3½-by-10-foot table lay enough 500 to regenerate a thousand acres of dying soil. A tiny lump of the blackish stuff between Alex's fingers, about an ounce and a quarter—enough for one acre—looked not like earth, nor putty, nor the heart of a newly-baked loaf of Russian black bread, but like a combination of all three, pliant to the touch and smelling sweet.

"To use a negative analogy, it's like a speck of matter in atomic fission," he said. "It can cause an enormous though much quieter explosion."

To activate it, Alex mixes a little over a pound of the 500 into fifty gallons of water, which he stirs automatically with a specially-built machine. He says that stirring, to be more effective, should be done under the open sky rather than under the roof of a barn or shed, that during the stirring a suctional force is created in the vat by the whirlpool-like vortex, which Alex explains as drawing in "cosmic power."

"It's like planets being drawn in centripetally by the force of the sun, which allows them to remain in their orbits and not drift off into outer space. As soon as a good vortex is created, the machine abruptly reverses itself, completely destroying the vortex, leaving a foamy confusion and chaos in the water before setting up another eddy whirling in the opposite direction."

Alex Podolinsky setting up an automatic stirring machine for BD 500 on a farm in eastern Australia.

Alex emphasized that the chaos between the alternating vortices was a most important factor. To the Greeks, he explained, chaos was the first state of the universe, from which cosmic order and harmony evolved. "Though we don't always recognize it, chaos applies to our own creative lives. Every creative artist is in a state of chaos before he brings whatever his creation will be into existence. He is thus, in relation to his environment, in an unbearable state in which something is given birth!" The chaotic change of state to which he refers can also be seen in such occurrences as the sudden collapse of a bridge spanning a river. Largely unpredictable, they lie outside the normal dictates of scientific lore.

The next day, in the early-morning sunshine, we set off in Alex's Peugeot 505 D on a nearly six-thousand-mile trek to visit, one by one, more than fifty of Alex's biodynamic farmer adherents on their widely

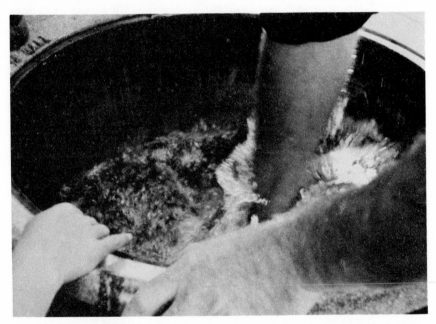

Hand stirring the BD 500.

differing farmsteads. The journey was the equivalent of a wintertime trek from northern Maine to southern Florida, with side trips cutting in and out of the Appalachian Mountains.

The trip was to take us from Gippsland in the southeastern quarter of Victoria to the valley of Australia's largest river, the Murray; over the Blue Mountains west of Sydney; up the coast to Port Macquarie; onto the four-thousand-foot-high New England plateau in New South Wales; back over the Great Dividing Range, which runs north-south along the eastern edge of the continent at an average distance of fifty miles from the sea, then up into tropical Queensland, more than double the size of Texas, with inhabitants twice as ornery.

On the first leg of the trek the road wound through a forest then out into open farming country toward the Poowong farmstead of Trevor Hatch, one of the very first members to come into the association, and the man whom Alex considered the one who would eventually lead the association.

In his mid-forties, lean and as strong as a wrestling champion, Hatch is owner of a 450-acre spread, a beef-cattle-raising and potato-growing operation, bought by his father in the early 1950s when the land was considered fourth class. The older man had cleared it almost completely of "bush"—meaning a heavy forestation—leaving only copses or clumps to dot the landscape. But the soil had been the poorest grade of whitish clay from less than an inch below the surface.

To show an example of this aboriginal dirt, Trevor walked to the base of a large eucalyptus tree, where he forced a sharp-bladed shovel into the ground. There was a crunchy sound even louder than at Alex's, and up came an example of totally lifeless soil. Trevor then headed for a large paddock—the Australian term for a fenced field—which sloped toward a creek at the bottom of which cattle were feeding. There he slid the shovel blade noiselessly into ten to twelve inches of soft, black, rich loam, teeming with earthworms as well as with white "feeder" roots. Picking up a small pea-shaped object, he held it in the palm of his hand, his lips curled into a satisfied smile. "Worm egg," he said, "millions of them."

In a glade of eucalyptus, where the bark had been stripped by cattle from the ground up to a point as high as the animals could reach, Trevor explained that before the land had been converted to biodynamic agriculture, the cattle had chewed the bark off trees to supplement something lacking in the pasture. "They needed no dietician or nutritionist," said Trevor, "to inform them that the grass was deficient."

Alex smiled. "You only have to look at the sheen on the coats of those animals, or even take notice how they stand and move, to realize that now they've been raised on biodynamic pasture."

As we scanned the landscape we could see where Trevor's pastures were still verdant, though it was the dead of Australia's winter, compared with the brownish-grayish pasture on the neighboring land, farmed in the orthodox Australian manner. Trevor led us down the straight driveway lined with yellow-flowering wattle trees, members of the acacia family, to a shed where he stored his 500 in containers exactly like Alex's.

There was also a machine he had specially designed for making the 501. From Queensland, he had brought back several hundred pounds of pure crystalline quartz for grinding to a flourlike powder; abraded for hours between two large, oblong crystal plates, it comes out as white as new-fallen snow.

"There is a light-bearing power in the substance of the crystal," said Alex, "but it's congealed, and has to be crushed and ground to be liberated. Buried in a cow horn throughout the summer, the powdered quartz absorbs, not sunlight, but 'light ether,' which is activated by being stirred like the 500."

Trevor showed off the equipment and machinery he had made for himself so he wouldn't need to get off his tractor either to put up hay three levels high in the shed, or feed it out to the cattle in the paddocks, enabling him to manage the whole farm by himself, his father being too old to do much work. Other implements invented and put together by Trevor in his own welding shop were a hay spreader to deliver winter feed to pastured cattle from huge eight-hundred-pound bales in the shape of fat jellyrolls, and a potato picker that, after clawing potatoes out of the ground, sorts and bags them by size, the whole mounted on a single vehicle.

"Our biodynamic farmers," said Alex, "have learned to design and make their own equipment out of scraps. It has freed them from debt and dependence on the manufacturers of heavy equipment that only destroys the land."

On the way north to the Goulburn Valley near the Murray River, Alex spoke of one of the main concepts he had had to drum into farmers, especially those inland: to make hay while the sun was shining. "In a good season they have a lot of feed going to waste. If anything, instead of conserving it, they'll put more stock on their pastures. Then, when a drought comes, they lose most of the herd—up to 90 percent. Those aren't really farmers," Alex added with a shake of his head. "They're miners!"

Six miles north of the Waranga Reservoir we came to our first night's stop, in Merrigum at the home of Farry Greenwood. Until his death in 1986 Greenwood owned and operated an enormous pear orchard, which is now in the capable hands of his son, Lynton. As we sat at dinner in the small kitchen of their modest bungalow Lynton was explicit: "I take a lot of fruit to the cannery, where I have to sit with other farmers while our produce is unloaded and inspected. I've had several growers ask me how on earth we manage without using lots of pesticides. But they never come over to our place to actually have a look or find out how we get by with no 'chemistry' at all."

"Mainly," said his father, "it's because they're scared that their *gross* profit—that big figure that seems to hypnotize them—will suffer if they abandon chemicals. And it will, to an extent, but what they don't understand is that their *net* profit will not suffer, and their soil will improve. The federal government's Agriculture Department has complete lists of all the BD farmers in the country, but they keep them hidden. They know all about BD, but they never let on. They see our method as a threat encroaching on their very reason for being. The only justification for their existence is to tell, or rather to order, farmers to use more and more of their destructive chemicals."

Lynton leaned forward to fix us with a steady gaze. "Around here all the orchardists have gone, their land burnt out by chemicals. Did you know that a pear tree should be productive for something short of a century, and peach trees should normally live forty to fifty years? The largest tomato growers in Australia used to be right here. Now they too have departed, having destroyed all the good land in the district."

Two hours down river in Walkool, in a mile-wide valley, Thory McDougal, a tall, sinewy fifth-generation dairy farmer of Scottish descent, had been facing ruin. Ten years ago, his land was overworked, his soils dependent on more and more fertilizer. Then he met Alex, and his farm was saved.

In 1984, he purchased another large parcel: rundown, farmed out, little more than a spread of oversalted bare dirt, abandoned by its previous owner. Today, operated as a large and flourishing farm, it is unrecognizable as the place McDougal bought—except for one corner.

"Before we applied 500," said McDougal, "the soil was so compacted that in some areas there were no plants, and in others the plants themselves were sickly. Now look! They're deep-rooted and healthy! They considered me mad when I bought this place. One man said to me: 'I thought you'd have more brains.' And it was true. In the sorry condition the property was in, it wouldn't have impressed a cat. But a couple of years ago the same man came over and said: 'I'll take back what I said. You've proved what can be done with your biodynamic method.' "

"On that farm," said Alex the next morning as we drove along the Murray River at first light, a silver moon still hanging in the sky, "McDougal has real BD rice, the only truly unpoisoned rice on earth! And now the Australian Rice Board has approached us because, internationally, it's very difficult for non-BD farmers to sell agricultural products. The prices are almost 'give-away.' But we can sell any amount of biodynamic rice at a very good price because there's a big demand for it in such countries as Switzerland, where baby-food manufacturers and other processors need unpoisoned food, if only to conform to new legislation setting much higher standards on quality than before. We export

BD grain overseas in twenty-ton containers, and we pay our farmers an extra premium per ton for BD-quality grains. Today they're getting $200 a ton, a price at which they can make a living. Thory gets about a ton more rice per acre than the average farmer, and he doesn't have to pay for any of the two dozen different kinds of chemicals the others want to make him use!"

After several months of effort Podolinsky even managed to persuade the Rice Growers Cooperative in Leeton, New South Wales, to store, mill, and market McDougal's first BD crop in complete isolation from rice grown by conventional means, according to the stringent guidelines put out by the marketing wing of the BD Association. Impressed with the prices the BD rice could command, both in Australia and abroad, a part of which is tithed back to it, the cooperative published a cover story in the May 1987 trade journal, *RCL Magazine:* "A Day with Alex Podolinsky."

Because rice growers as a whole were in such bad straits, the editor, Chris Black, announced in an introduction that the cooperative was keen to have more BD paddy, which was tantamount to admitting, for the first time, that rice farming with chemicals, as practiced for decades, was a "sunset" occupation, whereas BD rice production was its "sunrise" replacement.

"A decade or so ago," Alex told the farmers, "thousands of babies were deformed by Thalidomide, a drug to stop morning sickness in their pregnant mothers. Had its disastrous effects been known beforehand, it would never have been used! Right now, the same situation holds, not just for pregnant mothers but for all of us. We are all guinea pigs at this moment. The effect of hundreds of agricultural chemicals on future generations is as little understood scientifically as those of Thalidomide."

As the result of Podolinsky's speech a number of growers in the "Riverina"—as the rice area is called—began to strive for classification as BD farmers. Thirty years of work had broken down barriers, and farmers desperate to free themselves from the thralldom of expensive chemicals could see with their own eyes how biodynamics could provide a key to their release.

The Rice Cooperative's story was followed, six months later, by a full-page feature in the country's only nationally circulated newspaper, *The Australian,* entitled "Bio-Dynamic Man." Depicted with his left hand gripping his ever-present spade, his gaze directed heavenward, Podolinsky laid his central message on the line: "If it's true that you are what you eat, then at this moment, most of us and our livestock are a complicated chemical cocktail of insecticides, pesticides, fungicides, weedicides, and synthetic fertilizers."

Described as leading a lone cavalry charge against "the forces of chemical might," Alex was quoted as saying: "The worst poisons in use today do not even show a toxic physical symptom when they pass through the body. Almost impossible to detect, they nevertheless cause irreversible changes in the DNA pattern of cell reproduction. Even if an official inquiry were ordered into the harmful effects of pesticides, there is no laboratory equipped to perform it. And so far there have been hardly any funds allocated for that purpose, largely because universities —no longer free institutions—serve commercial interests instead of the public as they should."

Contrasting this situation with the one created by biodynamics, the newspaper article continued: "Podolinsky asks why, on B-D farms, plants are never subject to disease, cattle abdomens never bloat from digestive disturbances, and sheep never require drenching to rid them of intestinal parasites, as they are, and do, on all conventionally run farms. And if B-D practices were inferior, as officially claimed, then why was his association being inundated with orders for B-D barley, oats, wheat, rice, vegetables and other products? Do our customers know something that the national and provincial ministries of agriculture do not yet know?"

As we continued our trip toward New South Wales, Alex turned with a slight smile of contempt and a shake of his head: "The use of toxins in farming is all the more ironic because this continent of ours is otherwise the least polluted in the world and therefore ideally suited to profitably export huge quantities of agricultural products, were they nontoxic."

As he spoke, we crossed over the Murray and drove into New South Wales. Huge black-and-white birds with curved beaks and long legs were scrambling for their breakfast in an irrigation ditch. "They're ibises," said Alex, "and over there's a flock of swans, black like they all are in these parts. A little further on is where Kevin Twigg lives. But I don't think he's home."

The country flattened, with barely rolling hills. "Pity, because he's one of the most intelligent people you'd ever meet. He made a first-rate stirring machine. Many years ago when I first gave a lecture on stirring, Kevin was there. We had no machines in Australia at that time; we still did all the stirring by hand."

"I rang him right away," said Alex, "to say I'd drive on up to see his machine. When I arrived, there it was: the main pillar of the BD edifice. No man could have done a better job."

A road sign appeared, infrequent in the area, and Alex veered off the main road. "Before Kevin accomplished his feat," he said with satisfaction, "no one in the world had a decent stirring machine. I saw the one on Ehrenfried Pfeiffer's old farm at Loverendale in Holland. It only went one way round. It couldn't reverse to cause the 'chaos' between alternating whirlpool vortices. To produce a stirring machine that went only one way was just not a correct solution. Kevin adapted an ordinary electrical motor so it would automatically reverse the direction of the stirring at the proper interval, every fourteen or fifteen seconds. No electrical engineer would have thought of adapting an electrical motor the way Kevin did. They can't accept that the motor can be so used. I'll show you how it operates when we get to the grazing lands of the Stephens brothers in New England. They'll be trying out a machine for the first time and want me to check on its proper functioning."

Alex told of how he had learned from *Sensitive Chaos*, a book on water written by a Black Forest hydrological engineer, Theodore Schwenk, inventor of an optical method for testing water quality, that the right shape of a vortex was essential to the stirring procedure. Schwenk had discovered that only if the vortex is sufficiently open does it produce a correct vortical flow to make possible a renewing involution of the liquid in the outward-downward movement. And, because Schwenk's tests convincingly demonstrated that water freely moving in a brook or creek, constantly purling to form little whirlpools, is of a "liveness" not seen in city piped water, Podolinsky began to insist that only the highest quality of water available on a farm be used for stirring.

The Stephens brothers, powerful, rugged Aussie sheep and cattle graziers, had a sheep and cattle range outside the tiny New England hamlet of Emmaville in high country, bought in 1956 near the New South Wales–Queensland border.

"But the place wasn't big enough to support us all," said the younger brother, "so I went away for twelve years and when I had put a few dollars together sheep shearing I bought the place next door. In the course of shearing, I ran into a fellow, Max Thompson, another shearer, who told me about biodynamics. Alex here came right away and talked to us as if we were his own family."

Outside the house, on legs with rollers, stood a fifty-gallon cylindrical vat, already filled with water, enough to treat sixteen acres. Alex surveyed it with satisfaction, turning the switch to set it in motion. "To begin with, I look to see, as in hand-stirring in a bucket, that the vortex which starts at the rim is even. Then I break it and make a real chaos; that's what we want; but I start with the rim. There! It's finally reversing. But not soon enough. It should switch over earlier."

He paused to fiddle with a float mechanism, and the vortex changed direction more rapidly than before. As we watched the blades create the whirlpool, Alex was inspired: "There are certain things beyond predictability. An important one is a vortex. All weather works in vortices, and that's why with weather 'fronts' man can tell you *where* the front is but can never know what's developing *inside* the front. Here we have something that is independently new each time as it develops inside the vat. It's recently been confirmed in physics that vortices are not predictable. Each of these stirring machines is exactly identical; but each performs a bit differently. Each has to be adjusted just right."

As the stirrer picked up its rhythmic movement, Alex looked up with satisfaction out across the Stephenses' rolling countryside, which could have been the piedmont near the Big Horn Mountains west of Sheridan, Wyoming.

"What about the water?" asked one of the Stephens brothers. "Can we take it right from the river?"

"Fine!" said Alex. "Because your river's still *alive*. But make sure you pass it into the vat through hoses or pipes that are uncontaminatedly clean! When I go to farms I look for the most alive water to use for the 500 stirring. Just let samples be given me in a glass and I can tell which is better. It's a function of the light in the water. Dead water doesn't take in light. Living water does. Water gets rejuvenated by going deep down into healthy soil and coming up again through the plant. There is no better purification for water. Water that has gone through our soil cycle and is breathed out by plants is just perfect."

Alex tested the temperature of the water. "Lukewarm water is much easier to stir. I have always been able to adjust the flow of a mono-faucet to the proper temperature, hot or cold, just looking at the water, not feeling it, just looking. See how warm this water is? I know by sight when the temperature is right. When this is lukewarm, it stirs easier and quicker. Can you see the difference? Sense how this is like a great orchestra performing Beethoven's *Eroica*, while before it was just limping along. Watch this boiling action inside the center! That is real chaos."

He switched off the machine. "You'd normally keep this stirring going for exactly one hour. That's a German, not an Italian hour. And not a minute longer. Then be sure to have an eighteen-mesh sieve, copper, through which the activated water passes before it ever goes into your spraying unit. Otherwise you'll have fiber residue, which blocks up your sprayer nozzles. Some say you can use porous cloth, like pantyhose, but don't. If it bursts, you'll get fiber throughout your system."

Riding out onto the Stephenses' pasture through mobs of sheep crossing the dirt track in front of us, Graham Stephens explained that he had constructed a trailer to carry the new machine to any part of the property to allow tracts to be sprayed immediately following stirring.

"The most important thing of all," said Alex, "is to get the 500 sprayed onto the land within an hour after stirring. The longer you dally, the more of its power you lose. It's the short Australian moisture period that keeps the 500 going. If you drag the spraying out, then you'll come toward the end of your rainy season and the 500 will have much less chance to work in the ground than if you'd got it all out the first day. It works down as far as the roots go. So it will go on working even if the next year is dry, because there will still be moisture deeper down with the roots."

Alex waved his hand expansively: "That's the only thing that will get BD going on extended grazing land that makes up so much of Australian farming." Then he added with a friendly smile. "It could even benefit huge areas of your American West where the ranges have been brutally grazed."

On the last leg of our long journey, traveling up the coast from Brisbane to Townsville, to see Barry Ahearn, a third-generation sugar-cane grower, the weather became pleasantly hotter. It was like leaving the winter-bound mid-Atlantic seaboard in Delaware and reaching the warmth of the vegetable country near Homestead, south of Miami, Florida.

In Australia, sugar cane is raised, not as in Hawaii and other parts of the world, on huge monopolistic plantations, but by individual farmers on no more than 150 to 200 acres.

Alex said that Ahearn was one of his brightest BD adherents, the first cane grower to come into the association. Two brothers, also cane growers, were just beginning in biodynamics, and Alex was anxious to see what one or two sprayings of the 500 were doing to the sugar-cane soil.

"Barry only came into BD a little over a year ago," said Alex, "but in that time he's made leaps and bounds. His understanding of what it's all about is so sure I'm beginning to send him new people who contact me as a result of the ABC film. Some of them come down all the way from Atherton Tablelands in the mountains west of the seaport of Cairns, to talk with him, and now he is going to make trips to visit with them on their own farms way up north."

We were on the long middle stretch of the Pacific Highway set so far back from the coast there was no view of the ocean. Small resorts catering to "summer people" could be reached only on side roads. As we advanced through the flat countryside, Alex told of his first trip across the continent to Western Australia in the fall of 1986.

"I drove for two days across the wild Nullarbor Plain in the Great Desert. It was one of the most exhilarating experiences of my life as far as nature is concerned. Never have I seen so large a stretch of country anywhere on earth that is so totally and incredibly natural. Most people who decide to travel that route are told to expect nothing but sand, which, in fact, is all that most of them see. But the whole terrain is overgrown with a great variety of vegetation, and nowhere, for hundreds of miles, did I see a single sick plant. It was a wonder to behold thousands of lush plants growing with almost no rainfall at all, even in the hot April wind. It was a marvelous lesson on how nature can adapt, just as it has done in the tundra, which, though nearly always frozen, supports copious vegetation of another kind."

Alex paused as we crossed the Burdekin River, then continued: "Yet the population is little over one million, with the bulk of it in the port city of Perth. The farms in that dry country are enormous, a lot with thousands of acres each. Two thousand acres in Australia is a smallish property. The roads are arrow-straight, and very, very good, easily the best in Australia. You can go for miles on them at very high speed before you find a village. There's one man who grows sixty thousand acres of wheat alone! The biggest wheat grower on earth, they say. All that territory, so different from the 'old country' where biodynamics got its start, is largely sand, much of it now with a hardpan so dense you can only get through it with a crowbar."

In the northeast we were to see the solution; brand new rippers with steel blades as thick as a strong man's arm, able to go several feet into the ground to break up the compaction and aerate the soil.

A superb chauffeur, Alex took a curve with finesse, then added: "The 500 must be put on quickly so the plants can get their roots right down to finish off the job. In Australia BD has become an operation of large scale. We've got a BD farm of nearly ten thousand acres just starting. To get biodynamically operational, some farmers use a hundred-foot boom to put on 500 in swathes one hundred feet wide. Most of the Western Australians are new to the method. To get their 500 on just right will require a keen sense of timing because they have to catch the rain. Ninety percent of the bit of rain they get, in some areas only about ten inches a year, is subsequently blown away, leaving them with a bare two inches! But, once they get their soils in shape, they'll capture it all."

He wrinkled his forehead, as if perplexed. "So far none of them has made his own 500; they haven't enough suitable cattle for manure. They have some horns which they send east to us. It'll be a few years before they get enough experience. When I was out there for the first time in the fall of last year, I gave the longest lecture I've ever given, eleven hours in a single day to farmers who came from all over, some of them having driven hundreds of miles. Many camped out on the floor of Bob MacIntosh's big house, their total holdings representing 272,000 acres, over 400 square miles of farmland with all kinds of crops. They'd all seen the film and many of them had written to me via the broadcasting company. It was nearly midnight when I finished."

As we approached Ayr, in the heart of the cane country, Alex made a salient point. "You have to realize that out here in Australia we haven't developed just one subsidized model farm. That has been done in the past in Europe, in India, and elsewhere. What we've accomplished is the development of a whole system that can be learned by any able farmer who is interested in preserving and improving his land."

At Barry Ahearn's in front of a tree-shaded farmhouse we were met about noon by a short, muscular, suntanned fellow wearing an ancient Australian bush hat and an engaging grin.

"Barry has three brothers," said Alex with a subtle smile. "Two of them are sugar growers like himself, who are just beginning in biodynamics. The third is a salesman for an agricultural chemical company; so there's something of a rift in the family, as you might suspect."

Over a bowl of soup in an undecorated second-floor kitchen, Barry described the pressure being exerted by the agricultural advisers and chemical salesmen pushing and promoting all the farmers into greater production, whether of sugar cane, wheat, or whatever.

"We're all constantly pressured by the Ag Department to get the maximum out of each hundred acres. It doesn't seem to matter to their field advisers what becomes of the soil, so long as we're forcing the largest crop possible. And the big corporations are trying to get rid of all the little farmers and make huge agribusiness sugar plantations with the use of Japanese and other foreign money. It's all for the Australian dollar, which is not as big as it used to be!"

The edge of a field full of tall green sugar cane lay not a hundred yards from the house. When we reached it, Alex walked right into a thicket of plants and opened a hole in the soil to reveal good biodynamic structure.

Barry said he had planted his stalks further apart than other farmers because he had noticed that plants at the edge of a field did much better, obviously needing more room to develop. "The Ag people, of course, insist on more stalks per acre."

In another field, younger sugar-cane plants, a few inches high, were just getting started. In between the fairly widely spaced rows were zucchini, beans, melons, and other cash crops in rows a hundred yards long.

"Alex explained to me," said Barry, "how vegetables can be intersown with young cane plants as an extra-income crop, so long as enough space is allowed them, and that they'd generate money while I waited for the cane to mature. Add to this the fact that the vegetables are being raised biodynamically, which gives me a higher market price. It's all part of getting away from the insane monoculture of sugar cane, a system which only contributes to the degradation of the soil. Since getting into BD about two years ago, I've been taking an honest look at things, for the first time in my life. Now I know that all the weeds and 'rubbish' coming up in the fields all over the place—stuff you never would have seen years ago—is due to bad farming practices, for over a generation now, such as the continued uninterrupted growing of the same crop, forced by the greed factor. It's a system that actually suits the weeds."

"A lot of people," Barry added with a laugh, "told me I wouldn't be able to grow vegetables in this ground, that they'd rot with mildew, that pretty soon I'd go broke. But look at those flourishing rows! And as for going broke, when I got into BD farming my gross income was about $130,000. True, it has fallen to $90,000. That seems like a terrible drop in revenue until you realize that my *net* income has remained the same. Why on earth, I asked myself, had I been spending all that money to enrich the chemical companies all those years when I got no more cash out of my whole operation than I do now with no chemicals at all, and with less than I was planting before! And that's just the cash-flow side of it. As I go along now, my soil, which is any farmer's real capital, his money in the bank, so to speak, is going to increase in quality and therefore in value. My raising vegetables, which I've never done before, never even thought of before encountering BD, is helping to improve my land while it adds substantially to my net income. BD promises to make my land so much more richly productive, my crops so much healthier, and to relieve me completely of all that terrible pressure."

Looking out over the swaying sea of green, he sighed with satisfaction: "It's truly economical in every agricultural sense of the word!"

Barry's brother Kevin walked over to join us, so much taller, thinner, and darker than his brother, with luminous, fey eyes set in an unmistakably Irish Celtic physiognomy, that one would never have guessed they were related.

Picking up on our exchange, Kevin said he was following Barry's adoption of BD on his own sugar-cane acreage: "I think going any other way makes no sense. When I saw the ABC film I was amazed, and though I couldn't really understand what was going on, I saw that the main point was its improvement of the soil. And soil is just like your body. Without good soil, we'd all be dead."

As we took leave of Alex in a Townsville motel, it was clear that Steiner's system is being profitably applied in the great Down Under to over a million acres of land, some of the world's poorest, once dying, now flourishing, completely free of any chemical additive. With the help of the BD preparations, healthy plants, animals, and humans are being regenerated throughout Australia. Thanks to Alex Podolinsky and his thirty years of effort in applying Steiner's teachings, Australian biodynamic farmers are producing healthy crops, which can be sold at a profit, and in addition they are increasing their real capital—soil.

In the course of thirty years, Podolinsky has spread Steiner's method to perhaps twenty times as much acreage as is cultivated biodynamically in all the rest of the world, and as startlingly positive proof of its magic we brought home two small plastic bags, one containing a hard clayey substance that looked like off-white gravel, taken from a farm in the coastal region of New South Wales on which it extended from half an inch below the surface down as far as one could easily probe. In the other was rich, brown, fertile loam to which the infertile claylike material had been converted, down to a depth of twelve to fourteen inches, in only three years of BD spraying. Shown to American farmers, the two bags evoked comments of disbelief.

But one thought kept surfacing in our minds: if such a fruitful transformation could take place on the world's smallest continent, why not on the five more extensive ones? Was there any such prospect for the U.S.A.?

CHAPTER 7

It Can Be Done

SO SPARSELY TREED are America's flatlands in North Dakota that the state tree is mockingly featured as the telephone pole.

From the window of a Fairchild Metro-3 thirty-seat aircraft flying at twenty thousand feet the country east of Jamestown, North Dakota, resembles a vast, borderless checkerboard with mile by mile squares subdivided into patches of green, ocher, or chocolate brown. Each quadrant represents a 640-acre section of land—an amount considered by the first settlers to be adequate for the support and survival of two farm families. The whole "dry-land farming" expanse lying west of a former inland sea called the Red River Valley, and east of the Missouri Escarpment, which rises slowly up to the Badlands, depends on an annual rainfall of seventeen to nineteen inches. Pumping water from a 420-foot-deep underlying aquifer is too expensive for crop irrigation; the wells are strictly for household use.

Inside the Jamestown Aviation Service building an impressive bulletin board advised visitors they were at the center of one of America's farming heartlands, the Drift Prairie. But stapled and thumb-tacked to the board are not posters or photos of the amber waves of grain coming to head in early July fields that can be seen from level ground to stretch for thirty miles across the prairie's flatness in every direction, but a paper mosaic of flyers and broadsides advertising poisons for bugs and weeds, the brand name of one of which was as intellectually poisonous as its content: "You'll feel safer using MORE!"

In this once beautifully virgin land the chemical industry has clearly gained a raping *droit de seigneur.*

70

Up to the building drove an ancient Chevrolet Impala, out of which sprang Fred Kirschenmann, a blond giant in sturdy leather brogans, dark blue coveralls, and a red farmer's cap. What had attracted us to him, apart from his M.A. in history and two Ph.D.'s in political science and theology, was a feature article in *Agweek,* published by *The Herald* of Grand Forks, North Dakota, whose headline proclaimed: *"Switch to Organics Can Be Difficult But Profitable, North Dakota Farmer Says."* It reported that his poisonless methods worked: "His production costs are lower than his neighbors. . . . He gets premium prices. . . . His soil is rich and healthy."

One part of Kirschenmann's three-thousand-acre spread lies off U.S. Interstate 84—which runs all the way from Minneapolis–Saint Paul in Minnesota to Seattle-Tacoma in Washington. The farm is near the tiny hamlet of Windsor, which boasts no more than a tiny Roman Catholic church, a grain elevator, and a single public building—a cream-colored cubic blockhouse, self-advertised as "The Best Bar in Town!" An informal meeting place and "watering hole" for farmers from miles around, it serves as a haunt where they can compare notes and troubles over steins of beer, or noggins of harder brews, while sampling pickled hogs' hocks and chicken gizzards from fat stone crocks.

In the late 1970s local residents were amazed to see a strange house going up on, or rather under, a stretch of land a mile or so from Windsor. It was being set, like a bunker, into the side of a hill by Kirschenmann, born and raised on his father's farm at Streeter, thirty miles distant, and by his wife Janet, brought up in urban Braintree, Massachusetts. They had been inspired by an article in the Boston *Globe* about the New England architect, John Benard, to build one of his "earth homes," which seemed like an outsized root cellar. Naturally and air-conditionally cool in summer, the house is heated muffin-warm by a single small wood stove in winter, though when a snowstorm piles drifts in front of its doorway, to get out of the house to the roadway, Fred has to cut a tunnel with walls up to eight feet high.

The cozy kitchen is lit by a central thirteen-by-thirteen-foot shaft, or "well," driven into the center of the building from the surface of the ground, which provides more interior light than in many a windowed house.

As we sat near a big table strewn with an accumulation of farm publications, overloaded with huge encyclopedic volumes on organic gardening and natural insect and disease control published by the Rodale Press in Emmaus, Pennsylvania, Fred launched into the story of how he developed the largest biodynamic farm in the United States, growing wheat, buckwheat, rye, millet, oats, and sunflowers. It all happened, he explained, because of two graduate students at the University of Ohio whom he had taken under his wing. One of them, David Vetter, born and raised on a farm in Marquette, Nebraska, and committed to a farming career, had felt, after four years of soil-science studies for a B.S. at the University of Nebraska, that something vitally important had been left out of his farming curriculum, namely any references to loving or wise care for the soil. With Fred as his new and more liberal professor at Ohio, Vetter had been allowed to plan a private study course in the Philosophy of Agriculture, which included a reading program he hoped would broaden his perspectives on soil stewardship enough to tackle the crucial issues in what he saw was an ever-sickening American agriculture, and, he hoped, launch his "Ministry of the Soil."

One morning Vetter rushed into Kirschenmanns' office with an excited expression: "You've gotta read this," he said, handing his master a blue-covered compendium of the agricultural lectures delivered by Rudolf Steiner in Germany more than fifty years earlier.

"When I finally read the book," said Fred, "it seemed to me that it contained powerful material. But at the time I really couldn't fathom what Steiner was talking about. It was completely above my head. So I just filed it away in my memory drum, saying to myself that perhaps one day I'd get back to it."

The other important student in Fred's life was a young woman who had come on a scholarship in the 1960s to North Dakota's now-defunct church-sponsored Yankton College, where Fred was teaching theology. When she graduated and enrolled in a drama school in New York City, she was so appalled by the cutthroat competition alien to the cooperative "help-one-another" spirit among North Dakotans that she wrote a letter to her former professor seeking advice on where academically to pursue her overriding interests in philosophy and religion.

On a trip back East, Fred looked her up, and for the next several months she used her lunch hours at her telephone-company job to carry on a budding romance with cost-free long-distance calls, a perquisite of her employment. By 1968 she and Fred were married, and in 1972 moved to Massachusetts where Fred served as Dean of Students at Curry College in Milton, while Janet taught drama and theater arts.

Each summer the Kirschenmanns spent their vacations back in North Dakota helping Fred's father, Theodore, by then in his seventies, with the harvest on his Windsor acreage and on the older "home farm" thirty miles distant at Streeter, including six hundred acres of short-grass prairie still as virgin as in the time they had been roamed by thousands of bison hunted by Sioux Indians.

One evening in 1977, Theodore called to say he'd had a heart attack:

"It looks as if I'll have to sell the farm," he told Fred, "which isn't as good as it used to be, unless, of course, you want to take it over."

Across the dining-room table the couple looked at each other and wordlessly recognized their destiny was to return home to wrestle with deteriorating conditions on the family acres. With all their possessions, including their two-thousand-book library, they drove back to the home farm in Stutsman County, where the Kirschenmann parents, first cousins, had continued an agricultural heritage begun a generation earlier by their German forebears, immigrants, like many farmers in that area of North Dakota, from distant Mother Russia.

Partly inspired by the elder Kirschenmann's feeling that something was fearfully wrong with conventional agriculture as practiced under Extension Service advice for more than two decades, Fred and Janet knew that, whatever the difficulties, they intended to dispense with chemical additives and "go organic." They had been particularly impressed by the fact that the weight of a bushel of wheat grown chemically on Fred's father's and on many a neighbor's farm was mysteriously and inexplicably declining.

Unsure of exactly how to start, Fred called his old graduate student David Vetter, who drove up from Nebraska. Over several days he helped them work out a plan, based on his analysis of their soil conditions, for converting one-third, or about seven hundred acres, of their cultivated holdings to organic practice.

"David warned us to go slow," said Fred. "To convert seven hundred acres of tillage all at once was to court disaster. Instead, he advised beginning with a few control plots. But I wanted to do more than that right away; so, relenting, he suggested planting side-by-side companion fields, the ones farmed conventionally, the others organically."

Beginning with wheat and oats, the Kirschenmanns heeded Vetter's advice to supply fertility to their newly converted fields over several years to give their soils a chance to adapt to the new environment and get their nitrogen-fixing clover going. For this they imported a fish-and-seaweed liquid slurry from Texas via a Nebraskan distributor.

At the end of their first growing season they were elated to see that there was no difference whatsoever between the grain yields on their converted acreage and those on the conventionally chemically-farmed ground.

"Right then we decided to rush along, to go whole hog," said Fred. "We were eager to prove that organic agriculture was just as good, if not better than expensive chemical programs."

The following year, the Kirschenmanns cast aside all chemicals and treated the whole of their tillage with natural products, and, as Janet lamented: "We almost lost our shirts."

To their horror, wild oats, mustard, foxtail (locally known as "pigeon grass"), creeping Jennie, Canadian thistle, and other perennial and annual weeds that had never appeared after sowing the previous year burgeoned early and all over their planted ground to rush ahead of the sprouting grain crop. "It was as if a flock of sheep was being bruted away from a feeding trough by a herd of ravenous cattle." For the cereals, the competition was overwhelming.

"We hadn't realized," said Fred, "that the first year we'd been plain lucky. We had an unseasonably warm spring just like 1987. Temperatures rose in late March and early April into the upper eighties and low nineties. This anomalous hot weather was accompanied by a rash of thunderstorms with myriad bolts of lightning that somehow enhanced nitrogen supplies in the soil. The conditions were exactly right for us to get onto the land before planting and root out the weeds with disk harrows and rod weeders."

This ideal set of climatic conditions, rare in the Drift Prairie country, was supplanted the following year by a more usual cold, wet spring, which made tillage for weeds impossible until late July. Several fields produced only seventeen bushels an acre instead of a hoped-for fifty. Fields of hard, durum wheat, excellent for making spaghetti and other pastas, were heavily infested with pigeon grass; others hardly made any crop at all.

"We were really shocked and scared," Janet put in. "There were plenty of 'we told you so' remarks by our neighbors, who had warned us that organic agriculture was a thing of the past. But what could we do? We'd made a commitment, and it made absolutely no sense to retreat from it, to go back to chemicals. We were like the master of a ship in a storm forced to sail on until he can break out into clement weather. So we bit the bullet and forged ahead. It got so tight," she said, laughing sardonically, "we had a freezer full of meat from two of our slaughtered cattle, but hardly any toilet paper or laundry soap or other essential household supplies."

Then endurance paid off. During the third year of their organic operation, the Kirschenmanns were able to establish a system of crop rotation to supplant the monocultures—one crop and one crop only year after year—that had first become popular in corn-raising country in the Middle West and were being increasingly adopted in the wheat country of the Dakotas and neighboring states.

"That mentality of corn, corn, more corn, and corn only," said Fred, "accounted for why the massive use of herbicides first took hold in the Corn Belt. Weeds love and thrive in a monoculture environment, such as is being widely accepted in cereal-grain regions, even though it is wholly unnatural. Monoculture crept up here gradually when larger farmers were talked into getting rid of their cattle, plowing up their pasture land, cutting down all the windbreak trees so carefully planted after the 1930s dust bowl, and putting the whole of their acreage into cultivation, concentrating on one, or at the most two, main cash crops. This transition, fostered by Extension Service advisers, began to really take hold in the late 1960s and early 1970s. The advisers were telling producers that this was the only way they could survive. It was the same thing in forestry, with mixed forests being felled and single tree varieties being planted on thousands of acres, as you probably know."

But with a system of rotation, Fred explained, he and other organic farmers are able to keep weeds "off balance." It begins when a cold-weather early-spring crop of winter wheat or oats is planted in a field in mid-April of the first rotational year. The next year this is followed by sunflowers seeded in the ground in mid-May, or millet at the same month's end. It allows the weeds that have established themselves in the "wheat year" to come up before sunflower sowing and be killed off with tillage.

"This way," said Kirschenmann, "we wait later and later in successive years to catch another batch of weeds before sowing a grain crop and kill them as they are coming up with an implement called a rod weeder, a steel rod on the ground, turned by a chain in the opposite direction to the machine's travel, which yanks weeds out by their roots and spreads them on the surface to die in the hot sun, to nourish, not deplete the soil."

Fourth-year fields, allowed to lie fallow, can be kept free of weeds all summer. In the fall, a clover crop sown on the resting ground is disk-harrowed in to prepare for the original rotational cycle of wheat.

"You can't imagine how our soil improved," said Fred, smiling. "The worms just multiplied tremendously in our organic fields. There were so many they hung off the tines of my chisel-plow diggers whenever they lifted out of the ground." His eyes sparkled, then a shadow crossed them. "It's a crying shame what's been going on in these naturally rich Drift Prairie soils. On the home farm we began, like many others, with an average six to seven inches of black topsoil. Chemicals had been ruining it. The pity is that it was all completely unnecessary. A sane organic program would have required no chemicals at all. When Alex Podolinsky visited our North Dakota Natural Farmers group in 1984, after touring the countryside he told us that if an Australian farmer had suddenly woken up on this kind of land he'd think he'd died and gone to heaven!"

North Dakota, like other farm states all over the country, is now on the threshold of having to pay the high price for its "free ride" with chemicals, the binge it's been on, in terms of soil contamination. But there are a few new straws in the wind. Last year, South Dakota passed a law requiring cattle feedlots to prevent nitrates from running off their properties with fines up to $10,000 a day for each day the runoff remains uncorrected. Important consumer pressure is being put on State House legislators to clean up sources of ground water and food contamination.

"As the chickens of our previous practices come home to roost," said Fred, "farmers will have to foot long-overdue bills, and fundamental changes will be in the offing. The biggest of all is the one to be met on the question of depleted soils. That crunch would come much faster if our oil and gas resources run out. That won't happen soon, but if the price of hydrocarbons goes up again farmers simply will not be able to afford fossil fuels or the petrochemicals derived from them."

Fred paused to let us ponder the implications of his words, then went on, a little more sanguine. "The thing I sense happening is that, finally, conventional farmers are beginning to look sharp-eyedly at the bottom line. They know they've been at the brink of disaster for years and have read all the recent horror stories about increasing numbers of farmers going bankrupt. But the university extension people *still* keep telling them the *last* thing they should do is cut back on their chemical inputs, claiming that if they try to do so, they'll see their yields drop and get deeper and deeper in the hole . . . and into the red."

All this would, in his opinion, lead down the road to a food-production crisis of mammoth proportion, followed eventually, though unforeseeably—because of world-wide *overproduction* of food—by depressed prices that would put out of business the very farmer in a position to restore the health of the land. In the Drift Prairie area, costs for fertilizers and weed-bug killers run $60 to $70 an acre, which for a section of land means $40,000 or more. When that cost is compared to Fred Kirschenmann's input of $1.50 for clover seed per acre, plus $3 an acre for the biodynamic preps—a saving of more than $36,000 on his 2,100-acre spread, it is economically puzzling why farmers constantly faced with bankruptcy do not convert to organic or BD agriculture. The main stumbling block appears to be fear of a *single* year's failure, for which the banks could rapidly foreclose.

"I've talked to nearly a thousand farmers in these prairie states," said Fred, "and not a single one of them told me: 'Chemicals are terrific, just the thing we need for farming in the future.' What they told me almost to a man is that in their guts and hearts they know something is fearfully wrong about the way they've been advised to operate their farms. But they shrug helplessly, or stare at the ground and ask what they can do, which is nothing. Then they say sadly: 'That's how it is. One more bad year and I'm scheduled for service by the sheriff.' "

So trapped by the chemical system are the Drift Prairie farmers, says Fred, they can think only in terms of what chemical to use next. Many have now replaced ordinary NPK, that no longer supplies the "kick" it used to, with anhydrous ammonia, a gas knifed into the ground under pressure from a large tank, which initially, over the first three years, creates what Fred termed a "terrific response" in crops, but later is less and less effective while at the same time it inexorably turns farmland into the equivalent of a concrete airport runway.

One of the Kirschenmann neighbors managed to acquire eight thousand acres by cheaply buying up land from destitute smaller farmers. "He's a man who farms with huge eight-hundred-horsepower tractors and extra-wide planters," said Janet. "He can afford to leave more crop in the corners of his fields than many little farmers take off the whole of their land. This agricultural mogul was so pleased with first-year results of his 'anhydrous,' as it's called here for short, that he succumbed to the slogan 'More is better,' and greedily applied a double amount the following year. When the overdose killed every stalk of wheat planted fence row to fence row, to the exclusion of any other crop, he filed for a $4 million bankruptcy. But, having cleverly put the title to his acres and expensive farm implements in the names of relatives, he fought off his creditors and has gone right back to his chemically-dependent ways."

Fred nodded, and added: "A second neighbor, in his sixties, with three sons helping him farm another eight thousand acres, contrasted with the bankrupt by at least understanding the need to use as little chemicals as possible. Of all the conventional farmers around here, he's the best. He's been over to our place several times to carefully observe what I'm doing. In his heart I know he'd like to change over to the organic method we're using. Standing next to me in one of my fields, while admiring my soil structure, he said: 'What you're doing is fine for you, but it won't work for the rest of us unless a lot of us make the decision in common to convert together. Otherwise the banks'll pick us off, one by one.' "

"I gave him a hard stare," said Fred, "and told him: 'If you want to have a look at my accounting ledger, I'll prove to you in a few minutes that, going it all alone on my new program, I'm doing as good as you are acre for acre, or maybe better.' But he just sort of gazed at me slack-jawed without saying a word."

Used to growing only a single crop of wheat or sunflowers, such farmers are loath to adopt new measures of rotational cropping that require a process of "rethinking." "The Herculean task," said Fred, "is one of undermining the current mind set and status quo kept in place largely by a yearly avalanche of chemical-company propaganda. You simply wouldn't believe the amount of TV advertising for chemicals to which we are exposed. It starts early in the dead of winter on regular networks and cable TV to catch captive audiences sitting idly at home."

With new pest-control toxins appearing each year, in addition to advertising, farmers are beguiled by and rely upon poison salesmen rather than agricultural advisers for advice about which products to put on their fields. In winter, printed ads of the type posted at the Jamestown Aviation Service proliferate wherever farmers congregate to announce what amounts to "customer appreciation" meetings to be held at the local Ramada Inn or other motels where farmers, served free coffee, doughnuts, Danish pastries, and crullers, are briefed in detail by chemical-company representatives promising new poisons. The system is similar to the one in which pharmaceutical company "detail men" push new drugs on medical doctors, and the list of herbicides contains no fewer than seventy-eight separate products, most of them with macho names, such as Bronco, Roundup, and Blazer.

Flipping open a copy of the local Jamestown *Sun*, Fred pointed to a story in its business section by an Associated Press "Farm Writer" in which economist Lester C. Thurow, newly appointed Dean of MIT's Sloan School of Management, was quoted as dourly and direly recommending the government buy up surplus land and help farmers move on to other jobs wherever they might be available.

Fred waved the article: "How naïve and totally stupid can the Thurows of this world get? Of course that's been the policy for forty years, ever since it was made in a Washington, D.C., office by the Committee for Economic Development to get two million farmers off the land and into industrial servitude. But, as Wendell Berry in his book *The Unsettling of America: Culture and Agriculture*, asks: Who benefitted by so grotesque an agricultural policy? Certainly not the farmers or residents of rural communities. Did city folk benefit? Berry says no. So who did? What the Thurow-type proposals leave out of their narrowly conceived Cartesian mathematical-economic model is any idea of the amount of fossil-fuel energy, and concomitant inputs, spent or misspent for agriculture. *That* was the policy!"

The tragedy of this bleak philosophy, which recommends that farmers be sent to the cities, has already led in North Dakota to rich farmers buying out poorer ones or leasing their land after its takeover by banks or insurance companies. But the worst problem, according to Kirschenmann, is that the larger farmers, practicing the agribusiness "Get Big or Get Out" philosophies of city economists, are exactly the ones with no concept of "land stewardship" or concern for soil care. On the other hand many of the smaller farmers now bankrupted, or forced to sell, owned farms that were models of soil conservation, all of which have been or are being destroyed by the deliberately developed big proprietors.

As we set forth in Fred's battered car to tour his BD farm, we passed what seemed to be strange "artillery pieces" in a nearby field of knee-high sunflowers: special "cannons" timed to go off at intervals from twenty seconds to two minutes, to keep ravenous flocks of redwing and yellow-headed blackbirds from gorging on the sunflower seeds before they could be harvested.

On the way to the home farm in Streeter, south of Medina, fields were surrounded on every side with golden-yellow wild mustard that penetrated into corners and patches wherever farmers had not been able to lay down herbicides guaranteed to kill it. Other fields of nearly the same color on the always distant horizon turned out to be not mustard, but natural wild native pasture clover dominating all the other grasses with its early-July yellow blossoms.

Dotted over the Kirschenmann acreage, as indeed over all of the surrounding country, were ponds or small lake-sized sloughs, bodies of water that do not drain in the geologically adolescent overburden, the last in North America to be freed from tons upon tons of glacial ice. Related to the "10,000 lakes" advertised on Minnesota license plates, they teem in late March with ducks, geese, and swans on their way north to arctic breeding grounds and again in mid-October on the return trip, when they come under the guns of hunters from as far away as New York and California.

More anomalous were flocks of pelicans, which first mysteriously arrived in the 1950s after two years of incessant rain had filled the sloughs to the brim. Finding the climate to their liking, the pelicans have been returning ever since to the northern plains to pass their summer vacations in the company of sea gulls, which could be seen here and there in groups of three to six.

"Years ago, there were many more gulls drawn to this part of the country," said Fred, "but a quarter of a century of chemification has so decimated the worms that the sea gulls no longer follow the cultivators the way they used to. On my organic acres you can still see plenty of them at soil-tillage time."

We passed over a little levee cutting across the middle of a slough, where a lone mother Canada goose honked a warning to stay away from her nest with its new generation of goslings.

Up a rise on the other side of the slough Fred stopped the car and pointed to a field of wheat. "I don't normally like to put down what my neighbors are doing, but I want you to have a close look at that wheat field of mine. I planted it in hot weather on 10 April this past spring, when all the ground around here was so dry I knew it would be difficult for any of us to get good germination. Nevertheless, as you can see, my wheat seeds all sprouted and the plants are now nearly three feet tall."

A mile around two sides of a quarter section to where a neighbor had sown his own wheat at the same time as had Kirschenmann, some stands of varying height were barely heading out, others had no grain development at all. "It's going to be difficult for him to get a decent crop," said Fred. "When he harvests, he'll either have to lose the late-developing portion or risk waiting till it gets up, which may be too late. Note the empty little canyons running parallel through the stand in that field. That's where the weight of his tractor added additional compaction to already severely compacted soil to prevent good growth. Unable to retain moisture, his land dried out so fast that lots of his kernels didn't get a chance to sprout or, if they did, came up very weak. But the spongelike quality of my soil structure caught and held every drop of rain."

Approaching the home farm, we drove past a fenced expanse of luxuriant mixed grasses nearly three feet tall. "What you're looking at," said Fred, "is natural native short-grass prairie exactly as it was grazed by millions of bison. We have about nine hundred acres of it. It's loaded with thousands of varieties of grasses and wild flowers that bloom in a different set of hues each year. The quality of the forage is unsurpassable. We have learned to use the same practice recommended to Australian graziers by Podolinsky. We turn out our cattle to munch on one portion of it for twenty to thirty days, then transfer them to another portion to allow the first to regain its normal height. Bison, which never read any books on pasture management, followed this same practice, naturally moving from one unfenced range to another, never overgrazing any of their God-given territory."

At the top of a nearby slope was a hundred-head herd of cattle, basically Theodore's Angus-Hereford cross to which Fred has added genes from Tarantais bulls, a French upland breed known for its cold-weather-resisting stamina and the excellent mothering instincts of its females. Each year they drop some eighty feeder calves, sold when they reach 650 to 700 pounds, a few heifers being retained to replace cows too old to bear young. The steady income from the calf sales is more than outweighed by the manure from their mothers accumulated during winter months, the mainstay of biodynamic farming when injected with the preps.

In the farmyard to one side of Fred's parents' house was a half-acre feedlot, the herd's winter quarters: two sheds with open-sided fronts allow the animals shelter during bitterly cold nights or sudden savage blizzards. Amassed in the lot were seven to eight hundred tons of cow manure and straw, scraped with an old Case W-12 front-end loader from where cattle hooves had impacted it, piled into heaps four and a half feet high and twenty feet long. Each year Kirschenmann, using the same loader, turns the organic matter upside down and inside out to allow it to compost before preparing it and spreading it on 250 to 300 acres of his clover-fallow land.

"I don't do it exactly right," he admitted forthrightly. "I don't have time to run it through a manure spreader to break it up properly, as the purists preach. If I had to do that with this amount of manure, I'd be at that single task all year to the exclusion of anything else."

As a spotted-rump Appaloosa mare named Sally, as tame as a lap dog, nuzzled in our pockets for sugar, Fred related the saga of how he came to adopt the biodynamic agricultural system.

In 1979, well on his way to solving most conversion problems through "learn by one's mistakes" practice, he organized the North Dakota Natural Farmers Association with some twenty members who had like-mindedly begun to pioneer organic methods.

That winter, the newly-formed association came to the attention of Michael Marcolla, an energetic young man in his early thirties who had organized the Mercantile Food Company in Bridgeport, Connecticut, to supply organically-grown cereals and processed foods to enlightened North American and European consumers demanding something nutritionally more wholesome than what they could normally find on grocery shelves.

To assure that the products he was wholesaling were in fact of authentic quality, Marcolla organized Farm Verified Organic (FVO), a program defining and guaranteeing clearly stated organic standards.

When Marcolla visited the Kirschenmann's farm, he was as impressed with their management practices as Fred was by Marcolla's offering him a 15 percent premium above the price he was receiving for his grains at the local elevator.

"When I started in the organic program," Fred said with a smile, "I had no notion at all of markets anywhere that were willing to pay premiums. I'd begun in organics primarily because of my concern for our nation's soils. When Marcolla told me he was in the market for 15 percent protein organically grown wheat, I told him *my* wheat contained 15.5 percent. After having my farm checked out, he informed me, to my surprise, that I could get 50 cents more a bushel than the then fairly high $4.50 being paid at the time to farmers growing conventionally grown 'chemified' wheat. I was so amazed and excited that I made a trip to Bridgeport to talk with Marcolla, telling him I would grow anything, in addition to bread and flour wheat, and sunflowers, for which he might have a demand. He said I should immediately consider high-protein durum wheat, buckwheat, and millet, grains I had never thought about, let alone considered growing."

Marcolla surprised Fred even more when he told him that most of his customers, more sophisticatedly attuned to the value of organic produce than Americans, were located in Europe, where they were particularly eager for biodynamically-grown produce. "Do you know what biodynamics is?" Marcolla asked.

Fred's mind flashed back to his office at Ohio State ten years earlier where he had been handed Steiner's *Agricultural Lectures* by David Vetter. When he mentioned the incident, Marcolla pulled out his own copy of the same volume and Fred Kirschenmann experienced a *déjà vu*.

Fred asked what he should do to take steps to becoming not just an ordinary organic, but a biodynamic, farmer. Placing the book in his hands, the president of Mercantile Food soberly replied: "First read this. Then subscribe to back issues of the Biodynamic Association journal put out in Kimberton, Pennsylvania. Then call your midwestern neighbor Bob Steffen for help with specifics."

Bob Steffen, born and raised on a farm in a region of northeast Nebraska settled by Germans, studied with Pfeiffer for a year and a half, during which he was handicapped in his comprehension by the German's inability to speak good English and by his surprising lack of practical farming experience, at least so it seemed to the brash young native of a "cornhusker state."

"I wasn't yet all that sold on BD," Steffen told us. "Pfeiffer had given me what I thought was more theoretical than practical knowledge. The difficulty I had in becoming acquainted with him only began to be corrected when, with his help, we started to make large quantities of compost at Boys Town, Father Edward Flanagan's home for orphans near Omaha. With a seven-hundred-head herd of cattle we were turning out thousands of tons of composted manure each year. It could have been one of the largest, and probably the first operation of its kind in a truly biodynamic United States. But, when Pfeiffer and Father Flanagan both died I felt pretty much alone. Then I heard of Podolinsky's life work in Australia, and I said to myself: Maybe our BD people have been going about it wrong all these years! For one thing none of us has ever dared to mention the cow horns or the stirring. I guess that was the legacy of Pfeiffer, to adamantly avoid any reference to what he thought might be considered too arcane, indeed abandoning the preps in favor of his starters. And I understand his reticence. Even today I know very few of our BD leaders who can stand up and convincingly talk to a crowd of eager yet skeptical farmers about stuffing cow shit into cow horns or oak bark into sheep skulls, and I include myself in their number. Therein may be a clue to the failure of biodynamics to spread throughout America with either the speed or acclaim it achieved in Australia."

Despite these misgivings, two days after receiving the call from Fred, Steffen drove up to Windsor to tell the North Dakotan he could improve his compost by inserting the 502 to 507 biodynamic preparations. Steffen also explained that Fred would have to learn how to treat his fields and crops with 500 and 501, recommending that he visit Dennis Montgomery, a young farmer near Carrington, south of the Fort Totten Indian Reservation, who had a stirring machine given to him by Marcolla.

"So I started in," said Kirschenmann. "I called Josephine Porter in Stroudsburg, Pennsylvania, who was making all the Steiner preps and training a retired naval officer to take her place after her retirement, and she began shipping them up to me."

Steffen later told us he was at first dubious that the composting method Fred had been using would work properly; but when, a year later, his compost was as good as any Steffen had seen or smelled, with an ideal temperature, he believed that the successful result was due to the large amount of straw, with its high carbon content, incorporated into the piles.

Steffen then built for Kirschenmann a $4,000 stirring machine mounted on the back of a pickup truck, including two seventy-five-gallon wooden barrels, the water in which is simultaneously stirred by a single electric motor with a float mechanism to reverse its direction each time the vortex in one of the barrels reaches an optimal depth. For field application, Steffen told Fred he could use his father's old chemical sprayer, a long-chassis vehicle looking like a futuristic racing car, called a "Kirschenmann Coupe," made by an unrelated member of the Russian-German clan, installing on it a new tank and two sets of circulatory hoses and nozzles, one for 500, the other for 501.

In 1983, having decided to treat biodynamically one crop at a time in rotation, Fred started with buckwheat. Dennis Montgomery drove from Carrington to show him how to stir the 500 and 501 and apply it to his fields. As Fred watched the cone-shaped vortices forming and swirling simultaneously in the two big barrels, he wondered to himself how in the world anyone could have possibly dreamed up such a crazy scheme. He simply couldn't believe that the water, strained of all its residue, could provide any benefit at all to his land or anything growing on it. Only later, when he was introduced to the concept of homeopathy and its central tenet that a solution, potentized to the point where anything chemically analyzable was no longer present, could be effectively used in the treatment of disease, did he begin to appreciate what at first seemed to him to be an idiotic and obsoletely "alchemical" process.

Vortex *(above)* and chaos *(right)* obtained by Kirschenmann with the automatic stirring machine designed for him by Bob Steffen of Nebraska.
(Credit: Fred Kirschenmann)

Fred Kirschenmann in North Dakota preparing to transfer 150 gallons of BD 500 from his stirring machine to a special spraying machine.

"There have been some pretty funny rumors going around several counties," Fred said, grinning. "When we truck the stirring machine with those huge barrels from Streeter to the Windsor farm and back, people are sure we have a new still to make our own booze. And right up to this day several neighbors think my BD sprays are no more than witches brews! I let 'em kid me all they want because when, before we began, Marcolla told me I could get $4 a bushel for biodynamic wheat instead of the depressed going rate of $2.60, I was perfectly willing to humor anyone who considered my stirring and spraying to be so crazy. Now that we see what the method is doing for the land, I have great respect for those who prescribed it, even if it's difficult to rationally explain."

It was father Theodore who first saw signs that the BD treatment was having an effect. Walking out into one of the fields late one summer afternoon, he noticed that the leaves on buckwheat plants in low-lying wet areas, which had turned almost completely yellow, had become green again only three days after being sprayed with 501.

"About a year later my dad also said to me that he firmly believed the 500 spray was benefitting the structure of the soil, making it 'mellower.' That's the term he used. Of course, it wouldn't produce as dramatic a change on our soils, which had been in an organic program for six years, as it would in the really seriously compacted soils of average farmers in our area, who have hardpan problems going down as deep as they plow, if they do. A lot of them try to break up the compaction with products sold by the multi-level sales companies, such as Shaklee and Amway, but these are likely to have a detrimental effect on the capillary action of the soil."

Over the next three years Fred Kirschenmann gradually extended the BD program to his wheat, oat, millet, and sunflower rotations while adding still another crop, rye, for which, as Marcolla told him, he had a good market in Europe, though there was none in the United States.

"When we put in that rye," said Fred, "we were overjoyed to see that it had a strangely inexplicable, but most welcome, alleopathic effect on weeds in the sunflower crop which followed it. That's a natural process in which some plants exude substances from their roots to ward off others."

Fred Kirschenmann in a three-wheeled "Kirschenmann Coupe" (no relation), spraying BD 500 from two twenty-five-foot booms. Traveling at eight to ten miles per hour he can spray forty acres in one hour.

After considering two schools of thought, one recommending three gallons of Steiner spray to the acre, the other five gallons, Fred began treating 40 acres with the 120 gallons produced in two barrels, into which, during the last twenty minutes of each hour-long stirring, he mixes 1½ ounces of Hugh Courtney's barrel compost for each ounce of 500. The 501 leaf spray is put on in the early morning, the 500 root spray in the late afternoon and evening. In half of a long farming day, Fred can treat two hundred acres, spraying fifty-foot swaths with each pass.

Today, with all of his cultivated acreage under BD treatment, he buys 2,400 ounces of 500 in plastic bags, each labeled "1 gallon," and stores it in clay pots in a peat-moss-lined box. And so renowned is his success with his miraculous preps that in March of 1987 Kirschenmann was contacted by Stephen Gage, president of the Midwest Technology Development Institute, to help influence the agriculture committees in Congress. The institute was set up during a governors' conference by seven midwestern states to pool their resources in an attempt to obtain federal funding for new approaches and methods in agriculture. The hope was that the institute would be able to span a seemingly uncrossable chasm separating successful organic farmers and university agricultural faculties and get them to exchange views on problems.

Gage asked Fred to testify at a meeting of the House Committee for Agricultural Appropriations, which had done nothing to put teeth in a widely-reported Agricultural Productivity Act, passed in 1985, that was designed to support research in alternative agriculture by allocating funds to get it off paper and into action.

"At first we were allotted half an hour to address the committee," said Fred, "but just before we got to the Capitol, we found that our time had been cut to ten minutes, not for each, but for all of us. It was enough to make me wholly cynical about the political process. Here was this act that had just been lying on a shelf, unfunded for two long years, but the legislators running the hearings apparently thought so little of whatever we had to say about our problems that, with the time allotted, it was hardly worth the Midwest Institute's spending all that money on our travel. If our experience is typical, then the whole situation is ironic, to say the least. Very few people in Congress have any inkling of the larger economic or ecological issues related to agriculture as a whole. One has to wonder how our government gets anything at all done with any real wisdom."

After giving a short history of his farm conversion, Kirschenmann told the committee: "By now it is crystal clear to almost everyone that agriculture is in serious trouble not only in the United States, but in many other countries of the world. It would be a terrible mistake to think that only a little tinkering can correct its ills. The system is so flawed economically, agronomically, and environmentally that it needs no less than a full-scale overhaul."

Most crucial for farmers locked into the present chemical system, Kirschenmann told the politicians, is practical information on how to get out of it. The current agribusiness structure, with all the weight of forty years of research on capital and energy systems behind it, has given farmers the biased impression that it is the *only* viable agriculture available. This is patently false in that hundreds of small and large farmers around the world have demonstrated, with no help whatsoever from any national or international agricultural bureaucracy, that low-input systems are not only feasible but highly profitable.

The Agricultural Productivity Act, he went on, calls for exactly the kind of research needed to allow farmers, from whom he is receiving desperate letters and telephone calls almost daily, to consider transition to naturally healthy production with minimal risks. Some initiatives in this direction have already been taken by land-grant colleges, notably the University of Nebraska and Iowa State University. But those represent only a tiny step on a trail that could only be blazed with funds appropriated to relieve researchers from continually having to direct their efforts mainly to prove what the chemical companies, which supply the lions' share of funding in the form of research grants, want them to prove.

To bolster his appeal to the senators and representatives, and forcefully bring out to what depths purportedly objective agricultural research has fallen, Kirschenmann needed only to add how one land-grant-college professor told Charles Walters, Jr., editor of *Acres U.S.A.*, that if Walters, rather than the chemical companies, wanted to put up $100,000 to fund research, his department would supply data backing up *anything he wanted to prove.* "What's happened," Walters said wryly, "is that they've changed the Golden Rule from 'Do unto others . . .' to read: 'He who has the Gold makes the Rule.' "

Back in the kitchen where Janet was preparing a beef stew for supper, we asked Fred what impression his biodynamic practice was making on any of the agricultural schools or their extension agents.

"Good Lord," he replied, "I haven't even raised the question with them. We're still in an uphill struggle to get most of them to appreciate that ordinary organic farming is not only possible but profitable. I'm trying to get ready for the meeting of our association, now expanded into the Northern Plains Sustainable Agriculture Association, where I'll be giving the keynote speech, which will be attended by some twenty European representatives of Marcolla's FVO group as well as by our own members and new farmers desperate to convert to nonchemical methods. It will focus on the problem of transition from the old way to the new. We'll hold workshops in which our organic farmers who have made a successful go of it can explain to the new men who've already signed up for the conference all the problems they're going to face and get into the nitty-gritty of what to do about them, and tell them about the mistakes that have been made.

"It's a whole new ladder for them to climb, or even a tightrope to walk," Fred went on, "given their precarious financial conditions. One misstep and they'll plunge downward into an abyss, with no safety net to break their fall. Most of the new ones will be there out of desperation with what they've been forced to go along with all these years. They'll be both curious and afraid. It's a big responsibility. We have to go as slowly and carefully with them as any teacher instructing a child in swimming, horseback riding, or mountaineering."

Continuing this theme during dinner, Janet spoke of a "mile-wide" gap separating the day-to-day concerns of farmers from those of city folk represented by the majority of politicians who hold the lives of food producers in their hands. "As one brought up in an urban community who had to learn about a new culture, the culture of the land," she said, "I wonder if that gap, so painfully felt by farmers, will ever be bridged. To give you but one example, a Massachusetts congressman on a delegation to study farm problems here in North Dakota was invited to take part in a TV panel discussion including both members of the delegation and local farmers.

"The sharp contrast in thinking between the two groups, as striking as their dress, was unforgettably brought out when the congressman, attired in his natty suit and tie, turned to a dairyman in work clothes to ask in all sincerity: 'Do you have to milk your cows *every* day?' There must have been an explosion of laughter in kitchens and living rooms all over North Dakota. The dairyman, as he would to a child, quietly replied: 'Yes, sir, every day of the year, in fact twice a day.' You can't really blame that congressman. You should hear the same kind of naïve questions I get from my family and friends when I go back East. It's sad that all of them think agriculture can be run like a factory. A lot of them see milk running out of a tap, like water, and their packaged cereal grains as one more analogue of a mechanical production line turning out widgets."

The same mentality, said Janet, was illustrated when a New York TV morning show viewed by millions called one cold February morning to say they were doing a story on organic farming and wanted to feature the Kirschenmann farm as a model of that method. The TV crew would arrive in two days, the caller said, to get footage of the farm and especially the land.

"Don't you realize," Janet replied, unable to suppress a giggle, "that if you come now, all you'll get on your film will be thousands of square feet of snow as white as Antarctica?"

Silence ensued, broken by an embarrassed: "Oh, I see. . . . We never thought of that. Perhaps we'd better delay till summer." Then, with cheery abandon: "We just want you to know we feel you are true pioneers."

CHAPTER 8

Heaven on Earth

THAT ALL THIS chemical horror is as unnec-
essary as it is unnatural, that men and women can live into their second
century free of all disease, if fed on properly grown organic food, was
first proved by an early follower of Sir Albert Howard, a distinguished
Scots physician, Robert McCarrison, who did so with the aid of a crew
of city rats. As head of the Nutrition Research Agency for the Imperial
Government of India, and director of its Pasteur Institute at Coonoor,
McCarrison gained a knighthood and an appointment as personal physician to King George V. But it was his interlude with rats, while he was
still a young man attached to the Gilgit Agency in northern Pakistan,
that made his fame. It led to the discovery of the legendary health and
longevity of the inhabitants of Hunza, a hardy people living in a remote
and inaccessible valley, surrounded by the highest Himalayan peaks.

In the course of a comparative study of the dietary practices of people
from various regions of India, young McCarrison was surprised to find
that rats that ate the diets of Pathans and Sikhs increased their body
weight much faster and were much healthier than those ingesting the
daily fares of neighboring peoples such as the Kanarese or the Bengalis.

Even more extraordinary, when his rats were fed the same diet as
that of the Hunzas, a diet limited to grain, vegetables, fruits, and unpasteurized goats' milk and butter, the rodents appeared to McCarrison to
be the healthiest ever raised in his laboratory. They grew rapidly, never
seemed to be ill, mated with enthusiasm, and had healthy offspring.
Autopsies showed nothing whatsoever wrong with their organs.
Throughout their lifetimes these rats were gentle, affectionate, and play-
ful.

Major-General Sir Robert McCarrison,
C.I.E., M.A., M.D., D.Sc., LL.D., F.R.C.P.,
B.A.O. (From a bust by Lady Kennet)

Other rats contracted precisely the diseases of the people whose diets they were fed, and even seemed to adopt certain of the humans' nastier behavioral characteristics. Illnesses revealed at autopsy filled a whole page. All parts of the rats' bodies—skin, hair, blood, ovaries, and womb —and all their systems—respiratory, urinary, digestive, nervous, and cardiovascular—were afflicted. Many of the rats, snarling and vicious, had to be kept apart if they were not to tear each other to bits.

So extraordinary was this data that several explorers made the difficult and dangerous journey to the remote and inaccessible source to find out just what could be so miraculous about the Hunza diet.

In a narrow sunlit valley, dominated by the snow-capped Himalayan peaks, they found a real-life Shangri-La, inhabited by what they described as the healthiest, happiest mortals, many of them centenarians, subjects of the tiny, semi-independent kingdom of Hunza, legally a constituent of Pakistan.

Until very recently the only way to reach this oasis of life and health from a degenerate and dying West was via the outpost of Gilgit, a month's hike from the nearest Pakistani railhead at Rawalpindi, along a rough foot path over a 13,700-foot snowbound pass, open only three months of the year. This was the route taken by Marco Polo on his return from Cathay, over swollen rivers and icebound passes, past the rotting carcasses of pack animals that could not keep pace with the trains of silk and porcelain hauled from China to be bartered in India for gold, ivory, and the spices of the great subcontinent.

For centuries entrance to the land of Hunza has been over a treacherously swaying bridge supported by ropes of stranded goats' hair. Then, at an elevation of a mere eight thousand feet, there appeared before the traveler's enchanted eye a flowering seven-mile-long valley of green-gold orchards sparkling in springtime with a pastel carpet of apricot blossoms, sharply contrasted against the surrounding forbidding, snow-capped Karakoram range.

In the valley, small stone farmhouses dot a carefully cultivated landscape, whose endless terraced fields rise up the escarpment in geometric steps toward glacier streams which fertilize them with powder more precious than gold, a glacial dust, teeming with minerals ground from the raw living rock of the mountain.

Five hundred feet above the murky rushing waters of the Hunza River—which threads through the whole valley—stands the town of Baltit, dominated by an old fort, built while Canute still reigned in England. As the traveler approaches, the women of Hunza rise and place two welcoming fingers to their forehead, pretty children smile enchantingly, mustachioed gallants shout greetings in Burushaski, a native tongue of unknown origin.

Hunzakuts, as the natives choose to call themselves, differ from other races living in that distant corner of the world. With bronzed but Caucasian features, like southern Europeans, they love to boast that they are descended from three foot soldiers of the Macedonian army of Alexander the Great who took refuge with their Persian wives in this remote valley in the third century B.C. There they bred a warrior race of brigands who thrived on raiding the trade route to China, that is, until conquered by the British in 1891, and pacified by a domestic Ismaili ruler, known as the Mir, who had the wit to learn the rules of karma.

Visitors have written in great detail about the extraordinary health of the Hunzakuts. There is practically no plant or animal disease, and virtually none in humans: absolutely no cancer, no heart or intestinal trouble; and the people regularly live to be centenarians, singing, dancing, and making love to the edge of the grave. Visitors tell of seeing no cripples. Wounds are said to heal with remarkable speed, seldom becoming infected if rubbed with the local soil, rich in minerals, which somehow obviates sepsis. Hunza women are so healthy they need no assistance in delivering a child, whom they breast feed for two to three years, deliberately spacing their children so as to wean them one at a time. Children are invariably reported as growing up unneurotic and healthy, with none of the normal childhood diseases such as mumps, measles, and chicken pox. The girls' complexions are depicted as without acne or blemish, attributable in part to the application of oil of apricot seed. Nor is there any evidence of juvenile delinquency; visitors remark that one never hears a mother scold or bribe a child. Treated as integral members of society, trusted and given responsibility, the children are described as growing up healthy emotionally as well as physically.

Travelers who have been through northern India agree that the Hunzakuts are superior mentally and physically, excelling in grace, charm, and intelligence. They are frank, and have a fearless look on features that are chiseled and appealing. Strongly built men, with bold eyes and jovial expressions, have physiques that would delight a Rodin.

There is no moron or cretin among them, which contrasts sharply with the people in neighboring valleys, many of whom suffer from goiter and cretinism due to a lack of iodine in the water. Although surrounded by peoples afflicted with all kinds of degenerative and pestilential disease, the Hunzakuts, who lead a strenuous and spartan existence, do not contract them.

As mountain climbers they are superb and unequaled, with great powers of endurance. Agile as goats, they leap over boulders and icy streams, make holes in the ice in winter to plunge into glacier-fed streams, as swimmers second to none. They are also splendid horsemen, always ready to indulge in their national game of polo, played in every village on its own narrow polo ground with balls and mallets carved from bamboo roots.

Whether at work or at rest, or playing some game, the Hunzakut is always accompanied by a melodious tune which invites him to hum and to dance. The sound of flute and drum is everywhere carried on the air, drifting up to the perennially nourishing glaciers.

Unanimously described as having "rosy complexions, good features, and lovely eyes," the women of Hunza are treated as true partners in life, both at home and in the fields. Slim and erect, dark-haired and svelte, they walk with a light glide, easily climbing the thousand or fifteen hundred feet to the terraced gardens they lovingly tend with their menfolk as the source of their enduring health.

Every bit of Hunza soil has been hand-prepared with delicate care, often brought from afar, laid in narrow ribbons on terraced fields hacked from the raw face of the surrounding mountain and built up with stones without mortar or mud. Kept fertile with organic compost, the strips are watered by the silt-laden melt of glaciers brought in by man-made conduits of extraordinary engineering, constructed with nothing more than a wooden spade and an onyx-tipped pick. Without benefit of transit or theodolite, the channels go through tunnels, over aqueducts and even hang from the side of cliffs many hundreds of feet high, flowing through timber troughs supported by brackets.

And therein lies the secret of Hunza health: in the finely powdered detritus of rock made by the massive glaciers as they inexorably grind down the raw mountains to produce a silt containing all the mineral elements required by plants. The silt mixed with organic compost provides plant, animal, and human with every element they need for life. Every possible stream from a glacier is harnessed and the water channeled to the fields. In winter the channels are cleaned and the silt is spread on the fields to give a fresh layer of soil for the coming season.

In the Karakoram range lies the most immense accumulation of glaciers anywhere in the world, with the exception of the polar regions: millions of tons of ice and snow, hundreds or thousands of feet thick, are piled high on the mountain flanks. As these glaciers recede or advance, their slippery footing liquefied by pressure, the mass of ice scrapes off the face of the mountain. No mineral, no rock, no metal is strong enough to resist the weight of these millions of tons of ice.

A finely ground coat containing some or all of the elements moves down to be carried away in the *nullahs*, or glacial torrents. The water, a milky gray, spurts, splashes, and spills out of crevices, canyons, and any aperture in the mountain walls, where numerous *nullahs* are formed. These streams carry in colloidal suspension some, if not all, of the mineral constituents of the mountains, the necessary fertilizer for the Hunza fields.

In their manuring, the Hunzakuts return everything they can to the soil: all vegetable parts and pieces that will not serve as food for man or beast, including such fallen leaves as the cattle will not eat, mixed with their own seasoned excrement, plus dung and urine from their byres and barns. Children follow the cows through the fields with shoulder baskets ready to catch the precious substance as it emerges. To be sure that not an ounce is wasted at threshing time—when a mixed team of cows, donkeys, mules, and yaks, yoked abreast, are driven over the cut crop on the threshing floor—a person driving the team, usually a pretty young maiden, keeps a pan in her hand for a catch in mid-air, uncannily anticipating when and which of the animals is about to drop.

On mountainsides, children scour the fields to garner every last blade of grass for compost, every last errant speck of goat or sheep manure. Like their Chinese neighbors, the Hunzakuts save their own manure in special underground vats, clear of any contaminatable streams, there to be seasoned for a good six months. Everything that once had life is given new to life through loving hands.

For a thousand years the Hunzakuts have farmed with the same seed, and the result is reflected in the health of their plants. They have no central source of seed supply, yet the species all survive. Each farmer saves what's necessary for the next year's crops, or barters with his neighbor. No Hunzakut owns more than five acres, and many families have divided up their ownership into tiny lots, sometimes no bigger than a Persian carpet, every inch of which is cared for with affection and understanding.

Carefully cultivated strips of land in Hunza, every stone of the walls hand-laid. Raka peak, one of the highest in the Himalayas, in the background.

Yet in all these plots the produce flourishes, carefully tended by hand, with no other instruments than a pick of ibex horn and a wooden plow drawn by bullock or yak. Plowing is necessary to aerate the roots of the plants, which would otherwise suffer partial asphyxiation. But the Hunzakut does not plow deep, and would not, even if he had the modern moldboard plow. Deep plowing compacts the surface and submerges bacteria and other valuable organisms to levels where their numbers are reduced.

Rich in organic matter, Hunza soil is porous and spongy, encouraging the multiplication of earthworms, which make thousands of burrow holes into which water can easily seep. Each farmer applies his water personally, and when the right amount has seeped into his field the flow is stopped by maneuvering the requisite stones. Were overflows or run-offs allowed to develop they fear a leaching of the nutrients would end in an imbalance of elements, causing crop disease and bringing on insects. The terraces afford perfect drainage and allow for deeper penetration of air into the soil, all of which promotes the growth of micro-organisms, enabling both the bacteria and the roots to benefit.

Visitors to Hunza describe the vegetables as rich in flavor despite the fact that there is no seasoning, which is not missed when vegetables are grown in mineral-rich soil. Popeye's spinach-generated muscles may be more than Walt Disneyan malarkey; Hunzakuts eat large quantities of the green leaf at almost every meal, and the mineral-rich water it is steamed in is always saved and used.

But it is not just their spartan diet that accounts for the extraordinary health, amiability, and lack of neurosis among the Hunzakuts. The real secret appears to lie in the same mineral-rich glacial water that fertilizes their fields, brought to the villages in canals for domestic use. Cool and inviting, it is called "glacial milk" because of its pearly gray color. Every man, woman, and child drinks an abundance of this water, which is neither boiled nor filtered. And so does the Mir, though he has plenty of clear, "clean" water for unwise, finicky, foreign guests.

Sir Albert Howard once said he hoped someday some enterprising fellow would get himself samples of this Hunza water, have it analyzed, and find out exactly what it contains. "It should give us a clue," said Sir Albert, "to the health of these remarkable people."

John A. Tobe, an American traveler and adventurer, first to scientifically analyze the "glacial milk," says the Hunza minerals go into the soil in a colloidal state, which is described as the state of a solute when its molecules are not present as separate entities, as in a true solution, but are grouped together to form solute "particles." These particles, approximately one hundred-thousandth to one ten-millionth of a centimeter across, only detectable by means of an ultramicroscope, carry a resultant electric charge, generally of the same sign for all the particles, usually negative, and it is their colloidal state that enables the human body membranes directly to absorb essential mineral elements without their having first to be processed organically by plant and animal. Every cell of the human body is made of colloids arranged to perform specific functions. Colloidal particles are so small, and therefore have such a large surface area—a teaspoon of particles has a surface greater than a football field—that, according to Gustave Lebon in *Evolution of Energy*, they generate surface energies that have powerful effects on physical and chemical reactions.

Tobe believes that the Hunzakuts' "glacial milk," with which they water their fields and which they drink so copiously, supplies them with all the elements required by their bodies, in nutrient form, and that it is the explanation for how they can be so strong, so energetic, so unneurotic, and have such powers of endurance, both physical and sexual, on such a spartan diet. He believes it is why they always look content, why their gait is springy and full of vitality, why they look forty when they are seventy, why the men can sire children in their nineties, and why they can walk to Gilgit, sixty miles away, over the grimmest terrain, transact their business, and nonchalantly retrace their steps.

Harsh Hunza country and a few of its aged residents. Good health at over a hundred years of age is common among the Hunzakuts. (Courtesy *Acres U.S.A.*)

In a world of atomic threats, and the instant danger of an epidemic of AIDS that might make the Black Plague pale, knowledge of a group of people living in such complete peace, free of the major fears of mankind—disease and war—should have made an impact. You might think so, says J. I. Rodale, an early enthusiast, who wrote much about the Hunzas. You might think that demonstration of the fact that practically complete elimination of disease in an entire group could be effected by the mere eating of proper foods and drinking of vital water would create a tremendous stir in medical circles, would crystallize a demand that the mechanism be immediately created for carrying these findings into actual practice. "It didn't," he complains, "even produce a tiny ripple in the pond of medical inertia. The doctor is too much involved in the morasses of the disease and physic to be able to give much time to the question of mere health."

In the mid-1970s Senator Charles Percy, a member of the Senate Committee on Aging, visited Hunza to note the absence of heart attacks, cancer, and neurosis. Back home he made a laudable effort to awaken his colleagues in Congress to the amazing potential represented by the Hunzakut phenomenon. He might as well have dropped a pebble in the Hunza River.

Now the Pakistani army has built a military road through Hunza to the Sino-Soviet border. Where Polo's mule trains carefully trod, the treads of tanks grind hard, and in their train come sugared drinks, NPK, and the inevitable pesticides to raise the specter of disease and early death. Shangri-La is Shangri-La no more.

Yet the secret of the Hunza health may yet survive to save the soil and health of a sick and ravaged planet. A sample of Hunza water, brought back by an American traveler, the late Betty Lee Morales, has been thoroughly analyzed with sophisticated laboratory techniques by Patrick Flanagan, author of *Pyramid Power*. The results are as astonishing as they are encouraging. They explain not only the water's life-giving qualities, but vindicate Steiner's clairvoyant vision of the function of the vortex and of the intervening chaos in the life-giving stirring of his biodynamic preps, designed to revitalize a dying soil.

CHAPTER 9

Vortex of Life

SO STRANGE and mysterious are the properties of water, just ordinary tap water, that anything new about its behavior has become a landmark in the history of science. In the 1920s when the Romanian scientific genius Henry Coanda made the discovery, apparently banal, that a fluid flowing over any surface tends—as if it were alive—to cling to that surface, it was considered by physicists so important it was called the "Coanda Effect."

Intrigued by tales of the longevity of the Hunzakuts, in the early 1930s Coanda made the arduous journey to Hunzaland. As a water expert he was delighted to be told by the Mir that the secret to the Hunzakuts' longevity and health was hidden in the water they drink so freely.

Convinced that any health-improving properties of the Hunza water would be related to its molecular structure, Coanda set about analyzing a sample of it alongside "ordinary" water, using the facilities of the Huyck Research Laboratories in Connecticut, to which he was at that time a consultant. His novel method was to study water in its crystalline form as snowflakes, each of which, as they fall by the zillion, is uniquely designed and molded by unknown forces in its micro-environment, no two being alike, as far as anyone can tell.

With a "fluid amplifier," a device he invented that could make snow from water, Coanda found in the center of each snowflake a circulatory system composed of tiny tubes in which still unfrozen water circulates like sap in plants, or blood in animals—water which he considered to be what dowsers characterize as "living," to distinguish it from its stagnant, lifeless counterpart.

Dr. Henri Coanda (*right*), the father of fluid dynamics, and Patrick Flanagan at Huyck Research Laboratories in Stamford, Connecticut, in June 1963. Dr. Coanda was the scientist who discovered that the secret of long life was in the structure of water. He told Patrick: "We are what we drink." (Credit: G. Harry Stine)

By carefully timing the life span of snowflakes, which "die" when all the water in them becomes congealed, Coanda was able to establish an extraordinary and direct relationship between the duration of his snowflake water and the life span of people who regularly drink such water. The "living" fluid appeared to add more life to humans.

On far-flung travels Coanda found that the water that produces long-lasting snowflakes is the principal beverage not only of the Hunzas but of other long-lived peoples in Soviet Georgia, Ecuador, Peru, and on the mountainous Tibet-Mongolian border. Still he did not know why or how the special glacial water extended human life. Before returning to his native country to become president of the Romanian Academy of Sciences, Coanda entrusted his water research to a young collaborator at the Huyck laboratories, Patrick Flanagan, a prodigy who, at only seventeen, had been listed by *Life* magazine as one of America's top ten scientists.* "I think you are the only one I know," Coanda told him, "who can eventually come up with a system to make Hunza water available anywhere in the world."

Flanagan, a short man of forty, with a closely clipped mustache, his head as razed as Kojak's or Yul Brynner's, met us in a small motel next to the tiny Sedona airport atop a mesa in the mountains of central Arizona. All around us rose a magnificent series of red rock cliffs and spires, dominating the town in the canyon far below.

"I read everything I could on water," said Flanagan, at ease in his khaki clothes from the Banana Republic, "only to discover it was one of the world's most mysteriously anomalous substances."

Once a flamboyant adventurer who would savor the gall bladders of rattlesnakes and cobras to assume their powers and immunity, regularly earning thousands of dollars with far-flung lectures on such arcane subjects as Pyramid Power and Tantric Sex, Flanagan is now a quiet vegetarian, as reclusive and peace-loving as a Tibetan monk, healed and domesticated by his lovely wife, Gael, soulmate, so he claims, from an equally adventurous past, an expert on the structure, function, and properties of crystals. Like her husband, Gael received the degree of Doctor of Medicine from the Multi-Disciplinary World Medical Congress in Colombo, Sri Lanka, and now helps, in the seclusion of their Sedona laboratory deep in a pinewood forest, with research on the colloidal properties of water.

Oddly coincidental, or serendipitously propitious, the town of Sedona —lying in the coils of the twisting Oak Creek Canyon, surrounded by gnarled terra cotta peaks against a sky so blue its energy is tangible—is what Lyle Watson in his *The Romeo Error* describes as a unique "power spot," one where great concentrations of energy can be felt coming up from the earth through four separate telluric vortices.†

Visitors from all over the world, alerted to the existence of what they believe to be "psychic energy vortices," flock to Sedona to bask and meditate in the extraordinary energy emitted from the soil, and describe experiencing visions, telepathic communication, past-life regressions, precognitions, UFO sightings, enhanced automatic writing, spiritual healing, and other psychic phenomena, especially during the period of the full moon.

* For his invention of the *neurophone*, a hearing aid which bypassed the eighth cranial nerve associated with hearing and allowed the deaf to hear directly via the skin.

† At one such mystery spot in Southern California when two men were photographed facing each other across the "vortex," the man on the right appeared shorter in the developed picture, even when the two men changed places.

Two of the vortices are said to be magnetic, the third electric, and the fourth electromagnetic. The electric vortex, being charged with yang or male energy, is reputed to stimulate and elevate consciousness; the magnetic, or yin vortices, charged with female energy, are supposed to open one's psychic perception; and the electromagnetic vortex is credited with balancing both body and spirit, stimulating memories of past life experiences.

Author and lecturer Dick Sutphen, an aficionado of Sedona's vortices, has collected an anthology of visitors' experiences, ranging from intense spiritual visions to impressions of what may have taken place in the canyons long centuries ago, especially visions of refugees from what they believe to be the cataclysmic destruction of Churchward's legendary pacific continent of Lemuria, the claim being that the nearby native Hopi Indians are descendants of ancient Lemurians. Many a visitor attests to sensing the presence of great crystals buried beneath the existing town, to which they attribute the emission of intense radiating energy, the source, they say, of the vortices' power in the area.

But, whether or not this is just the bright dream of a successful real-estate developer, happy to welcome little old ladies with pendulums, Ouija boards, and Tarot decks to dabble in the vortices, it is a fact that magnetic anomalies do exist on the planet, places where cars mysteriously roll backward up a hill as gravity turns to levity. Indeed, at Sedona the trees in the vortex in the canyon by the airport mesa do *not* grow perpendicular, but lean toward the center of the so-called vortex.

Our purpose—not with these anomalies—was with the secret of the Hunza water and its relation to the stirred BD 500, a field in which Pat Flanagan turned out to be an expert.

For years he has been collecting samples of water from all over the United States. Considered no more than an amalgam of oxygen and hydrogen, water is, in fact, far from banal. Constituting 90 percent of the human brain, it may, says Flanagan, be the most important substance on the planet, perhaps in the universe. With thirty-six distinct isotopes, each possessing different properties, it is the universal solvent of chemistry, capable, with time, of dissolving any and all the elements, even gold. Among its odd attributes, such as growing lighter instead of heavier as it freezes, water, as Flanagan explains the matter, has what is known as surface tension, a force that causes it to stick to itself, to form a sphere, the shape with the least amount of surface for its volume, requiring the least amount of energy to maintain itself. Yet its potential strength is ominous. Were all the extraneous gases to be removed from an inch-thick column of water, that column would become harder than steel.

In a bathtub, Flanagan elaborated, bulk water is composed of a small number of liquid crystals and a very large number of chaotically random molecules. "In theory, liquid water, even when boiling, has microscopically tiny 'icebergs' of crystalline water within it, liquid crystals that retain their set structure, whereas the rest of the water is all randomly oriented, vibrating vigorously. Cooling water automatically creates more of the crystals until nearly the whole mass becomes crystalline ice."

The point he wished to make was that when a living organism, such as a plant or an animal, takes in water it structures it into a composition with a high percentage of octagonal liquid crystals and a very low percentage of unorganized molecules. This, said Flanagan, is done by means of high-energy colloids—particles in suspension or solution too small to be viewed accurately with an ordinary microscope. "Colloidal particles act as tiny 'seeds' of energy, charged to attract freely-roving water molecules, and thus form the nuclei of liquid crystals. But to do this the colloids, normally of unstable charge, require a high electrical charge. In living systems, they retain this charge by being *protected* by a coating of such materials as gelatin, albumin, or collagen."

To illustrate his meaning with an analogy, Flanagan explained that similar colloids, not found in nature, are artificially manufactured by the detergent industry. These, we learned, have two poles, one lyophilic (liking water), the other lipophilic (liking oil), the latter on the inside facing the colloid, the former on the outside facing the fluid in which the colloid is suspended. This structuring allows the water to penetrate dirty clothing.

Another force causes water molecules to form long complex structures known as hydrogen bonds. These enable water to *wet* substances such as glass, clothes, powders, or one's hands, a force that can be strengthened or weakened by structuring the water's internal composition.

It was these two properties of water that showed Flanagan the way to duplicate the Hunza water, and thereby resolve part of the riddle of what takes place in a bucket of Steiner's BD 500 being "potentized" as it is stirred into vortex and chaos.‡

‡ The mystery of why water *stirred*, as in the preparation of the BD preps, 500 and 501, or *succussed*—violently shaken—as in the preparation of homeopathic remedies was finally given world exposure at the end of June 1988, when the British bellweather scientific journal *Nature* published a report its editors said they could not believe. According to deputy editor Peter Newmark, if the results of experiments run by four reputable researchers from France, Canada, Israel, and Italy turn out to be true, "we will have to abandon two centuries of observation and rational thinking about biology because this can't be explained by ordinary physical laws."

The work implied that antibodies in the immune system can function even when the solution they are in is so diluted that no antibody molecules are left in it. There is no known

In 1974, Flanagan found that crystals of all kinds, such as quartz and precious gemstones, have a marked effect on water surface tension, a characteristic known to ancient Tibetan physicians, who applied it to make crystal-affected water potions for their patients. If poured on wheat, mung-bean, soy, alfalfa, or radish seeds, the "crystal-affected" water produces more vigorous growth, and sprouts much larger and tastier than normal.

But where, Flanagan asked himself, did the crystals get the energy with which to affect the surface tension of water? It seemed to him they must be "resonators of cosmic energy impulses generated by super-novas, and other deep-space influences." To detect such forces he basi-cally thought to be gravitational in nature he constructed a device to pick up cosmic gravitational waves. These he converted and amplified so as to be recorded on a chart recorder, heard through a loudspeaker, or displayed on an oscilloscope.

Testing showed that, whereas ordinary tap water has a surface tension of 75 dynes per centimeter, the cloudy Hunza water that appears "dirty" when held up to a light source has a much lower one of 68, and is negatively charged.

Hunza water was also revealed, through spectrum analysis, to contain almost every known mineral element, with an especially high content of silver. To Flanagan the most interesting feature of the minerals was their being not in ionic but in *colloidal* form—meaning that the minute min-eralized particles, though microscopically small, do not dissolve in the water, as salt does, to become ionic sodium and ionic chlorine, but remain in suspension as tiny electrically-negatively-charged and there-fore self-repelling stable particles.

Pure water melting from the glaciers in the mountains of Hunzaland, Flanagan realized, would be devoid of minerals. But pressure from the millions of tons of ice—enough to grind to powder four inches of surface every century—would scrape and carry minerals to the valley at great speed to spurt into the *nullahs* or vertiginous torrents. And such is the peculiar nature of water that each time its speed is doubled it can carry sixty-four more times the amount of matter in suspension. In the fast

---

physical basis for such an action. It would mean that the solution was able to "remember" the presence of the antibody molecules and act *as if* they were still there.

Conducted in seventy separate trials in several different laboratories for a year, with numerous special controls designed to weed out errors and erase the results, the basic experiment held up under various stratagems to prove it wrong. "This is why we feel," wrote Newmark, "that it would be unfair *not* to publish the report."

Particularly objectionable to Newmark was the fact that the experiment tended to sym-bolically support homeopathy, or what he called "a generally discredited practice of using herbs and oils 'supposedly attuned' to organs in the body, to cure ailments in them, by dilution and vigorous shaking of the remedy."

nullahs, water thickens up with sediment.

"This colloid matter seemed so important," said Flanagan, "that I pursued the idea that it was these stable colloidal minerals that gave the Hunza water its special structure."

For years Flanagan tried without success to duplicate the colloidal minerals in his laboratory. "I tried all kinds, many of them clays. But none had sufficient electrical charge to lower the surface tension of water down to the sixty-eight figure characteristic of Hunza water."

Flanagan next discovered that the mineral particles in Hunza water had a fatty or oily organic acid around them, derived, he assumed, from old strata through which they had traveled, which, in his eyes, had to be made up of petrified forests or something equally ancient. What was necessary, he saw, was to find a way to make the nonsoluble minerals *colloidal* by artificially putting an electric charge on them.

It flashed to him that such a charge could be produced in the Hunza water by a vortex, or whirlpool, such as exist by the hundreds in the fast-flowing glacial torrents, an idea he had extracted, as had Podolinsky, from Theodore Schwenk's *Sensitive Chaos*. Therein he learned that all flowing water, though it appears to be uniform, is actually divided into extensive inner surfaces, or layers, moving against one another. Any

Patrick Flanagan in his forties in 1988 with his vortex machine in Sedona, Arizona.

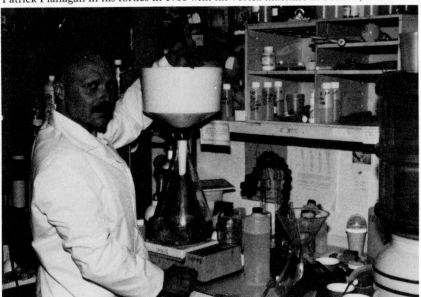

obstruction will cause these inner surfaces to flow at different rates of speed and form spirals or vortices, which separate from the rest of the fluid and generate electric current. In flowing rills, brooks, streams, and rivers, millions of vortices form when water rushes over or against stones and other obstructions. It was the vortex, Flanagan realized, that could put an electric charge on the particles of matter suspended in the water and thus render them "colloidal," each tiny element forced to be distinct, repelled from its neighbor by a similar charge. If he could find a way to duplicate this charging of the microscopic matter, he might produce an effective facsimile of the Hunza water.

From careful observation, Flanagan found that vortices have a special cadence of their own. They shrink in diameter and extend in length at one moment, then expand in diameter and contract in length at the next, continuing this oscillation in a periodic manner just like a pendulum or the mainspring of a watch.

To easily view the separate parts of a vortex, Flanagan added a little glycerin to the water and poured it into a clear cylindrical vessel—similar to, but much smaller than, the BD 500 stirring vat—with a hole bored through the bottom from which the water emerges in vortical form. When the shape of the container is altered, the closer its curvature comes to the ideal for sustaining a vortex, the smaller is the hole required. The perfect container, Flanagan found, was an egg-shaped ellipsoid whose length-to-width ratio was $1:\sqrt{2}$.

If a few drops of food coloring are added, the whole vortex appears to come alive. Not only can its rhythmic pulsation be observed, but layers of internal formative surfaces can be seen spinning much more rapidly than external ones, which themselves form corkscrew patterns reminiscent of the spirals inside conch shells or on the surface of various African antelope horns, all of which are built to scale on the basis of the Fibonacci series: 1, 2, 3, 5, 8, 11 . . .

In the vortical flows of water, says Flanagan, reside the secrets of its great sensitivity to cosmic force and its power as a bearer of formative living processes.

"When you read Schwenk's book," said Flanagan, "you realize the organs of every living thing are parts of frozen vortices. Schwenk gives example after example of vortexial formative processes in nature and comes to the conclusion that the vortex formation is tuned to the warp and woof of the Universal matrix. This accords with ancient Vedic texts written millennia ago in the Indian subcontinent that indicate the shape of the Universe as ellipsoidal."

In a monumental ten-volume work, *Wave Theory: Discovery of the Cause of Gravitation*, published in 1943, T. J. J. See, an American pro-

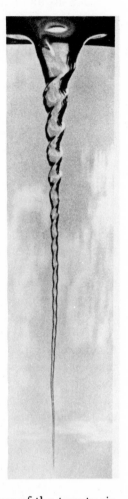

Vortex photographed underwater reveals the
spiraling surface between the water and the air which
is being sucked in. (Credit: *Sensitive Chaos*,
Theodor Schwenk)

fessor of mathematics who in the 1930s was in charge of the twenty-six-inch Equatorial Telescope at the U.S. Naval Observatory in Washington, D.C., showed that the entire physical universe revolves around a geometric figure known as a rectangular hyperbola, which also defines the curve of a water vortex. This basic curve, says Flanagan, was discovered by See to represent many phenomena, including the inverse-square law of electromagnetics; the laws of magnetism, gravity, and planetary motion; the temperature of the sun at any given point from the center outward; and—most important to the subject at hand—the surface-to-volume relationships and the structuring forces binding all matter.

When Wilhelm Reich, once considered by Sigmund Freud his most brilliant disciple, nevertheless broke with the psychoanalytic movement to make his momentous discovery of a life energy, "orgone," related by some to Steiner's "chemical ether," he found it to be made of *Kreisel-welle*, or spiral waves. In his book *Cosmic Superimposition* Reich describes the creation of matter from the throat of cosmic vortices, such as nebulae.

In his laboratory, Flanagan demonstrates the cosmic properties of a vortex. Closer examination eerily reveals that the circulating water in its laws of movement is a miniature copy of the solar system, and, on a larger scale, is reflected in the great stellar nebulae, as Reich already showed in *Cosmic Superimposition*.

Schwenk pointed to other properties of the vortex, suggesting cosmic connections. One such connection can be revealed by affixing a pointer, like a compass needle, to a wood chip. As it circles, the pointer always points in the direction in which it was originally aimed when the chip hit the water. Like the compass needle, it is constantly directed to a single point in infinite space. According to Schwenk, this is a clear indication of how a vortex is oriented as if it were held in place by mysterious cosmic threads.

Fascinated by Kolisko's experiments, which showed changes produced in crystallization of mineral salts by planetary transits, Flanagan realized the same cosmic energies could be captured in water undergoing turbulent chaotic movement or vortical flow, and—most important— that these energies would remain in the water after the flow ceased until the water was again agitated violently enough to disrupt them.

When the velocity of a vortex increases, the diameter of its throat gets smaller. In a perfect vortex, as the diameter approaches zero, the velocity approaches infinity. Because infinite velocity in the physical universe is impossible, something has to give. In the case of water, its molecules begin to dissociate into vapor, releasing electric charge. And here comes the first clue to what could be so mysteriously charging the Steinerian BD 500 with electric energy.

What happens, Flanagan explained, is that the hydrogen bonds of water molecules subjected to such stress are stretched like rubber bands; at which point they absorb planetary forces, which, as the molecule snaps back into shape are retained and energize the molecule. This goes a long way toward explaining Steiner's and Kolisko's planetary forces energizing the BD preps.

To find out just how much charge is generated by a vortex four inches in diameter, Flanagan lowered a thin specially shielded wire electrode into the center of its vortical throat, being careful not to allow the wire to touch the water. By means of another electrode touching the water he was able, when the vortex was moving at approximately one thousand revolutions per minute, to record a charge of more than ten thousand volts emitted from its swirling water: quite a boost from the cosmos.

As for the telluric forces that affect the BD preps while they are buried, in the 1930s an intrepid German physicist, Paul E. Dobler, showed that water moving in constricted underground passages radiates energy in what was then called an X-band of the electromagnetic spectrum, X because nobody then had the spectrometric equipment to differentiate any signal in this part of the infrared band of radiation. As set forth in the first of two books—most copies of which were destroyed by the Nazis—Dobler was able to demonstrate mysterious earth rays, such as are said to radiate from the Sedona vortices, by using a highly polished enameled aluminum plate inscribed with the words *Unterirdische Wasserader*, or "Underground Watercourse."

Combined with X-ray film and positioned on the surface of the ground over where a vein of water was known to be flowing, the film became exposed to earth rays via the aluminum, and when developed gave a clear picture of the words "Underground Watercourse." Physicists repeated Dobler's experiment, but were so perplexed by the results they rejected them because "physics recognizes no such radiation."

Continuing to work on the problem of vortices, Flanagan found that the vessel best suited to contain a vortex was one designed as a mathematical complement to See's rectangular hyperbola: a type of ellipsoid. His device, called a *vortex tangential amplifier*, was put to work in 1983 to create what Flanagan describes as a "perfect vortex," allowing newly-formed colloids containing all the ingredients found in both human mother's milk and in fresh fruits and berries to be fed into the vortex, where they became subjected to forces he says are creatable in no other way. These forces lower the surface tension of water treated with the colloidal mix—all the way down to an all-time low of 26 dynes per centimeter, the same as for ethyl alcohol.

"But very low tensions are not necessarily good," he explained, "in that when they get too low they are far from equilibrium. Over a period of time they lose energy and revert to water with a normal surface tension. We found that if we produced a water with a tension of 38—lower than the 45 created by washing-machine detergents—it has a stability that could last for years, perhaps as long as a century."

Flanagan based his statement on the fact that colloidal chemistry has discovered that large-sized colloids tend to "bounce around" and lose their electric charge, but tiny ones retain a charge, called a *Zeta*-potential, that is optimally long-lasting.

Thomas Riddick, a pioneer colloid chemist who formed his own Zeta-Meter company in New York, says that the *Zeta*-potential is a basic law of nature; it plays a vital role in plant and animal life to maintain a discreteness among billions of circulating cells that nourish the organism. The whole human body is made up of colloids, and all its flows are based on electric attractions. Blood cells have a protective coating of albumen, which keeps them charged, stable, and uncoagulated. Wrong foods, says Flanagan, tend to destroy the electric charges on your blood cells, which then coagulate and get sluggish, eventually dying. But, if you are able to take in highly charged colloids from fresh food—or from Hunza water—they help enhance the overall negative electric charge on the blood cells.

By adding one ounce of his newly-made colloidal mixture with its 38

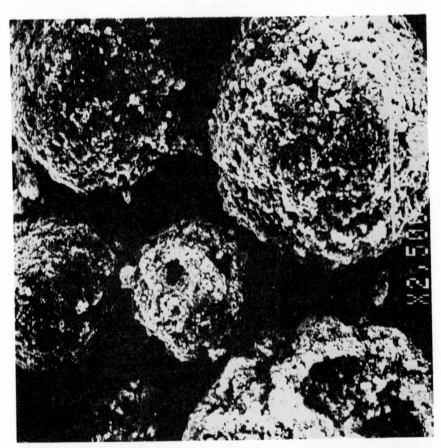

Professor Kenneth Reed, head of the University of Minnesota's department of mineral engineering, examining the colloidal particles in Flanagan's water with an electron microscope found them to be "a hundred times more active and stronger than the particles in any known surfactant due to a much greater dispersal capability in suspension."

surface tension to a gallon of distilled water, Flanagan has been able to create a product with a surface tension of 55 to 65 that he feels may have the same positive biological effects on living organisms as does the Hunza water.*

To find out, he allowed his thirteen-year-old bitch, "Wishes," a 115-pound cross between a St. Bernard and a Great Dane, to drink nothing but "Hunza" water imbued with what he calls his "Crystal Energy Concentrate." Within just three weeks, the enfeebled dog, which previously had to be lifted over logs on the trail in the forest behind the house, was able to leap them by herself. In Ohio, one of Flanagan's friends gave the treated water to a valuable thoroughbred stallion so old he could only get a live foal once in three breedings. Soon after being put on the water, the horse, whose coat went from dull to shiny, was impregnating all his mares without a miss. Given the same water, the mares also had such easy births that often the stable men would arrive to find newborn foals on their feet, though no signs of labor had been evident in their dams the previous evening. Race horses known as "bleeders"—because their lungs are so weak they emit blood from their nostrils as they run—showed no more bloodletting symptoms after a regime on Flanagan's water.

More important to humans, and especially to secretaries, authors, and all those obliged to face a computer terminal for hours on end, is the risk, according to specialist John Ott, of having their red blood corpuscles coagulate or clump into aggregated masses the French call *rouleaux*. Flanagan videotapes, taken through a microscope in his laboratory, show such masses disaggregating in a matter of minutes after ingestion of his water.

While these extraordinary biological results are as yet no more than empirical, strict "double-blind" tests not having yet been attempted, laboratory trials run in Minnesota proved incontrovertibly that Flanagan water has amazing structuring properties.

In this instance the test subject was neither animal nor human, but plain old cement or—mixed with other aggregates such as sand or gravel—concrete. Made with ordinary water, a block of cement, thoroughly dried, has a resistance, prior to cracking or sundering, of 8,400 pounds per square inch. Add only a thimbleful of Flanagan's water to each gallon and a half of ordinary water in the mixer, and the same cement has a resistance of 12,300 psi.

* By June 1988, a professor at the University of Minnesota, Kenneth Reed, had found, through the use of an electron microscope scanner, that photos of Flanagan's colloids taken at magnifications of 25,000x and with a resolution of 1/10 micron were able to show that, unlike silica particles that have angular shapes, the colloids are spherical in shape. But they are so small it has not yet been possible to measure their exact dimensions.

In another trial, called a *slump test*—basically a measure of air content in partially liquid cement—it was found that the normally "entrapped" air, which makes cement weaker, was only 30 percent compared to 70 percent with ordinary water. Remarkably, the treated cement contained less water: with greater plasticity or flow capacity it is also 30 percent *lighter* than the untreated and, because much *less acid*, less corrosive to steel embedded in concrete. It also does not burn the hands and arms of construction workers who use it in such jobs as stuccoing.†

Flanagan says the electric charge on his colloids is so well protected they do not become unstable even if frozen, boiled, autoclaved, microwaved, irradiated by gamma rays, or immersed in powerful cationic electrolytes. He found this out, after exposing them to such influences, by measuring their electrical charge in a U-shaped glass tube, a type of electrophoresis cell in which colloids, if overcharged positively or negatively, will migrate to one of two poles.

Using this simple system, Flanagan established a direct relationship between *Zeta*-potential, surface tension, and the structuring effect on water.

When Flanagan viewed the Australian film on Podolinsky that shows the 500 stirring process, he at once saw the connection with his own research, and his explanation of what occurs is at last comprehensible in lay terms. "Steiner's idea that an energy enters the water each time the direction of stirring is reversed is right on the mark," said Patrick with excitement. "I, too, have run tests with reversing vortexes but, in my case, I have an instrument that creates instantaneous reversal and I've measured the electromagnetic charges involved."

Flanagan further cleared up the mechanics of how the vortex is actually created. "What is first formed in the Australian BD stirring machine is not a true vortex, but a *parabolic* curve called a *paraboloid of rotation*. Actually an egg shape, it relates to Steiner's idea that eggs and other ovoid forms such as walnuts or pecans are receptacles for the life energy suggested by Reich."

It is only when the machine stops and reverses that, if one looks closely, the parabola is seen to collapse instantaneously, and only just preceding the onset of what Podolinsky calls "chaos" does it then form a perfect vortex.

"The wide, deep center of the paraboloid collapses," said Flanagan. "It virtually *implodes* toward the center, at the time of reversal. It then

---

† Extended testing of the new cement was undertaken at the Lafarge Concrete Company, largest of its kind in Canada, where a Finnish-born expert and consultant declared he could hardly believe the results produced by Flanagan's water. So hard is the ensuing concrete that it is being considered as the best possible casing for highly toxic waste.

creates for an instant a tiny vortex throat, which disappears into foam and chaos. It looks like chaos, but you must know that any time it's created in water, it's filled with millions and millions, if not billions, of small vortexes. Also we have a certain amount of *energy* involved here, the energy of *revolution* rotating this tremendous fluid mass in one direction. When the direction abruptly reverses, where does that energy go? It has to be *absorbed into the hydrogen bonds* of the water and be absorbed *onto the particles* of the 500, rendering them *colloidal*, and readily ingestible by both the microorganisms and the hungry single-celled root hairs of plants."

That Steiner was onto a phenomenon somehow familiar to peasants in the old country is clear from the testimony of Viktor Schauberger, a forester who spent a lifetime observing the behavior of water in virgin tracts of forest in his native Austria and also in Bavaria. In a biographical book on *Living Water*, Swedish author Olof Alexanderson includes Schauberger's description of how liquid vortices are associated with strange energy discharges such as halos, ball lightning, and a levitational force capable of bringing heavy egg-shaped stones up from the bottom of a stream to its surface, where they float on top of the vortices. And *implosion* was the name given by Schauberger to the act of vortical collapse. It is the levitational energy, said Schauberger, that allows fish to leap over high waterfalls from the agitated pools formed by the cascading water below them.

Schauberger was the first to refer to the creation in nature of *Edelwasser*, or "live water," the artificial production of which he effected in his laboratory with egg-shaped "vortex reaction chambers," also called "implosion chambers," because the energy developed within them was centripetal (acting toward a center) instead of centrifugal (acting away from a center). "Live water" is the name given by dowsers to the flowing water they detect in underground veins that emits the radiation discovered by Dobler not detectable by dowsers in water coursing through an ordinary pipe. Schauberger maintained that the *implosive*, centripetal energy was the basis for life, whereas its opposite was the cause of decay and destruction—patently exemplified by atomic *explosions*. His cosmological outlook insisted that all our technology should be harmonizingly "going *with* the flow of Nature" rather than forcing actions contrary to natural motions.

Brought up, like Steiner, in the countryside, Schauberger took as much inspiration from the observations and practices of simple peasants as from the academically-schooled who hardly ever left the city.

In an essay, *Natural Farming*, he recounted a visit made to an old farmer whose harvests were superior to any in the region, but who was considered a little touched in the head by everyone in the community.

"It happened one day, as darkness was coming on," wrote Schauberger, "that I came to the farmer's house. In the courtyard I met the somewhat unsympathetic son and inquired after his father. 'The old one is in the back of the house,' the young man answered with an unfriendly gesture. 'Shout loud enough and he will come.'"

Schauberger found the old father standing in front of a wooden barrel as large as three or four buckets, singing in a weird voice. At the same time, he was stirring the contents of the barrel with a huge wooden spoon. It was not really a song he was singing, but rather a musical scale, rich in tone, ranging from falsetto to double bass. This the old man did as he bent over the barrel, singing loudly down into it. As he went up the scale, he rotated the spoon in an anti-clockwise motion. When his voice deepened, he changed the direction of the rotation of the spoon.

Schauberger says he thought to himself there had to be some reason for all this.

The farmer did not hear me coming, and after I had watched him for a considerable time, I was curious as to what he was stirring. Unnoticed, I came up to the barrel and glanced inside; there was nothing there except clear water. Eventually the old man noticed me, nodded in reply to my greeting, and continued to stir without pause.

My glance alternated between the farmer and the contents of the barrel. With a flick of his hand, he would throw bits of loamy soil into the barrel as he continued to stir the liquid first to the right and then to the left. At the same time he sang quite loudly and not altogether pleasantly into the open container. His stirring the water finished, the farmer loaded his barrel onto an ox-drawn cart and, proceeding into his fields, would dip a palm-frond into it and sprinkle the water accumulated on it onto the ground in very fine distillations, not unlike a priest who sprinkles water on his communicants. Eventually the water evaporated leaving exceedingly fine crystals which carried a negative charge. These crystals attracted rays from all directions and then gave them out again.

Schauberger noted that the farmer had, through trial and error, learned at what pitch his voice would set up a resonance with the shape of the barrel that would stimulate the molecular vibration of its contents. The practice, called in German *Tonsingen*—"singing to clay"—was done at specific times, principally immediately after planting and firming of seed in the soil, at just this or that side of Eastertide.

When Flanagan read Schauberger's account he was so struck by the apparent similarity of the old peasant's stirring treatment of his water with a colloidal substance and the one shown in the Australian film that he wondered whether Rudolf Steiner, who was also raised in the Austrian countryside, might not have derived his knowledge from the same kind of *Tonsinger*—"clay singer"—as the one described by Viktor Schauberger.

"It is also interesting," said Flanagan, "that the earth travels through space in a vortical manner. Like the other planets, it circles the sun at about thirty kilometers per second. Constantly and rectilinearly, it is also moving in space toward the constellation of Hercules at twenty kilometers per second. The combination of these two movements produces a helicoid trajectory, as beautifully shown by the Italian scientist Giorgio Piccardi, whose animated model revealing it was set up at the Brussels World's Fair in 1958."

A single calculation shows that the speed of the earth's spiral trajectory reaches a maximum in March and a minimum in September, at approximately the time the cow horns are disinterred, depending on whether they are buried in the earth's Northern or Southern Hemisphere. Could this, Flanagan wondered, somehow be related to the energy-charging effect in the cow horns?

Flanagan further related the 500 stirring process to the work of the Russo-Belgian scientist Ilya Prigogine, author of *Order Out of Chaos*, who won a Nobel Prize for showing that, in a particular type of chemical reaction, chaotic forces in open systems create highly structured order. The kind of environment in which chaos leads to order is called "dissipative structure" which, as Flanagan points out, is characterized by any crystalline structure.

"The energy of chaos," says Flanagan, "transmitted into the water's liquid crystals, and absorbed by them, could only, *à la* Prigogine, destroy the crystalline structures or bring them into a higher state of order."

Steiner, with his vortex, his chaos, and his stirring, may well have been toying, knowingly or not, with the vortical and ellipsoidal sources of life.

CHAPTER 10

Claws of Chelation

HAD THE Great October Revolution never happened to transform "all the Russias" of the Tsars into the Union of Soviet Socialist Republics, Alex Podolinsky might never have made it to Australia, and Dr. Albert Schatz, a second-generation Russian-American, might never have made one of the most important discoveries about soil formation, the basis for which was laid by Russian science. Love for the dark earth of Mother Russia had seduced her scientists, both Tsarist and Soviet, into a sensitive analysis of her constituent elements and functions, more intuitively delicate than the grossly venal approach of the West to its virginal prairies.

Born in Connecticut, just after World War I, Schatz spent his early years on a 140-acre farm acquired in Yantic by his Russian grandfather, a former resident of the Jewish "pale" who had never been allowed to own a single sod of the precious soil of Tsarist Russia.

In a row house on a tree-canopied street in the Mount Airy section of Philadelphia, Schatz told us: "The land in Connecticut was marginal, even submarginal, but for my grandfather it was a real step up, the fulfillment of a dream, to finally own his own land and work in nature." Boyhood work on the Yantic farm instilled in Schatz the urge to devote his life to agriculture. New Jersey's Rutgers University gave him a degree in soil chemistry, and a Ph.D. in soil microbiology.

"The Soils Department was unique," said Schatz, self-effacing yet intense. "It was under the direction of Jacob Lipman, a Russian-born professor who had immigrated to the United States and later became the Dean of the Rutgers College of Agriculture. The whole field of *pedology* —from the Greek *pedon*, 'earth,' or 'soil science'—now in disuse thanks

116

to the proliferation of chemical fertilizers, was originated and developed in Tsarist Russia, beginning with Vassily Vassilyevich Dokuchayev, a seminarian turned geologist."

Albert Schatz, Ph.D., discovered the wonder drug streptomycin and developed the concept of chelation as a major chemical mechanism in the formation and fertility of soils.

Dokuchayev, who lived from 1846 to 1903, developed the singular view that soil was a "living body" as independently distinct in nature as any birch or oak tree, any heron or finch. His outlook was founded on a lifetime of work that took him all over European Russia to study soils, laying the foundation for a science that differentiated between living and dead branches of nature. At the age of thirty-seven, Dokuchayev published *Russkii Chernozem* *(The Black Soil of Russia)*, the first extensive treatise on the incredibly productive soils of the Ukraine. Three years later, he put out another volume, illustrated with maps, the first to present the concept that soils can be classified by "geographic zones."

"Prior to Dokuchayev's research," said Schatz, "and unfortunately subsequent to it, soil was and still is largely viewed, from the geological angle, simply as the upper, weathered crust of the earth, or, due to its

* The word *chernozem*, meaning "black earth" in Russian, has been incorporated into English and other world languages to describe humus-laden rich soils, also found in an estimated 560,000 square miles of land stretching from North Dakota to Texas and Minnesota to Oklahoma.

agricultural applications, solely as a medium for plant growth, as something in which a plant takes root."

Dokuchayev's follower Konstantin Dmitrievich Glinka (related to the
opera composer) extended the scope of his master's work with his massive
The Great Soil Groups of the World and Their Development. Translated
into German in 1914, it had a profound influence on the penetration of
Dokuchayev's views into Western Europe. Glinka's election to the presidency of the new International Society of Soil Science in 1927 led to the
book's translation into English by a Missourian, Curtis F. Marbut, head
of the U.S. Department of Agriculture's Division of Soil Survey, who
was so impressed with Glinka that he taught himself Russian at age
seventy to become acquainted firsthand with the literature of the expanding "Dokuchayev school," mastering the language well enough to
converse with Glinka and his fellow Russians in their native tongue. A
year before his death, Marbut wrote an introduction to Joffe's seminal
book *Pedology*, published in 1936,† which made use of hundreds of Russian articles scattered in dozens of journals, wherein he recommended
that all American agriculturists familiarize themselves with at least the
broad lines of the Russian development of the science.

"The Russian work," wrote Joffe, "brought the study of soils out of
the chaos and confusion of the geologic, agronomic and chemical (Liebig) points of view. For the first time, it plainly showed that soil is not
just physical but is intimately related to the biological bodies within it"—
the myriads of insects, worms, bacteria and fungi that give the soil its
life, and vice versa.

Marbut stressed that, at a time when Western Europe was still engaged in the futile assertion that soil *per se* is dominated by the materials
out of which it has been built, the Russian workers had long since shown
that soil is the product of a process rather than of material, and is,
therefore, not a static, but a developing body. They allied soil to *life*
rather than to *death*. In Marbut's view, Dokuchayev occupies the same
position in soil studies as Sir Charles Lyell in geology, Linnaeus in botany, and Mendeleyev in organizing chemical elements into his famous
Periodic Chart.

Marbut lamented the fact that soils in the United States were looked
upon solely as producers of crops, and

> no discussion of soil development worthy of the name was attempted in any
> book dealing with their study. The treatment of soils as such is not considered
> "practical," but the subject must be made practical at all hazards. It is outra-
> geous that the whole field of the significance of the soil as a body in plant

† Revised and greatly expanded in 1949.

production remains wholly uncultivated because of the dominance, even in the third decade of the 20th century, of the ideas of Justus von Liebig that completely overlook the fact that the soil lies in the twilight of life, as a connecting link between the living and the non-living, between material animated by vital forces and material subjected to physical forces.

Schatz, who had learned to speak Russian and could read the language, said this advantage was the one that led him into pedology. "I call myself a pedologist in tribute to my Russian teachers because people who say they are 'soil scientists' these days are largely taken up, not with pedology as such, but with the unnatural effects of chemical fertilizers as they apply to plant growth. And soil science here is concerned with, and equated to, agronomy, which focuses on the large-scale production of field crops such as grains, and others. My perspective has always been inspired by the original Russian definition."

In 1966, Schatz was invited to write an article for *Compost Science*, published by the Rodale Press in Emmaus, Pennsylvania, to explain why soil research, subsidized almost completely by the U.S. fertilizer industry, was leading to a decreasing interest in humus and soil organic matter in the United States to the point of near extinction.

The central problem, to Schatz, was that the grants to colleges of agriculture provide continual financial support to an army of graduate students while at the same time their professors are raking off thousands of dollars as consultants to the same fertilizer moguls who are supporting the students' research.

The research problems pursued are predominantly those from which the fertilizer companies can make more profit, while the indoctrinated graduate students go on to work along the same lines after receiving their degrees, and those receiving teaching appointments train new students in the same chemically-oriented philosophy.

"Thus," wrote Schatz, "with each passing generation, humus and soil organic matter become more and more abstract textbook topics which gradually lose their identity as real objects of vital interest. This decreasing attention dramatically illustrates how private and selfish economic interest can distortedly influence the direction of scientific research in a major discipline."

This neglect has resulted in a corresponding decline in *soil microbiology*. Graduates originally trained in the discipline were herd-mindedly rushing to accept lucrative appointments in pharmaceutical, fermentation, and fertilizer industries, where financial remuneration was steeply higher than in university faculties. All of which echoed Alex Podolinsky's complaint that any presentation of soil science, of whatever stamp, has been all but dropped from the curricula of Australia's agricultural learning centers.

In eastern Europe, on the other hand, scientists, while pursuing investigations on chemical fertilizers, insecticides, and weed killers, have not abandoned research into the organic content of the soil. In 1961, a collection of long essays, *Soil Organic Matter: Its Nature, and Its Role in Soil Formation and in Soil Fertility*, was edited by M. M. Kononova, a member of the U.S.S.R. Academy of Sciences and a senior researcher at the Dokuchayev Soil Institute. It was made available in English by the Pergamon Press in London. Three years later, an illustrated book, *Humus and Its Significance*, authored by S. Prat, was published in Czech by the Czechoslovak Academy of Sciences in Prague; it presented a complete history of humus research as it applies both to the natural and artificial fertility of soils. It went into detail on the definition, classification, and characteristics of all humus substances, their transformation under natural conditions, and their physiological effects on germination, morphology, anatomy, chemical composition, photosynthesis, assimilation, respiration, nutrition, specific enzyme systems, plant growth, and general development, as all these pertain both to algae and higher plants.

"But even in its Russian heartland," said Schatz, "this important humus research could not significantly influence the turn-around from Liebig-style artificial agriculture, so deeply entrenched in the Soviet and East European Ministry of Agriculture thinking."

Drafted at age twenty-two into the U.S. Army, Schatz worked in Florida military hospitals, where he collected microorganisms in the state's soils and swamps, and the oceans off its coastlines, to test them for antibiotic properties. At Rutgers, as the only researcher who dared to come personally in contact with the tuberculosis bacillus, he went on, alone, in a basement laboratory, to isolate a nontoxic antibiotic, which he called *streptomycin*—because the spores from which it is made appear in chains, *strepto* being the Greek for "twisted," and *mykae* for "fungi."

Possessed of an unusually broad antibacterial spectrum, and effective against tuberculosis, it surprisingly came from two independent strains of bacteria, one taken from cultured field soil, the other from the throat of a chicken, thus revealing the power of a family of microbes to exist in the earth itself as well as in creatures walking upon it. The discovery of this life-saving drug, made by a twenty-three-year-old student earning only $40 a week, was awarded the Nobel Prize. The recipient, however, was not Schatz but his professor, who coauthored the patent.

Unabashed by such injustice, Schatz's curiosity was next attracted to the problem of what exactly was happening chemically in the formation of soil as rock is disintegrated into smaller and smaller particles.

The question that puzzled Schatz was: how was it that lower forms of plant life, such as lichens, could extract mineral nutrients from the bare rock surface on which they grow.

A solution to the problem of how the obvious rock-weathering properties of lichens actually work seemed to Schatz to have important practical implications in agriculture, especially as it related to the total amount of minerals in soil on the one hand, and their availability on the other.

In 1924, E. J. Fry had suggested, in the *Annals of Botany*, a purely mechanical explanation for the action of lichens. And even in the 1950s the potential role of lichens in soil formation was subject to doubt by American soil specialists, two of whom concluded that the evidence for their importance was "exaggerated." They were, however, completely unaware, as Schatz was not, of important work by Soviet scientists on the pedogenic role of lichens. Russian studies indicated that though soils were essentially products of climate—of sun, wind, and ice—microbes and plants also contributed, like heat and frost, to the "weathering" of rocks, not only by penetrating deep into cracks and crevices to exert tremendous splitting pressures with their expanding cells, but, more importantly, by disintegrating rock through a mysterious chemical action known as "chelation."

Taken from the Greek word *chela*, meaning "claw" or "pincer," it is the property that allows the formation of a ring-shaped chemical structure, based on six carbon atoms, that enables lichens to clamp onto free floating metal ions. The ensuing more-complex compounds can then be absorbed by the lichen; once within their bodies, the metals are released to serve whatever function is required. The chemical explanation for this canny phenomenon is that in chelation one or more electrons can be *shared* between two elements, dragging the metals along to be released as the compound is soluble in water.

Chelating substances, present in up to 36 percent of the dry weight of lichens, give them the power to dissolve iron and other metallic minerals, grab them, and suck them up, enabling them to feast directly on the hard bare rock.‡

‡ Lichens, marvelously resistant to both heat and cold, considered by many to be the biblical "manna" of the Israelites, and eaten with relish in both Japan and China, are of unusual interest in that their thalli—or vegetative bodies, which, unlike leaves, stems, or roots, lack differentiation and do not grow from an apical point—are compositely made up of interdependently growing blue-green *algae* and filaments of higher *fungi*.
Here is a truly remarkable symbiotic relationship between two members of quite different species. Algae, which contain chlorophyll, range from microscopic unicellular plants to giant sea kelp a hundred feet long. Fungi, which also run from microscopic entities to twenty-pound puffball mushrooms, are now classed as neither plant nor animal, in a cate-

In 1954, Schatz published in the *Proceedings of the Pennsylvania Academy of Sciences* the first of a series of papers: "Chelation as a Biological Weathering Factor in Pedogenesis." As he continued to work on the problem, it came to him that if the chelating process could explain the predisposition of lichens to dine on no more appetizing a menu than a rock surface it could also be importantly involved in the creation of soil fertility. Were the plants not creating soil just as the soil created plants? And was soil formation not an ongoing process that *maintained* fertility?

"It was when I found out how lichens were chelating rock to extract minerals," said Schatz, "that I said to myself that something in the soil had to be working in the same way to keep on forming it from solid rock. One couldn't think that just a certain amount of weathering produces soil, after which the rest of the minerals remain intact in it forevermore. It had to be a case of a *continuing* chemical weathering process, which *continually* released minerals for plant growth, as the very basis for natural soil fertility. It was then that I concluded that there had to be a chelating agent in humus."

The clue—or claw—with which to unravel the mysterious mechanism of chelation came to Schatz when he ran across an article published 160 years ago in a Boston newspaper, *The New England Farmer*. It was the amazing story of a traveler to the ruins of the celebrated Temple of the Sun, in the high Andean town of Cuzco, Peru, who reported finding stones so finely cut and grooved he could not insert a needle between them. The art behind this feat of masonry had been lost with the demise of the Incas, who were said to have used the juice of a certain herb to soften stones prior to their emplacement.

Schatz immediately saw a possible connection between this lost art of masonry and the power of chelation. Delving into the literature on the civilization of the ancient Andean Incan population, he became aware of that people's finely experienced agricultural knowledge, perhaps the world's most sophisticated, then or since, as dramatically reported in *Lost City of the Incas*, by Hiram Bingham (later a U.S. Senator), who discovered the Inca city of Machu Picchu, one of the sites where Coanda found that the natives' water enabled them to live as long and healthy lives as the Hunza.

---

gory of their own. One member of this "odd couple," the alga, supplies carbohydrates through photosynthesis; the other, the fungus, supplies water storage and chemical salts. Alone on rock neither member could survive.

It is also generally accepted that, to break down rock, lichens use acids secreted by filamentous threads making up the bulk of the thalli, which are often submerged in soil or organic matter, the tissues of plants, or rocks. These acids are deposited on the outside of the thread cells as minute yellow, orange, red, or colorless specks.

To the Incas [wrote Bingham] the art of agriculture was of supreme interest. They carried it to a remarkable extreme, attaching more importance to it than we do today. They not only developed many different plants for food and medicinal purposes, but they understood thoroughly the cultivation of the soil, the art of proper drainage, correct methods of irrigation, land soil conservation by means of terraces constructed at great expense of labor. Most of the agricultural fields in the Peruvian Andes are not natural. The soil has been assembled, put in place artificially, and still remains fertile after centuries of use.*

Bingham considered Cuzco, the ancient capital of the Empire of the Incas, one of the most interesting places in the world. In the days of the Spanish conquest of Peru it was the largest city in America. On a hill back of the city stands an old fortress, whose northern wall is perhaps the most extraordinary structure built by ancient man in the Western Hemisphere. Wrote Bingham:

As an achievement in engineering it stands without parallel in American antiquity. The smaller blocks in the wall weigh 10 or 20 tons. Larger blocks are estimated to weigh 200 tons. A few of the largest weigh 300 tons! And yet they are fitted accurately together. The gigantic polygonal blocks cling so closely together that it is impossible to insert the point of a knife between them. The whole thing staggers the imagination.

Schatz's reading of Bingham convinced him that people who could develop an agriculture revealing an innate knowledge of the importance of organic material for soil health—which has nowhere been resurrected since its destruction at the hands of Spanish conquistadors—could

---

* Over a hundred years after Bingham's prophetic passage, American archeologists have discovered an advanced system of agriculture practiced by a pre-Inca civilization more than three thousand years ago in the Peruvian Andes. Using canals and three-foot-high raised beds, thirteen to thirty-three feet wide, and three hundred feet long, prehistoric farmers were able to reap bumper crops in the face of flood, drought, and killing frosts; with no chemical fertilizers, herbicides, or pesticides they were able to outproduce modern agricultural technologies.

An article in the Science section of the New York Times of November 22, 1988, describes how modern Peruvians, using nothing but ancient instruments and reconstructed pre-Inca platforms have reproduced an agriculture so hardy and so inexpensive as to form the basis for a new and healthier Green Revolution. The cost is minimal, amounting to no more than the human labor involved. Sediment in the canals from nitrogen-rich green algae and plant and animal remains provides natural fertilizer that in tests far outstripped chemically fertilized fields.

Millions more abandoned platforms have been found throughout Latin America. Dr. Clark Erickson of the University of Pennsylvania's Museum of Archeology/Anthropology, responsible for the discovery, hopes the old Inca system can be reintroduced as a replacement for the uneconomic capital-intensive systems so dependent on expensive machinery and fertilizers.

equally well have discovered the use of organic chelating material as it applied to stonework that made them into construction engineers of rare ingenuity, skill, and competence.

Why couldn't the Incas, he asked himself, have used the principle of chelation to soften a stone surface to the point where, like mortar, it could be leveled to requisite shape or flatness? Another clue to this possibility was presented when he came across the Inca tale of a tiny bird, the *pito*, living high in the Andes, that used the juice of a plant to dissolve niches in hard rock as nesting places.

A nineteenth-century English explorer, P. M. Fawcett, who disappeared without a trace in the Amazon jungle, left an account of how the birds themselves made the holes. A native who spent a quarter of a century in the mountains told him that the birds, arriving at a selected spot with a certain species of leaves in their bills, fastened themselves to the rock, like woodpeckers on a tree, and rubbed the leaves in a circular motion, constantly flying away for more leaves with which to continue the process. After three or four repetitions, the birds discarded the leaves and began to peck at the rock, quickly forming a circular depression. The process took only a few days, at the end of which the holes were sufficiently large to serve as nests.

"I have climbed up to see the nests," said the native, "and, believe me, a man couldn't drill them more perfectly. The birds don't peck away the solid rock with their beaks. Whoever has watched them at their work can see that these birds know about a leaf which has juice that softens the rock and leaves it like wet clay."

At first, Fawcett accepted the native's description only as a "fanciful tale," but when many others all over the country offered similar accounts, he began to wonder whether it could be true. Finally an Englishman, about whose veracity Fawcett had not the slightest doubt, told him a story that clinched his belief in the original tale.

My cousin was in the region of Chuncho," said the Briton, "near the Pyrene River in Peru. His horse became lame, so he dismounted and took a short cut through part of the jungle where he had never penetrated before. He was wearing riding pants, high boots, and spurs, not the English kind but those large Mexican ones four inches long with center discs as large as a half crown, and these spurs were practically new. When he arrived at his neighbor's farm, after a hot and difficult climb through tangled thickets, he was astonished to discover that his beautiful spurs had been destroyed, eroded in some manner and reduced to black nails one-eighth of an inch long. He couldn't understand what had happened until the owner of the neighboring farm asked him if he had walked over certain plants about one foot high and with dark red leaves. "This," said his neighbor, "is what corroded your spurs; it is the substance which the Incas used to work their rocks because the juice of that plant softened their rocks and converted them into a paste.

At last it was clear to Schatz that lichens produce and exude a chelating chemical that enables them to soften and sieze from raw rock the elements they need to survive. When he included this story in his *Teaching Science with Soil,*† a delightful handbook for high-school students, which describes many simple experiments to demonstrate visually what is happening in the earth, Schatz received a newspaper clipping from Chile, where he had taught as a professor at the University of Santiago. The Chilean newspaper *Mercurio* reported that a Peruvian priest, Jorge Lira, in one of a series of archeological expeditions carried out over forty years, had finally found a plant which the Incas could have used to soften rocks. Two more years went by before Schatz could locate Father Lira, who sent him a letter from Cuzco to say the plant's name was *harakkeh'ama* in Quechua, the Incan language still so widely spoken by many Andean Indians that Radio Moscow includes it in the several dozen languages broadcasting Soviet propaganda.

The remarkable ability of a plant to disintegrate iron is not surprising, according to Schatz, in light of 1964–68 research proving that chemical constituents of plants can attack even the hardest steel at a surprisingly rapid rate, as seen in the dulling and blunting of steel cutting tools in the lumber industry. These chemical compounds, also present in humus and various forms of composted organic matter, are among the substances that react upon minerals and rock particles in the soil, converting iron, manganese, copper, zinc, and other metals into water-soluble complexes, thereby making the trace elements available to plants.

Consequently, one of the most—perhaps *the* most—important properties of well-composted material, says Schatz, is its ability to react on soil minerals, just as lichens react on rocks. The compounds in the now rare Andean plant reveal clearly how soil organic matter solubilizes soil minerals.

It can be shown in a simple experiment that chelating acids are also excreted by many bacteria, which equally dissolve minerals in soil. It is their ability, with acids like those excreted by lichens, to put minerals into a colloidal solution that makes them especially important to the creation of soil fertility because plants can use such minerals *only* when they are chelated.

Knowing that bacteria are sparsely, if at all, present in dying soils, Schatz wondered which other chemical compounds might play the

† A companion volume, *Teaching Science with Garbage,* appeared a year earlier.

metal-grabbing and metal-releasing role comparable to that played in the much simpler lichen-rock association. More specifically, which chelating-effective compounds would be present in the soil, in *quantity*. After several years of detailed analysis of chemical structure of humus—performed mostly in Russia and Eastern Europe—Schatz concluded that only *humus* filled the bill.

Schatz also found that his quest was a lonely one in that no fund-granting institution responded favorably to his requests for support. The only interest shown in his work was that of the Kearny Foundation, which offered him a lectureship in soil science that allowed him to present his material orally to various academic audiences.

At the University of California's College of Agriculture, Schatz gave a lecture on "The Importance of Metal-Binding Phenomena in the Chemistry and Microbiology of the Soil," in which he presented all his own experimental results. But when he found, to his amazement, that no U.S. agricultural journal would publish his lecture, he was saved by an Indian publication in New Delhi, *Advancing Frontiers of Plant Sciences*, which accepted it in 1963.

"No one in the United States seemed to care about the subject," said Schatz, "though it is vitally connected with health, not only that of plants, but our own. The difference between soil richly endowed with humus and one deprived of it is the difference between a well-nourished citizen in a developed country and a man suffering from starvation in the Third World. Just as the poorly fed soil will produce weak plants so the poorly nourished Third World native will produce sickly children."

Chelation goes on not only in the soil and in microbes but in the cells of plants and in the bodies of animals and humans. How closely plants and humans are related can be explained by the extraordinary fact that both depend on a chelating chemical compound basic to their physiology. In man it is the deep red *heme* that transports in the blood the oxygen liberated by plants, which themselves have a compound, green-colored *chlorophyll*, that is so similar to heme that, to depict its chemical formula, it is necessary only to substitute an iron atom for one of magnesium. "It is one of nature's miracles," says Schatz, "that it could so simply modify a key life-compound, one way for animals, and another way for plants."

Asked why the feeder roots of plants prefer the humus packed in a below-ground open jar to the soil around it, as demonstrated in Podolinsky's experiment, Schatz replied: "First, because the trace minerals prized by plants are more readily available in humus since it has already acted as a chelating agent to solubilize them. Secondly, the far greater microbial activity in the jar breaks down the humus itself so that the roots can absorb the organic products coming out of it."

With respect to the colloidal nature of humus and compost, Schatz pointed out that the whole of humus is not made of colloid substances but only that part which is chemically constituted to act as a *chelate*. The reason colloids remain in a liquid suspension, he added, is that their surface-to-mass ratio is enormous. To illustrate his point, Schatz put a large wad of ordinary steel wool in a tinned coffee can and applied a match to it. The wool exploded into a white-hot flame. "If you try to ignite an iron nail with a match," he said, "nothing happens. But if you convert the iron in it to very fine steel wool, the surface-to-mass ratio, as in colloids, becomes very large. If you weigh the wool before and after burning you'll find the weight has increased because the final product is no longer just the iron, but iron plus oxygen, or iron oxide. The same thing happens with an iron nail if you allow it to rust in water, the difference being that in the burning of steel wool the oxidation effects are far more rapid.

"To relate this to the *chelating* and *colloidal* properties of humus means that their combination brings about a faster and greater chemical effect than if they were both not acting in conjunction. Nutritionally speaking, whereas a pig, or human being, can't eat a nail, it can easily ingest chelated iron."

Thus Schatz provided for his students a series of seemingly separate links, including studies of chlorophyll, blood, lichens, humus, Incan masonry, the *pito* bird, and a plant called *harakkeh'ama*. Forged into a chain they help disclose the chemical secrets of creation of soil and humus, from which all of life derives. But what the motivating forces might be behind this miraculous process of chelation remained a mystery. What was it that was moving the atoms of elements to put out claws with which to grasp other elements? Science, even with its most sophisticated instruments, could for the moment probe no further. Some finer form of vision was required. Hence Rudolf Steiner's clairvoyant gift, with which he was able to depict extraordinary cosmic formative forces at play and at work in the soil: forces that could motivate elements to search out and bond to each other by putting out claws of chelation.

Details of Steiner's background and of his development of the Anthroposophical Society, to whose members he gave his famous lectures on agriculture in the summer of 1924, can be found in the appendix to this volume, along with some of Steiner's more arcane explanations of what is actually occurring.

CHAPTER 11

Sonic Bloom

PLANTS, SAYS Steiner, can only be under-
stood when considered in connection with all that is circling, weaving,
and living around them. In spring and autumn, when swallows produce
vibrations as they flock in a body of air, causing currents with their wing
beats, these and birdsong, says Steiner, have a powerful effect on the
flowering and fruiting of plants. Remove the winged creatures, he warns,
and there would be a stunting of vegetation. His point was well made for
us in Florida.

A bird's-eye view across country south and east of La Belle, midway
between the great Lake Okeechobee and Sanibel Island, reveals an ocean
of citrus orchards cut by a skein of dusty "sea lanes," extending for miles
toward the shores of the Gulf of Mexico, once a paradise for seashell
hunters until ravaged by pollution.

Any bird overflying this greensward in the mid-1980s would have been
perplexed by the lack of avian fellows among millions of orange trees
growing in the confines of Gerber Grove, saturated by a fog of chemicals
laid down to ward off swarms of insects—except in Section I. There a
multitude of feathered fauna darted among the trees or perched singing
in their branches.

To this oasis the birds had been attracted, not by a natural concert of
their colleagues, but by a sonic diapason closely resembling birdsong,
which to human ears—incapable of distinguishing its varied harmonics
—recalls the chirping of a chorus of outsized crickets.

This sonic symphony was being emitted from a series of black loudspeaker boxes set atop twenty-foot poles, each resounding over an oval
of about forty acres. Its purpose was not so much to attract birds as to

increase the size and total yield of a crop of fruit, "hung," as they say in Florida parlance, on trees as if it were a collection of decorative balls at Yuletide.

"I have hung oranges the size of peas, shooter marbles, golf balls, and tennis balls, some still green, others fully ripe, all on the same tree, all at the same time," said Roy McClurg, a former Union City, Indiana, department-store magnate, part owner of the Gerber Grove.

We had driven down at dawn to his 320-acre holding, where two young field hands, brothers-in-law, each with a tractor and a trailer tank of foliar feed had started off between two long rows of trees, dousing them with an aerosol mist from top to bottom while a speaker, similar to the ones on the poles, tuned to maximum volume, shrieked a whistling pulse easily audible above the roar of the tractor motors.

Pointing to one of his many trees, McClurg raised his voice: "This is the typical fruit I'm getting with this brand-new method called Sonic Bloom. It synchronously combines a spraying of the leaves of any plants, from tiny sprouts to mature trees, with a broadcast of that special sound. With that process, simple but scientifically unexplained, I've been able for the first time to get fruit all over the inner branches of my orange trees, greatly adding to the 'umbrella'-type set which is everywhere the norm. And that's not all. I want to show you something far more impressive, even fantastic."

On a portion of McClurg's plantation three immature trees with more than half their branches withered or dead were being treated with a solution similar to the one in the trailer tanks, but delivered from plastic bags via tubes and needles punctured into the bark just above the ground.

"What you're looking at," said McClurg, "are three specimens afflicted with a mysterious disease called 'Young Tree Decline' or YTD. It has taken out at least one of every ten citrus trees in groves all over the state of Florida, which has spent more than $50 million, so far in vain, to try to come up with a cure."

Closer inspection of the sickly trees showed them, after only ten days of treatment, to be putting forth a host of brand-new sprouts all over their limbs, sure proof that their root systems, known to have been withering and sloughing root hairs, was recovering.

"YTD trees such as these," explained McClurg, "sicken at between eight and ten years of age, before they've begun to bear. It's as if a disease were striking young children. A healthy orange tree, like a human, can live up to eighty or ninety years. Most incredible, as you can see, these little trees are also beginning to put out a heavy bloom, trying to repropagate."

Back in his pleasantly refurbished clapboard house, oldest in the county, McClurg took from his refrigerator a dozen oranges the size of small grapefruit. "These were picked at my grove yesterday," he explained. "Ordinarily oranges as big as these would be pithy and woody inside, with very little juice." Slicing four of them with a razor-sharp butcher's cleaver, McClurg held up several of the hemispheres dripping with juice to show off rinds no thicker than an eighth of an inch. An electric juicer processed three of them to nearly fill a pint-sized glass.

"Oranges like these," said McClurg, "will give me a crop with at least a 30 percent increase in yield and a marked rise in 'pounds solid.' Add to that the fact that the Garvey Center for the Improvement of Human Functioning, a medically-pioneering research group in Wichita, Kansas, has tested the juice to show an increase of 121 percent in natural vitamin C over normal oranges, and you can understand that this new 'Sonic Bloom' discovery we're talking about not only improves quantity, but also quality. I've run blindfold tests with scores of ordinary people who have compared the taste of my juice with that of oranges from many other groves, and they all selected mine as the most lip-smackingly superior."

While McClurg was happily harvesting his oranges, Harold Aungst, a dairy farmer milking a two-hundred-head herd of Holsteins in McVeytown, Pennsylvania, was equally happily applying the Sonic Bloom method to a hundred-acre field of alfalfa, the deep-rooted leguminous plant grown for hay, brought to Spain in the eighth century by invading Moors and since spread to create agricultural wealth all over the world. Nor did his animals have any difficulty distinguishing the high-quality fodder sprayed with Sonic Bloom.

That year Aungst took off five cuttings, one shoulder-high and so thick he had to gear his tractor down to low-low to pull his cutter through it. With this harvest Aungst won the Pennsylvania State five-acre alfalfa-growing contest over ninety-three other contestants by producing an unheard-of 7.6 tons per acre as against a state average of 3.3 tons.

To dairyman Aungst, the size of his harvest was not its most important characteristic. Hay from this alfalfa fed to his herd that winter allowed the cows to step up milk production from 6,800 to 7,300 pounds per hundred-weight of cow, yet eat one quarter less feed. "I could hardly believe it," said the usually peppery Aungst, third-generation owner of his property. "My cows were devouring the alfalfa, stems and all. Other years they'd leave the stems just lay. A cow's nose is the very best barometer to tell how good your crop is. Cows are really finicky about what they eat. I threw down hay from another of my fields alongside this record-breaking alfalfa and the cattle first went for the feed exposed to that funny sound every time, changing over to the other only when the good stuff was all gone."

In the cellar of his house, Aungst showed us two dried alfalfa plants, one from his farm, another from his neighbor's. The Sonic Bloom specimen, twice as long as the other, was much greener and had a far thicker root mass.

"Let me show you something," said Aungst, holding the neighbor's plant by its root and flailing it against a bare table until the top was littered with dry leaves. Sweeping them off with his hand, Aungst slapped the Sonic Bloom plant down against the surface. Hardly a leaf fell off.

"There!" he said, speaking emphatically. "That should tell you something about the inherent quality of those two plants. When you have to move or ship the Sonic Bloom hay, it doesn't lose a lot of its bulk the way the other does."

One clue to the cows' preference was revealed in a test run on protein analysis by an "infrared scanner" at the Pennsylvania State University "Ag-Days" exhibition and fair. Aungst's sound-exposed hay scored a record 29 percent for protein and an extremely high 80 percent for "Total Digestible Nutrients" (TDNs). At the fair the same test showed similar percentages for Aungst's soybeans.

Across the United States in the Tiwa Indian pueblo of San Juan, New Mexico, twenty minutes' drive northwest of Santa Fe, the highly alkaline desert soils, composed of playa clay called *adobe*, best suited when mixed with straw to make cheap building blocks for houses, can be as hard-packed and impenetrable as a New York sidewalk. Yet a garden under the ministration of the same aurally-spiced nutrition as used in McVeytown and in Florida was growing as if in Eden.

Alongside more than fifty kinds of herbs, vegetables were flourishing, including tomatoes and carrots never before grown in that arid region at the confluence of the Chama and Rio Grande rivers.

To Gabriel Howearth, a bearded, pony-tailed master gardener employed by the tribe, veteran of several years' working with Maya Indian farmers in Mexico's Yucatán peninsula, Sonic Bloom was as miraculous in its results as was the Mayas' ability to grow crops with no chemical additives by simply mentally communicating with them in some mysteriously hermetic way long part of their ethos.

"As you can see," said Gabriel, parting the purplish-green leaves of a German beet to cup his hands around the top hemisphere of a swollen mauve-maroon root much larger than a softball, "I can't get my hands completely around it. All these beets, which normally scale off at no more than four pounds, will weigh at least nine, possibly ten."

With the steely features of a conquistador overlaid by the gentle traits of a Comanche, Howearth uprooted the giant and sliced it open with his Mexican machete. Like McClurg's oranges it had no spongy core. "Pure beet throughout," said Howearth. "Shows every sign of overwintering well. One of these will last a pueblo family a whole week."

He was also growing *quinoa*, favorite grain of the Aztecs, and *amaranth*, the prized staple of the Incas, both richer in the amino acids necessary to a body than any temperate-climate cereal. With Sonic Bloom, he had achieved a crop of both grains many times larger than any brought in by the Costa Rican *Centro de Agricultura Tropical y Ensenaza*, which had pioneered their cultivation at lower altitudes for over a decade and a half. "The remarkable thing about these crops," said Gabriel, "is their ability under this special treatment to adapt to altitudes much lower than those of their native climes. Like the beets and the rest of the herbs and vegetables, they're in fine balance. With this Sonic Bloom our sorry soils seem to be 'alchemized' into getting softer through the plants' transferring nutrients into the ground itself. You can test this by smelling, even tasting, the soil, feeling the 'crumb' of its structure, and noting how earthworms proliferate in it."

One of the native pueblo administrators scuffed the earth with his boot and said wonderingly: "I can't imagine what would happen if poor people like myself, working with bad soils all over the world, were widely afforded this remarkable method. It could help them grow a great part of everything they need to support their families, and on just a tiny plot of land."

Halfway up the vast arc which connects New Mexico to Pennsylvania, customers at the St. Paul Farmers Market were meanwhile raving about the taste of tomatoes, cucumbers, sweet corn, zucchini, squash, and other vegetables grown with Sonic Bloom, displayed there every Friday afternoon and Saturday morning. As one older buyer put it, as if speaking for the rest, "This produce tastes like it used to taste when I was a boy!"

The vegetables had been trucked to a special booth by William Krantz, a former successful Twin Cities stockbroker, sick of the financial rat race, who had bought a plot of ground in River Falls, Wisconsin, on the left bank of the St. Croix River, separating the state from Minnesota. On his two-acre vegetable plot, not much larger than those to which the Tiwa Indian had referred, Krantz saw cherry tomatoes, less than four feet tall, each bearing six hundred to eight hundred fruits per plant, *Cucumis sativus* vines sprouting three to six cucumbers at each leaf node instead of the normal one or two, sweet corn growing three stalks, each

Dr. Webster with tallest sweet corn on record (sixteen feet high), grown with Sonic Bloom.

with two to three ears, all from a single grain. In one corner, a lone viny plant occupied nine square yards of ground, mothering in the autumn sunshine thirteen huge saffron-colored pumpkins.

All of this produce had been treated by the same method used by McClurg, Aungst, and Howearth, obviating the need for any artificial fertilizer; it cost no more than $50 per acre to spray with Sonic Bloom. The same treatment has been experimentally applied to crops ranging from potatoes, broccoli, cauliflower, carrots, wheat, barley, and soybeans, to such exotic tropical ones as papayas, mangoes, avocados, and macadamias, in all fifty states, with results as startlingly impressive as those obtained in La Belle, McVeytown, San Juan Pueblo, and Three Rivers.

The idea was seeded in the mind of its developer one bitter cold winter day in 1960 in the Demilitarized Zone between North and South Korea. Dan Carlson, a young Minnesota recruit serving with the U.S. Army motor pool, happened to see a young Korean mother deliberately crush the legs of her four-year-old child beneath the back wheel of a reversing two-ton GMC truck. Tearfully the woman explained in distraught and incoherent English that, with two more children starving at home, only by crippling her oldest boy could she beg enough food in the city to feed her entire family.

There and then, Carlson decided he would single-mindedly devote the rest of his life to finding an innovative and cheaper way to grow food, accessible to anyone with even the smallest and poorest plot of land. Back home in Minnesota, he enrolled in the University's Experimental College. Like David Vetter at Ohio, he was allowed to design his own curriculum and reading program in horticulture and agriculture.

Soon he concluded that in poor soils, if plants could be appropriately fed, not through their roots, but through their leaves via the minute mouthlike openings called *stomata*—which plants constantly use to exchange gaseous aerosols and mists with the surrounding atmosphere—they might flourish and even grow rapidly in soils that were acidulous, alkalinely salty, arid, desert, or otherwise deprived of balanced nutrients.

But some motive force, he soon realized, was needed to awaken the stomata to action. Puzzling as to what this might be, Carlson stumbled on a record called "Growing Plants Successfully in the Home" devised by George Milstein, a retired dental surgeon who had won prizes for growing colorful bromeliads, members of an extended plant family as diverse as the pineapple and Spanish moss. Milstein's innovative idea had been to get a recording company, Pip Records, to amalgamate into a popular tune the pure sound frequencies broadcast by University of Ottawa researchers to increase wheat yields, which he had read about in *The Secret Life of Plants*.

Picking up where Milstein left off, Carlson focused on finding frequencies that would motivate the stomata to open and imbibe. Though he did not at first suspect a tie with the sound that caused the birds to flock to McClurg's orange grove, he managed through a stroke of spiritual insight to hit upon a combination of frequencies and harmonics exactly accordant with the predawn bird concerts that continue past sunup into morning.

To help create a new cassette tape of popular music into which his nonmusical sonics could be imbedded for inclusion in a Sonic Bloom home kit for use in small backyard gardens and greenhouses, and on indoor plants, Carlson enlisted the technical expertise of a Minneapolis music teacher, Michael Holtz. Within seconds of hearing Carlson's "cricket chirping" oscillating out of a speaker, Holtz realized its pitch was consonant with the early-morning treetop concert of birds outside his bedroom window.

The first cassette, using Hindu melodies called *ragas*, suitable to an Indian ear, and apparently delightful to both bird and plant, induced stomata to imbibe more than seven times the amount of foliar-fed nutrients, and even absorb invisible water vapor in the atmosphere that exists, unseen and unfelt, in the driest of climatic conditions. But the sound proved irritating to American horticulturalists and farmers, especially women, apart from those few whose tastes for the exotic accepted *ragas* as in vogue.

Looking for western music in the range of Carlson's highest frequencies, the ones which in Hindu experiments had shown the best bumper crops of corn, Holtz culled several baroque selections from the *Dictionary of Musical Themes*, settling on the first movement of Antonio Lucio Vivaldi's *The Seasons*, appropriately called "Spring." "Listening to it time and again," said Holtz, "I realized that Vivaldi, in his day, must have known all about birdsong, which he tried to imitate in his long violin passages."

Holtz also realized that the violin music dominant in "Spring" reflected Johann Sebastian Bach's violin sonatas broadcast by the Ottawa University researchers to a wheatfield, which had obtained remarkable crops 66 percent greater than average, with larger and heavier seeds. Accordingly, Holtz selected Bach's E-major concerto for violin for inclusion in the tape. "I chose that particular concerto," explained Holtz, "because it has many repetitious but varying notes. Bach was such a musical genius he could change his harmonic rhythm at nearly every other beat, with his chords going from E to B to G-sharp and so on, whereas Vivaldi would frequently keep to one chord for as long as four measures. That's why Bach is considered the greatest composer that ever lived. I chose Bach's string concerto, rather than his more popular organ music, because the timbre of the violin, its harmonic structure, is far richer than that of the organ."

Holtz next delved into what for him was a whole new world of bird melodies. In the 1930s, Aretas Saunders, author of *Guide to Bird Songs*, had developed a method of visually representing, through a newly devised audio-spectrogram, the arias of singing birds that can neither be described in words nor adequately shown with any accuracy on a musical staff.

Refined at Cornell University's Laboratory of Ornithology into "sonograms" which show electronic frequencies and amplitudes rather than musical notes, they were first popularly used in 1966 in a field guide, *Birds of North America*, where they are printed next to most of the individual descriptions of 645 species of birds representing 75 families that live north of the Mexican border.

A few songs of particularly high pitch—from 6,000 to 12,000 cycles per second or cps—such as that of the shy Tennessee warbler, whose protectively-colored bright green back blends into the leaves at the tops of trees, are as inaudible to many older people as high-frequency dog whistles. They are distinguishable in the guide because they have to be represented on an extra-large sonogram.

Soon Holtz came to see where the various predominating pitches in birdsongs could be calibrated by reference points on the musical scale and their harmonics. Don Carlson had instinctively hit upon frequencies that were the ideal electronic analogues for a bird choir. "It was thrilling," said Holtz, "to make that connection. I began to feel that God had created the birds for more than just freely flying about and warbling. Their very singing must somehow be intimately linked to the mysteries of seed germination and plant growth."

During visits back to the Iowa farm country of his birth, Holtz learned that there had once been literally thousands of songbirds all over the countryside. His Aunt Alice particularly missed the lyrically beautiful and extended flutelike trilling of various spotted-bellied thrushes, the high, thin whistling of the black-and-white warbler, and the buzzy five-note song of its cousin the blue-winged warbler, recognizable by its bright yellow head, throat, breast, and underparts. Most, if not all, of these songsters were long since gone from the landscape.

"I guess Rachel Carson was right," Holtz said nostalgically. "The spring season down on the farms is much more silent than ever before. DDT killed off many birds and others never seem to have taken their place. Who knows what magical effect a bird like the wood thrush might have on its environment, singing three separate notes all at the same time, warbling two of them and sustaining the others!"

One morning while Holtz was mentally bemoaning all the species of birds that had vanished from Iowa, a yellow warbler, looking for all the world like a canary, flew, as if reading his mind, to perch on the top of a tree outside his bedroom window and, as if cued by his band maestro's baton, burst into song. Holtz grabbed his tape recorder and managed to register an aria that went on and on for nine to ten minutes. In the field guide he found that the little bird registers a high 8,000 cps. Drawn deeper into the subject, Holtz consulted books that detail the structure of birdsong, such as *Vocal Communication in Birds, Born to Sing,* and *Bird Sounds and Their Meanings.* He also consulted biological texts to find that tiny *villi,* minute shaggy hairlike tufts in the cochlea of the human inner ear, vibrate to certain "window" frequencies.

"What I was trying to figure out with Dan Carlson was what exactly we were oscillating in plants," Holtz explained.

Looking at drawings of a cell, Holtz further discovered the representation of a subcellular structure within the cytoplasm known as a *mitochondrion.* Pointing to the enlarged drawing of one of them he asked: "Of what does their shape remind you?"

A glance suggested the form of the wooden-bodied sound box of a violin or viola.

"That's right!" Holtz exulted. "And I found it more than of passing interest that the resonant frequency of mitochondria is 25 cps, which, if interpolated upward gets to a harmonic of 5,000 cps, the same frequency used by Dr. Pearl Weinberger to grow winter wheat two and a half times larger than normal with four times the average number of shoots, as reported in Dorothy Retallack's *The Sound of Music and Plants.* It could be that the frequencies he used vibrated not only the mitochondria in the wheat seeds, but the water surrounding them, increasing the surface tension and thus enhancing penetrability through the cell wall."

Holtz connected this to Retallack's having also discovered that the transpiration rate rose, indicating greater growth activity in her experimental plants when they "listened" to Bach, 1920s jazz, or the Indian strains of Ravi Shankar's sitar; whereas exposed to hard rock, with the same rate nearly tripled, within two weeks the plants were dead.

"I believe such frenetic music," said Holtz, "was too much for their overall systems. The intense, grindingly monotonous energy in that rock sound could have virtually blown the cells apart! Young volunteers for the U.S. Navy who have listened to that type of music since childhood have been rejected because of partial deafness, even before reaching the age of twenty."

Asked if one could simply play the recording of a crescendo involving all of a symphony orchestra's instruments with their hundreds of frequencies and harmonics and allow plants to select those best suited for their needs, Holtz replied: "You have to take into account a law of diminishing returns. Too big a dose of anything is not necessarily of greater benefit than just a little or even a tiny dose."

It seemed significant that Holtz, the musicologist, could say this with-out any knowledge of homeopathic "potentizing."

Carlson, whom we met in Kansas City at one of Charlie Walters's annual eco-agriculture conferences, explained his approach with lively enthusiasm. "What I've tried all along to do with the *sonic* part of Sonic Bloom," he expostulated, his jet-black hair and pirate beard reflecting the hue of the Western-cut suit he wears for public lectures, giving him the air of an Amish elder, "is to stay within boundaries set by nature. I think there are certain cosmic forces which can account, however 'un-scientifically,' for much of our success. Properly adapted they will get plants to grow better, perhaps get cows to give more milk, or even inspire people to relate to one another more harmoniously. There's plenty of evidence that various frequencies of both sound and color can be cura-tive. But 'hard rock' is not consonant with nature's own harmonics. I believe birds exposed to it for long periods would fall ill and die, just as Retallack's plants withered away."

He waved his hands like an evangelist. "I get over a hundred calls a year, from people experimenting with my broadcasts. Most of them say that when the sound is turned on plants actually turn away from the sun to grow toward the speakers! Always! To me that means the sound is as important to plants as whatever we understand about photosynthesis. Perhaps that's what Rachel Carson meant when she intimated that 'spring' might one day be silent without Vivaldi's violins."

With a cold Minnesota winter coming on, and limited space in which to carry on his early experiments in a VHA-financed home, Carlson took a big step: he spent eighty-eight cents on a tropical *Gynura aurantiaca* or purple passion vine. Known also as a velvet plant, native to the Indone-sian island of Java, its fleshy teardrop leaves are densely covered with violet veins and hairs, and its yellow-orange dishlike flowers exude a nasty smell. But to Carlson this was his cherished baby. Once a month with a cotton swab he applied doses of nutrient to the tip of his vegetal pet, almost homeopathically weak doses, while simultaneously getting it to "listen" to his sonics. The swabbing turned the tip a withering brown, but quickly a new sprout burgeoned forth one leaf below the dead tip to grow at an accelerated rate. Within a few days, the original tip had completely recovered and was spurting rapidly ahead, both shoots ex-hibiting thick, healthy stalks and exceptionally large leaves.

As the vine crawled upward out of its pot, Carlson screwed teacup hooks into the wall of his kitchen, six inches apart, to support it; and so fast did the vine race for the hooks he had to add half a dozen every week.

At which point he made another startling discovery. If he snipped the growing tips with a scissors, the Javanese plant, far from daunted, put out a new shoot at the first leaf node below the cut.

As novel as this seemed to Carlson, he was even more puzzled by his pet's growing not only the teardrop leaves characteristic of its species, but also saw-toothed ones typical of its Indian cousin *Gynura sarmentosa*, along with completely alien split leaves previously never seen on any purple passion plant. The sound-plus-solution treatment appeared to be strangely affecting something to do with his vine's genetic qualities even as it grew.

In a paper on his experiment submitted to his professor, Carlson presciently asked: "Does one cell of a plant genus contain *all* the characteristics of *all* the species of that genus? If not, why has my plant, grown from a *Gynura aurantiaca* cutting, developed leaves, over 90 percent of its length, peculiar to the *Gynura sarmentosa* and, at the same time, exhibited an entirely new split-leaf form? Could the combined application of nutrient and audio energy result in such rapid growth rate that the very process of evolution is condensed? Have I enabled my plant to adapt more quickly to its environment? Is this the reason for the different leaf characteristics appearing on one plant? If any of these questions can be answered 'yes,' can this knowledge be applied to other plants? Could food crops be treated to achieve more rapid growth and better adaptability to their own or alien environments?"

As winter wore into spring, and summer into fall, Carlson noticed another oddity: his plant had bloomed not the usual once, but twice. Even more fantastic was its incredibly extending length. In only the first three months, the vine, which normally never exceeds a length of 18 to 24 inches, had grown a total stem of 150 feet. During the rest of the year it pushed on at the same rate, out of the kitchen through an inch-and-a-half hole bored in the wall leading to the living room, where it boustrophedonly roved back and forth along the ceiling on wires strung eighteen inches apart, to attain a length of over a tenth of a mile.

During the next year Carlson began snipping four-inch shoots from his vine, which he started in small plastic pots. Four hundred of these, labeled with his address and phone number and a request to call him for a replacement should the shoots die, he took to a flea market, where they rapidly sold for $4 apiece.

"I had many calls," he reminisced, "but none were to complain about sick or dying plants. Instead the callers wanted to know why the offshoots from my mother plant were growing twenty, thirty, forty, fifty feet long, and even more. I at once thought that this unheard-of development might give rise to the possibility of whole new strains of hardier superflora."

Despite this achievement, worthy of Luther Burbank, when Carlson, in happy excitement, asked members of his university committee to come to his house to see for themselves what he had done, their only reaction amounted to a yawn.

Didn't he realize, they asked, that, because his results had been obtained on a nonedible house plant, they were of no commercial value or interest?

"I was dumbfounded," said Carlson. "I could hardly believe this reaction. Here was the first time in their lives they had heard of sound being able to enhance the uptake of nutrients to produce the kind of growth I was getting, and they cast the result aside as worthless."

Desperate to get anything into the public record that would substantiate his achievement, Carlson wrote to Guinness Superlatives Limited in Middlesex, England, publisher of the now famous *Guinness Book of World Records*, which sent to Minnesota to check his claim "specialists in the matter of freaks in the plant kingdom."

Carefully measuring his plant's stem, inch by inch over its entire length, the freak specialists congratulated Carlson. That same autumn the new edition of the record book had an entry on page 113 extolling his find. To counter the notion that his new method was commercially valueless, Carlson next began to supply portable sonic equipment and nutrient mix to backyard gardeners who had called him after the Minneapolis *Star* ran a huge photo of the Carlson family standing under the passion plant, its leaves intertwined in the supporting chain of a chandelier before proceeding, through additional holes in the wall, into his children's bedrooms.

Not to be outdone, the St. Paul *Dispatch*, describing his African violets, with more than four hundred blooms in a full spectrum of colors, and his morning glories, purple, blue, white, red, and pink, as enveloping his house from its foundation to its roof eaves, quoted Carlson as foreseeing a Jack-and-the-Beanstalk world with gigantic flora capable of feeding multitudes while their stomata increased the earth's supply of life-giving oxygen.

Though he did not inform the reporter that the multicolored, oldfashioned trumpet-shaped morning glories had come from an ancient seed packet found by one of his mother's friends when she was cleaning out her attic, it did occur to Carlson that if Luther Burbank could coax a spiny cactus into losing its thorns, not through crossbreeding but by informing the plant that it no longer needed them because he would "protect it," he too might get his climbing plants to adapt to human desires.

"I subscribed to Burbank's idea," Carlson told us, "that at the highest level plants are capable of creating what is in the mind of man as a means of assuring their survival into future generations. I did not discount the many stories about trees which had borne no flowers or fruits for years, suddenly blossoming and bearing when threatened with an axe or a chain saw."

One spring, as he collected the seeds from his morning glories for successive annual planting, Carlson and his twelve-year-old daughter, Justine, meditated on how to get the vines to respond to their lovingly felt desires by focusing on their favorite hues, purple for Dan, pink for Justine. "We believed," said Carlson, "that the *plants* might respond to the colors we favored and draw closer to us as *we* were mentally and emotionally drawing closer to them." By late summer when the vines were putting out the usual mixed spectrum of blooms over most of Dan's house, he found massed all around his daughter's bedroom window nothing but pink flowers and around his own bedroom window only purple ones.

"This confirmed to me," he said, "that we can, in some still undefined way, communicate with plant life, which is even capable of altering the colors of flowers and the shapes of leaves. It must somehow be based on trust. The plants must feel your intent and realize that if they respond you'll save their seeds to assure their flourishing continuance."

Even more intriguing was Carlson's belief that his method would allow him to determine the very likes and dislikes of plants. By exposing them to a varied menu of nutrients hitherto unavailable to them, he aimed, through their reactions, to find out which selections they might prefer, instead of just forcing them, like human babies plied with distasteful turnips or liver, to accept what their parents believed, usually mistakenly, to be good for them.

This he hoped might ultimately lead to the elimination of deficiencies resulting in bad-tasting fruit or vegetables, the eradication of plant disease, and even, with their exposure to spice-laden aerosols such as mint, cinnamon, or nutmeg, the creation of apples with mint, cinnamon, or other flavors, right on the tree instead of in the pie.

"What I began to realize," said Carlson, "was that my method was *challenging the seeds' potential,* a potential maximized with the right number of Sonic Bloom sprays—which have turned out to be five—put on two weeks apart." Striking a massive fist on the table for emphasis, he added: "I believe I've come across a new principle that can be called *indeterminate growth!* It shatters the idea that plants are genetically limited to a given particular size or yield."

This belief in a lack of limitation led Carlson to another principle: *geometric progression*. "We began regularly to discover that plants treated during one growing season would pass along whatever changes were taking place in them, and create, right through their seeds, a successive generation 50 percent larger and more fruitful, even when the newly generating plants remained untreated with Sonic Bloom. I also call this *genetic elasticity*, the latent ability of plants to exhibit characteristics hidden in their gene pools, pulling out advantageous ones that may have been hidden for hundreds of years. This is connected to the ever-bearing trait brought out in McClurg's oranges."

Suggesting that the potentials in plants to respond to human wishes should be closely examined, he lamented that botanists, plant breeders, and genetic engineers have failed to understand the problem. "Scientists are rushing headlong into tampering with plants, monstrously slicing and splicing genes with as much surgical fervor as the ghouls who cut and maim animals in laboratories. This has led some of them to proudly announce that in order to produce a leaner grade of pork, they have developed a cross-eyed hog that staggers pathetically on legs that can hardly hold it up." He looked up and away with the firm yet benevolent gaze of a committed soul. "We should *tender* plants and animals, not distort God-given gifts still unrevealed in his creatures, but coax these gifts and learn to live cooperatively with all God's creatures."

He paused to allow the emotion in his words to simmer away. "Some people, with particular philosophies," he added, hardening his tone, "might accuse me of torturing plants, abusing their delicate nature. This is not so. I would challenge anyone to look at the model gardens I've set up, examine the radiant health of the plants, witness the remarkable fructification, and taste of this fruit. It is all done with nothing but affection, natural nutrients, and sound."

But perhaps the most encouraging prospect for fulfilling Carlson's dream of growing large quantities of food on very small plots of ground in a very simple manner is the marriage of his system with one developed by Ron Johnston of Mississippi, an amateur farmer in his thirties who doubles as a night nurse in a hospital in Memphis, Tennessee.

In a mixture of nothing but sawdust and sand in long rectangular boxes ten inches high, Johnston has been growing a staggering amount of delicious healthy produce. With discarded lumber from the sawmill, plus two pickup trucks full of free sawdust and one of sand, each box requires no more than a few hours of labor to build; and by Johnston's conservative figures a box eight feet wide by sixteen feet long can produce as many as 800 cantaloupes or 5,000 pounds of tomatoes—many times more than could ever be grown on the same size plot of ground.

Ron Johnston spraying Sonic Bloom onto tomato plants grown in boxes of sawdust and sand. Each plant, seven inches apart from its neighbor, can produce upward of sixteen pounds of fruit for a total harvest from a box eight feet by sixteen of over a thousand pounds of tomatoes.

"It all came together for me," says Johnston, "three years ago. Before that, I couldn't grow a thing down here on the dead soil of Mississippi. Then I got hold of a tape of Dan Carlson and I ran into a farmer using microbes. I also read about the French intensive method and that gave me the idea for the boxes. The system eliminates plowing, cultivating, and weeding. A daily watering can be automated and extremely economic. My water bill has gone up only a few dollars since I started; and during the drought of 1988, while my neighbors were cropless, my plants were a jungle of healthy green."

With a mere expenditure of $150, Johnston added a frame and plastic hothouse to his first box of sawdust and sand to produce tomatoes two months before his neighbors. Each tomato plant, planted seven inches apart, and producing twenty-five to thirty blossoms, gave as many as sixteen pounds of fruit per plant, some individual specimens weighing as much as a pound and a half. The chlorophyll content of the leaves was almost doubled, and they contained so much sugar that insects nibbling

Johnston's tomatoes, growing in boxes of sawdust and sand, matured weeks earlier than his neighbors' and produced individual tomatoes weighing as much as a pound and a half. Such tomatoes, planted in shallow ten-inch-deep boxes, can be grown on the terrace of a Manhattan penthouse or even on a sidewalk. (Credit: Ron Johnston)

on them were killed by an overdose of alcohol. Johnston uses no insecticides.

Two hundred strawberry plants in a narrower box produced two hundred quarts of strawberries with double the normal sugar content. And just one normal box of bean plants alone is enough to feed a family of four for a year. With cantaloupes clipped onto strings and climbing toward the rafters of the greenhouse, Johnston is able to hang twenty full-sized fruits from each plant.

Sawdust and sand form a fluffy consistency that allows plenty of essential air and water to reach the roots. But the real heroes of the system are forty-seven strains of microorganisms that Johnston obtains from a cultivator out in California. "I call them piranhas," said Johnston, only half joking. "They devour whatever nutrients are in the air and turn into healthy plant food whatever fertilizer I put into the boxes, transmuting potentially toxic salts into a balanced diet for each specific type of plant, providing them with a continuous flow of nutrients."

One teaspoon of microbes is added to a gallon of water and sprinkled around the plant stems; there they proliferate at the rate of 200,000 a minute, dying off individually every thirty minutes, but lasting, as a strain, as long as there is food for them to feed on. "The microbes," says Johnston, "eat any cheap fertilizer I provide them, and switch the elements around. They can turn potassium into sulphur, or whatever is excess into whatever is scarce. And my microbes feed the plants just what they need, just when they need it, providing them with a variety of minerals, the more of which the plants can get the better they taste and the longer is their shelf life."

Like camels, says Johnston, his microbes absorb a great deal more water than they need, which they then relinquish to the plants in moments of drought. Well fed, they proliferate down into the soil below the boxes to a depth of several feet, turning it to humus.

But all of this is only half of Johnston's story. The rest is provided by Dan Carlson's Sonic Bloom.* Every morning Johnston plays the enchanter sound to his plants, enabling them to suck in element-laden moisture from the air; and once a week he saturates their leaves with Carlson's liquid nutrient. "It all works in concert," says Johnston. "Sand and sawdust; microbes and fertilizer; Sonic Bloom and sound. Each by itself will not give the same results."

The whole system, as Johnston explains it, started as a hobby, then turned into a driving force. He's now determined to teach people anywhere in the world to grow a garden in their own backyard or terrace that will feed an entire family and leave a marketable surplus. "At first," he says, "people will find it hard to believe; but they'll be amazed when they find they can grow cantaloupe in quantity on the small penthouse terrace of a skyscraper in mid-Manhattan, or when a peasant in the Third World learns that with only a tiny corner of land and a little labor he can flourish as never before."

Ron Johnston paused to survey a whole acre of his boxes laden with produce, worth potentially a couple of hundred thousand dollars. "What I'd really like," he added with a winning smile, "is to help turn this planet back to what it was before the 'original sin' of desecrating the soil of Mother Earth."

* By the autumn of 1988 Sonic Bloom was being experimentally used in foreign countries with startling success. Representative of several reports was one, dated 10 September 1988 from Mohammed Azhar Khan, technical advisor to the Northwest Frontier Provinces' Farmers Association in Pakistan, who wrote to Carlson: "Using your method I have been able to increase potato yields by 150 percent over the national average and increase corn harvests by 85 percent over the national average."

CHAPTER 12

Seeds for Survival

To MICHAEL HOLTZ, the mystery of Dan Carlson's success with sound may be explainable in terms of the basic philosophy of India. For thousands of years its sacred texts have taught that sound, as the integrating phenomenon of life, holds the key to the mysteries of the universe, to the creation and sustaining of the physical world.

In Indian metaphysics there can be sound without vibration, even without the usual medium of conveyance such as air, water, or so-called solid matter, sound being the cause and not the effect of vibration. This "soundless" sound is said to be the source of cohesion, of electricity, of magnetism, of all that is. And it was a universal ancient belief that God, or a Divine Being, created the universe by means of a vibrating emanation, referred to by early Christians as *Logos*, "The Word." And musicologists point out that the whole-number ratios of musical harmonics—octave, third, fifth, fourth—correspond to an underlying numerical framework which exists from astronomy to atomic physics, through chemistry to crystallography, and even to botany.

In the Indian view of the cosmos, sound precedes light, and the whole universe is viewed as an ocean of sound followed by light of varying degrees of density and luminosity. Strangely, much of this wisdom of ancient India has survived among the native tribes of "Indians" in America.

On an Indian reservation in the arid northeastern corner of Arizona, at high noon of a scorching July day, with a hundred-degree temperature hanging heavy over land parched by an extended drought, John Kimmey restlessly awoke from his siesta in a cool stone house.

A schoolteacher and founder of New Mexico's innovative Community School in Santa Fe, Kimmey was a guest of David Monongye, a traditional religious leader and elder of the Indian tribe, the Hopi, settled for centuries atop a trio of side-by-side mesas. In their language the appellation *Hopi* means "peaceful" or "good."

As most of the tribe dozed away the midday rest hour imposed by Grandfather Sun, Kimmey let himself out through the screen door into the glare of the *Kisnovi*, the central plaza of the village of Hotevilla, fit at that hour only for mad dogs or Englishmen. Drawn by an urge for which he could not account, Kimmey ambled to the edge of the third mesa and down a dusty trail, guided, as if from the spirit realm, by an iridescent lizard across the countryside to a pile of stony rubble at the base of a cliff. Pausing, Kimmey became aware of an eerie song wafting through the arid surroundings. Beyond the last lithic outcropping, he gazed upon an expanse of planted corn so vast it didn't seem Hopi. The singing became clearer, coming from a soft, powerful voice, though Kimmey could not see the singer. To his astonishment he noticed that every one of the thousand or more clumps of waist-high Hopi cornstalks, each with a dozen or more ripening ears, was growing as lushly as if it had been planted in rain-blessed Iowa, the whole field contrasting sharply with the brown, wilted crops scattered on parched land all around the village.

Tiptoeing through the cornstalks, Kimmey spotted the grizzled white pate of an old Indian seated on the ground, his closed eyes impervious to the presence of any passing creature, including Kimmey, the atmosphere about him indicative of deep communication with the cosmos.

Gingerly moving away, Kimmey returned to his quarters, where his Hopi host smiled: "So you came upon old Titus! He still keeps the ancient Hopi way, which unfortunately has all but died out among the present generation. It's not the water, it's the *Navoti* that keeps his plants alive. He knows by heart the right airs to chant to his corn children, for whom he offers prayers at planting time. Most important, he knows that he should never worry, like most farmers, about his crop, because anxious thought is as damaging as extended drought. Instead of worrying, he goes to his children in the high heat of noon to impart courage to them with his generations-old songs."

"Surely," Kimmey protested, "all the other corn farmers must see what a difference the songs make to his crop. Why don't they sing to theirs also?"

The old Hopi sighed: "It's too late. *Navoti* is no longer alive in other men's seeds."

Returning to his home in Taos, where for fifteen years he had apprenticed in native ceremonialism with the Tewa Indian elders, Kimmey noticed for the first time in several years how little of the farmland around each of the nineteen Indian pueblos he passed was under cultivation for food or even animal fodder. Listening in his mind to Titus's haunting refrains, he felt he heard the old man's seeds calling to him that their *Navoti* power might still lie trapped in other seeds long stored in half-forgotten caches, in clay pots, old coffee cans, lard buckets tucked away in the dark corners of tool sheds, in the walls of adobe houses, or in long braids of dried corn woven into *ristras*.

What was calling him, he felt, was the old seed, gathered years, decades, even a century before the commercially-sold seed packets began to appear each spring on general-store and supermarket racks, old seed that could still retain its inner vitality, the ancient force imbued into it during an era when men still sang to their plants. The old seed chorused its desperation that it be found and lovingly put into the ground before it disappeared into oblivion.

When he told his eighty-two-year-old adopted Indian father of his experience, the old man's face seemed to reflect a soul swimming back through time to happier days. Arising slowly from his chair he went into a side room to emerge with three small buckets full of what sparkled to Kimmey like bright sapphires. As he stared at the dark blue corn seeds, he heard a choir similar to the one that had come from Titus's throat.

His Indian father explained that only the week before he had found the buckets stashed in an old trunk in his sister-in-law's house among ancient tools, strands of rawhide, assorted turquoises, and other tribal mementos.

Planted the following spring to Kimmey's own singing, the seed grew into nine-foot-tall plants, to the amazement of the pueblo's elders, who could not remember seeing anything similar since their childhood.

It made Kimmey feel that he had at last discharged a duty to Kokopelli, the mythological Hopi spirit most closely associated with plant fertility and seed germination. Known to non-Hopi Indian tribes as "the hunchbacked flute player," his unique image is painted on countless pots, or etched in stone, throughout the Americas over many centuries, the hump on his back recalling a sack of seeds to be scattered as he moves about, his Pan-pipe flute said to be the source of the spirit breathed into the seed.

Through contact with small seed companies, Kimmey went on to discover the truth of his Hopi agricultural lesson. And he also discovered a serious problem that has developed in the arbitrarily enforced and unhealthy domination of the seed market by modern hybrids that are not as disease resistant or as nutritious as the older, nonhybridized seeds.

As Pat Roy Mooney, a leader in the effort to promote a wide diversity of seed for the world of farmers, put it cogently in his 1979 book, *Seeds of the Earth:* "It should alarm no one that some genetic material, which might ultimately have been of use to major crops, has been, and is, vanishing. What should really cause concern is the massive, wholesale eradication of irreplaceable breeding material over thousands of square miles of arable land."

So fast is the prairie fire of genetic erosion spreading that biologist Thomas Lovejoy estimates that the world will have lost one-sixth of all living plant species by the year 2000. And Mooney himself concludes that existing wheat cultivars cannot survive twenty years without a constant influx of fresh genes from plants to allow them to stave off new pests and diseases.

The situation is more than precarious. Based on the rediscovery of Mendel's laws of genetics, a new science of plant breeding has been erected, which, within only a few decades—an "eye blink" in plant, if not human, history—has been engineered to meet the technological requirements of machine harvesting, grain milling, beer brewing, bread baking, and many other food-processing industries. In the process, genetic strength has been sorely weakened. As new varieties of hybrid seeds are produced to meet ever new demands, the risk increases of huge crop losses through disease. The only remedy is to have recourse to the ancient centers of plant diversity.

As genetic uniformity of the world's main staple crops increases, breeders are forced to rove deeper and deeper into half-lost valleys and forests in search of new—or rather older—genetic material.

The real heart of the problem, says Mooney, is the so-called "Green Revolution," for which Norman Borlaug won a Nobel Prize in 1970. Its basic impetus derived from the idea that "High-Response" non-self-perpetuating hybrids be exclusively relied upon. While the Green Revolution has been a plague on genetic resources—because it has led to a galloping erosion of native plant varieties in favor of highly inbred imports—it has also been a boon to the world's seed industry. Cost-free, these industries have raked over the genetic riches of poor countries to breed new varieties whose high yields are assured only by massive additions of artificial fertilizers and pesticides, sold, obviously, by the same companies, with their built-in bias for industrialized agriculture.

At the beginning of the seventies, according to Mooney, a number of factors propelled chemical firms into seeds. They had money to spend on acquisitions, and the economic environment was conducive to mergers. Second, the costs of bringing new drugs and pesticides onto the market were skyrocketing, owing to increased public-health and environ-

mental-protection concerns. The easy victories of the fifties and sixties
were ebbing and the struggle to find new compounds was growing
harder. Third, many of the major patents won in the fifties were dying
out and some top companies were buying other firms in fear of losing
their profitability.

With these concerns in mind, and the prospect of having to double
food production by the end of the century, chemical firms bought out
large areas of farmland, not only in North America but on every other
continent, and moved aggressively into produce marketing. In November 1970, *Business Week* asked, "What has household bleach to do with
lettuce patches? What have man-made fibers to do with raising hogs, or
chlorinated solvents with catfish farming? Not much. But a growing list
of U.S. companies based in chemicals and allied products are venturing
back to the good earth."

Yet American business journals failed to note how the seed business
was coming under the domination of giant conglomerates. The growing
dependency on chemicals of monopolized seed production has led to
what Mooney calls a highly unsettling trend: increased buy-up and control of the world seed industry by large chemical and agribusiness corporations.

"The effect in the course of fifteen years," says Mooney, "has been
to utterly transform the seed industry. Where once it was small and
family-based, it is now large and highly corporate. In 1972, Royal Dutch
Shell had no involvement whatsoever in plant breeding. Now, having
acquired thirty major seed houses, it is the largest seed company in the
world, with sales of more than half a billion dollars, and is seeking still
more acquisitions. Every other company with an interest in agricultural
chemicals—Atlantic-Richfield, British Petroleum, Ciba-Geigy, Monsanto, Stauffer, Upjohn, ITT, Occidental, Sandoz—is moving into
seeds."

Sandoz, the $3-billion-a-year Swiss multinational, has spent more
than $300 million acquiring American seed and agri-chemical compa-
nies. Along with DeKalb, Pioneer, and Ciba-Geigy, Sandoz accounts for
about two-thirds of all seed sales in the U.S. for corn, America's biggest
crop, and about 60 percent of the sorghum seed. As seed prices rise,
costing farmers hundreds of millions of dollars, the companies, instead
of delivering a self-perpetuating variety, can repeatedly milk the farmer
with a nonreproductive hybrid for years and years to come. U.S. farmers
and ranchers spend nearly $4 billion each year on seed; world-wide, the
market for agricultural crop seed is $45 billion. And, along with Upjohn,
Lubrizol, Limagrain, LaFarge Coppee, and Arco, Sandoz has also gar-
nered more than 230 vegetable-seed patents.

Other corporations hope the seed industry will open the way to expanded agricultural-products divisions currently manufacturing and marketing fertilizers, pesticides, and other agricultural chemicals.

"The intent is to become a full-service operation," says Bob Skaggs of the California Department of Food and Agriculture. "Not only can disease and pest-resistance and earlier and greater yield potentials be transmitted to crops through new seeds, but these can also deliver agricultural chemicals to the fields through 'controlled delivery systems' such as seed encapsulation processes."

Companies such as the Celanese Corp, which owns the Harris-Moran Seed Company, based in Modesto, and Celprit Industries in Manteca, California, have been experimenting with polymeric seed coatings. This process, which has been applied to seeds in the past, makes the seeds the same size and weight, ensuring uniform planting. Celanese also has been developing chemical "additives" for seeds, such as fungicides, to "protect" plants from soil organisms, pests, and bad weather. Mooney foresees the budding chemical-seed partnerships pushing their wares in favor of developing new plant varieties.

Many of the multinationals venturing into the seed business hope to capitalize on the huge, and no longer so distant, promise of genetic engineering. According to Ray Rodriquez, a molecular geneticist at the University of California at Davis, they see an annual market for biotech agricultural products that may be as much as ten times as large as the potential market for biotech-medical products, with a market potential of up to $100 billion by the year 2000.

With gene splicing, the entire plant kingdom becomes one open-ended gene pool, in which genes can be moved about freely from one species or genus to another. In higher plants, upward of ten million genes—more than ten times the number found in man—control the way the plant performs. With the modern ability to screen millions of plant cells in the lab in a matter of hours to move the required genetic characteristics from one cell to another, new varieties and breeds can be created in a matter of days, avoiding the old painstaking process of cross-breeding, selecting, and further breeding that might take up to ten years. The results can now be spread around the world in a matter of weeks through normal trading and shipping.

The Japanese consider this biotechnology the major technological revolution of the century, launching a race comparable to the space race, with impressive possibilities. Were the Soviets—aggressively pursuing biotechnology—to develop wheat resistant to even a few degrees of frost, they could greatly reduce their dependence on foreign imports.

In this new biotechnology, almost anything can be envisaged, including, says Jack Doyle in his *Altered Harvest*, a dairy cow as big as an elephant capable of producing up to 45,000 pounds of milk a year. The Carnation Company, known for its dairy products, is already genetically manipulating and selling frozen cattle embryos, which, split and re-seated, can produce a herd of a hundred clones. And the futuristically outlandish plans of the new plant monopolists to use the newly-born and fast-growing techniques of genetic engineering bid fair *not* to dispense with chemicals but to foster a vast expansion of their sales.

As seeds remain the primary ingredient of agriculture, Kimmey found corroboration of Mooney's concerns about their plight as he visited the chronically-underfunded seed-storage facilities of the National Plant Genetics Resources Board in Fort Collins, Colorado—healthily located halfway between a nuclear power plant and the Rocky Flats plutonium facility—which operates for some 240,000 seed collections the equivalent of a Fort Knox for gold. So crowded are the facilities that seeds were piled on the floor in brown cardboard cartons and sacks, uncatalogued, some having lost their ability to germinate, or with viruses in their germ plasm, the equivalent of a biological time bomb. Not that Kimmey found the collecting effort any better operated by the United Nations Food and Agriculture Organization in Rome, Italy.

In search of remedies, Kimmey came across a network of individual seed savers, principal among whom was Kent Whealy, a young midwesterner. Having organized in Princeton, Missouri, his Seed Savers Exchange (SSE), Whealy publicly appealed for help in collecting heirloom seeds for propagation and offered instruction for their storage and planting in gardens.

Invited to lecture at the Fort Collins seed bank to a prestigious audience of plant breeders and genetic preservationists in October 1986, Whealy was dismayed but not surprised to hear staffers tell him that their way of storing seeds was leading to a decline in the rate of their germination after a period of only a few years. Whealy's invitation had been in tribute not to any scientific prowess, but to what a single citizen had been able to accomplish in twelve years of effort in the area of seed preservation in the United States.

Born into one of the forty million American families who grow part of their own food, and cultivating a vegetable garden of his own almost from the time he could walk, Whealy was entrusted at the deathbed of his wife's grandfather with seed from three garden plants the old man's family had brought over from Bavaria four generations earlier.

Growing them out to ensure their multiplication, and their preservation through distribution, Whealy wondered how many other home gardeners might also have cached heirloom seeds in attics or cellars. So he

Kent Whealy examining airtight bottles that hold the Seed Savers Exchange collection of 2,200 varieties of beans. Most are heirlooms passed down from generation to generation since the *Mayflower* landed at what is now Plymouth, Massachusetts, in 1620. Though some seeds have been known to germinate after thousands of years of hiberation, all these beans have to be regrown every four years or they will die.

decided, in 1974, to mount a campaign to seek out their locations. As a result of his forays, he discovered to his surprise that while old vegetable and fruit seeds are not common, neither are they all that rare, if one knows where to find them. Whenever he and his associates went into isolated areas, such as the rugged backwoods of the midwestern Ozarks or hidden hollows along the blue ridges of the Appalachians, they were offered seed riches no longer seen in towns or cities or advertised in catalogues.

A whole heritage of plant spawn traded at country general stores or over backyard garden fences could lay the basis for breeding programs to produce a vast wealth of new garden varieties or reanimate the old ones.

Particularly solicitous about seed saving were religious sects that have maintained uninterrupted ties to the land, such as the Amish, Mennonites, Hutterites, and Dunkards. As Whealy built trust in his exchange enterprise, he was honored to attract to his effort members of American Indian tribes, long reluctant to share with the perfidious white man seeds which, as the sparks of life, they considered sacred.

In 1986, Whealy published a 250-page eleventh edition of his *Winter*

Kent Whealy, Director of the Seed Savers Exchange, working in its Preservation Gardens where some 1,200 varieties of endangered plants are being regrown from seed.

Yearbook, containing lists of 630 members offering to trade seeds or seeking to find varieties half-remembered from childhood. Over a decade the publication had grown to the point where his exchange membership was collectively offering 4,000 different strains, to allow for over 300,000 plantings of truck-garden and orchard crops, unavailable from any commercial source and mostly on the edge of extinction.

"The material we have been able to locate is really just the tip of an iceberg," says Whealy. "The unique heritage of seeds in our country has never been systematically collected because most government-sponsored plant exploration has focused on foreign countries. Our next task is to organize the fielding of several hundred local plant explorers, region by region, and state by state. Many professional plant breeders are becoming excited about what we have already uncovered, because it's material they've never seen before, much less been able to work with."

With these rare seeds the breeders produce, not hybrids, but openpollinated, self-reproducing stock. Typical of breeding programs to which Whealy referred is one at the University of Wisconsin in Madison, directed by Dr. Fred Bliss, who concentrates on beans to create *upright* plants twenty-four inches tall that can be machine-harvested. He has married into their hereditary font the genes of older varieties with tough, rank stems in what Whealy terms an architectural application.

One of many old bean varieties being brought back from the edge of extinction is the tepary, discovered by Dr. Gary Nabhan, Director of the Native Seed/Search program of the University of Arizona. With the remarkable ability to grow lushly green, even in normally plant-withering conditions—perhaps because it, too, was sung to over several generations—teparies were once widely cultivated in northern Mexico and the southwestern United States, but today are found only around the Arizona homes of a few Pima and Papago Indians. "I've seen the tepary growing successfully in air temperatures between 115 and 120 degrees Fahrenheit, and ground temperatures over 170 degrees, where pinto and other beans completely failed," said Nabhan. "I've seen it survive on less than two and a half inches of rainfall under extremely arid conditions. It's just plain ornery—it refuses to die."

Other crops highly resistant to heat include teff, a low-yielding but high-protein extraordinarily drought-resistant Ethiopian grain that western scientists, knowing little about, have recommended be replaced with corn or wheat. Pat Mooney has seen fields of it flourishing next to African corn so drought-stricken as to resemble fields of withered onions. "A main reason," chides Mooney, "why people are dying of famine on that continent is because of rotten western agricultural advice. We do it with the best of intentions—not a mean bone in our bodies—but not much humility either. We need to recognize the humanity of the people who are starving. If we don't have the means to end world hunger, they do! And if we don't have the capacity, they do! Together, we might have the combined will to solve the problem."

Concerned with another aspect of the disappearing seeds once known to parents or grandparents, Whealy began systematically to collect every seed mail-order catalogue he could locate, no matter how small or obscure. By 1982, he had put out a 448-page *Garden Seed Inventory* compiled from 239 separate catalogues listing over six thousand non-hybrid varieties still offered throughout the United States and Canada. The greatest value of his compendium is its indicating which varieties are in the most danger of extinction, so they can be planted and redistributed before they are dropped completely.

"As the *Garden Seed Inventory* grew toward completion," said Whealy, "it became not only more and more fascinating, largely because the little-known diversity and extraordinary quality of the garden varieties offered is almost beyond belief, but also more frightening, because it became apparent that nearly *half* of all nonhybrid garden seeds were being offered, as randomly as haphazardly, by single companies. And a lot of those have either been dropped in the interim, or will be as soon as remaining supplies are sold. Our efforts aim at reversing this ominous trend."

In Colorado, Whealy was shocked to learn that the computer print-outs of the National Seed Storage Laboratory showed that it was stocking only 3 percent of the food plants listed in W. W. Tracy, Jr.'s *American Varieties of Vegetables* for the years 1901 and 1902, published by the United States Department of Agriculture in 1903. "It is depressing," he said, "to see those enormous lists of garden varieties available at the turn of the century and to realize almost all of them have been lost forever. Imagine the tremendous amount of unique material that would still be available to breeders today if that early USDA inventory had only been updated annually, and endangered varieties systematically procured and maintained."

Since 1950 most of the losses have been due to economic pressure, which focused, not on the needs of home gardeners, but on the interests of commercial growers, who demanded such features as ease of harvesting, extended storability, and shelf life. To this kowtowing by plant breeders to the needs of large-scale agribusiness is due not only the loss of irreplaceable heirlooms but of what Whealy calls the *"best garden varieties we will ever see."*

Even with his main tool, the *Garden Seed Inventory*, Whealy is just picking up the pieces. "But we must at least do that," he explained, "and do it quickly, updating the inventory every couple of years. One good example, among many, of material that is available right now, but probably won't be for long, are the Chinese Cabbages. There is a tremendous amount of Oriental material currently flowing into this country. But most of it is coming through just a few small specialty companies, and in many cases these varieties are listed for only a single year."

Because 80 percent of the rare, commercially-unavailable seeds sent him by his members had not been permanently "adopted" by his Growers' Network, in 1985 Whealy rented five acres of rich bottomland to grow out two thousand varieties—including three hundred beans, tomatoes, and squash, and one hundred varieties each of potatoes, corn, muskmelons, watermelons, peppers, with smaller amounts of peas, lettuce, and many other crops.

"All of the corns and all of the cucurbits were hand-pollinated," he said, "and we have been protectively caging all of the peppers after reading some studies from New Mexico which reported that, under some conditions, populations of peppers can cross as much as 80 percent, thus mixing genes and rendering individual varieties unpure. Our garden projects have been funded these last two summers by Pioneer Hi-Bred International, and I say that thankfully."

To Whealy the most impressive thing about his garden is the effect it has on visitors. When he takes a tour out onto land where there are two thousand different varieties growing, the excitement it generates in those gardeners is little short of amazing. Time and again he has seen people walking along its rows, their mouths falling open. "They just look and look, as if they could never get enough of it," he said excitedly. "We've talked incessantly about the loss of genetic diversity, but the concept is so *abstract* that most people, failing to even recognize it as a threat, block their ears or turn off their hearing aids. But if we take those same people into our garden, show them hundreds of unique varieties, and tell them that most would be extinct except for our protecting them—then they light up with understanding."

After a two-and-a-half-year search, during which he looked at dozens of properties, none of which fulfilled his criteria, Whealy at last located the ideal location, a former fifty-seven-acre Arabian-horse stud farm near Decorah, Iowa, with a large barn and a one-acre spring-fed pond with a continuing flow so strong it never freezes over. There his Seed Savers Exchange is in the process of setting up an exemplary seed-preservation center. "In our beautiful new setting, we will expand our huge garden and build, if we can, a system of special greenhouses and large underground root cellars. We also intend to plant large apple orchards of up to two thousand older historic varieties, including those for cider, hoping to focus on some that might fall through the cracks as existing collections are transferred into the new clonal repositories. The work will make possible a massive exchange of scionwood—detached living portions of apple trees for grafting—to spread the apple heritage across the land, just as Johnny Appleseed did in the early nineteenth century—now as mythical to orchardists as Paul Bunyan is to lumbermen or Pecos Pete to cowboys.

"Imagine," said Whealy, "a place where the public could come for taste tests at harvest time and our telling them: 'You probably have only known the taste of Red Delicious, Golden Delicious, or McIntosh, picked green months before ripening. Now try these tree-ripened Winesaps, Chenango Strawberries, Winter Sweet Paradises, Yellow Bellflowers, Westfield-Seek-No-Furthers, or one of Thomas Jefferson's favorites, the Esopus Spitzenbergs.' "

Several years from now Whealy envisions adding representations of rare poultry and minor livestock breeds as a way of supporting organizations that focus on preservation in those areas. He wants to attract young apprentices to come during the summer to learn and to work in the garden and, eventually, see a network of such preservation farms, each maintaining plant materials in different climates and sharing data via computers.

Over the past decade several pioneers have been working to fulfill Whealy's networking dream. In the 1970s a former Rockefeller University molecular biology and life sciences Ph.D., Alan Kapuler, who had taught at his alma mater as well as at Yale and at the universities of Connecticut, Illinois, and Wisconsin, sick, at age thirty, of urban life and the academic rat race, founded, with others, Earth Star Botanicals, since become Peace Seeds, a "planetary gene pool service," in Corvallis, Oregon.

In his 1987 *Catalog and Research Journal*, written together with a botanist and alchemist, Olafur Brentmar, Kapuler lays out long seed lists arranged according to a computer-based model for classifying plants devised by Professor Rolf Dahlgren. Because all the known plants are traced schematically to their near or distant relatives, the system makes clear in evolutionary terms which of them are rarest or unique. Thus, one can at a glance see how lettuces, the highest source of silica in common diets, including such varieties as Imperial Winter, the Italian ruffled minihead Lollo Rosso, the French Barcarole, or the purple-leafed Red Sails, are related to such medicinal herbs as common burdock, or to chrysanthemums, marigolds, zinnias, sunflowers, and Jerusalem artichokes.

The seeds of no fewer than thirty-six varieties of tomatoes are offered, including the orange and yellow Marvel, hatched with red stripes. Their relations to peppers, tobacco, huckleberries, purple Peruvian potatoes, and the Chinese *matrimony shrub* are spelled out in the catalogue.

Waxing poetic in its enthusiasm, Kapuler's *Organic Seeds, the Gene-Pool and Planetary Peace* (1987) touches on the fundamentally spiritual role of seeding:

> In our work of growing paradise we have broken from mechanistic rectilinear constraints to begin an exploration of earth sculpture and cosmic relationships to maximize niches for all kinds of vegetal creatures providing love as well as food, shelter, and happiness. We come to you in the spirit of avatars like Lao-Tse, Krishna, Krystos, Black Elk, Mahatma Gandhi; of musical virtuosi like Mozart, Schubert and Pablo Casals; of poets like Henry David Thoreau; of martyrs like Giordano Bruno; and of plant innovators like Luther Burbank and George Washington Carver.

A far cry from the monotonous cloning of white-coated biotechnologists in agribusiness. And, as a measure of the growing revolt against such autocratic standardization is *Guerilla Gardening*, a book by Washington State University English professor John Adams, in which he tells

Gabriel Howearth and his totally unpolluted farm in southwest New Mexico. With contributions from around the world he has created a unique seed bank to save precious stock from extinction.

would-be heirloom-seed collectors how to begin collecting rare plants and seeds. With the use of step-by-step illustrations, he provides lucid instructions for growing, grafting, budding, and other propagating techniques.

"Time and a sense of adventure are essential for success," says Adams, "but anyone who gets into planting rare seeds will be more than rewarded, if only because their fruits and vegetables will offer them extraordinary taste experiences they've never had before. The beans I've grown for staples and for chili are head and shoulders above red, pinto, and other varieties found in supermarkets that are grown from commercial seed. There are a dozen different varieties of beans, going back long before Columbus was absurdly given credit for having discovered the 'New World'! If you grow them you'll find out what the pre-Columbian natives were enjoying. The taste will thrill you."

For beginners, Adams recommends starting small. "When growing heirlooms, mark out your patch of garden so you know what is where, and mark the sturdiest plants at full bloom for future seed use. By following this 'First Commandment' of heirloom seed saving you'll wind up with your very own hearty strain."

To provide such strains to people all over the world is a central mission of the Abundant Life Seed Foundation, situated across the state in Port Townsend, a small town on a peninsula, which the Seattle ferry passes at the midway point of its shuttle route to Victoria on the island of Vancouver. In addition to a wide spectrum of vegetable seeds, it proffers a range of annual herbs, from heat-loving cumin to lemon and licorice, basils to fenugreek—also known as "Greek hay"; as well as biennial and perennial herbs as diverse as copper fennel, feverfew, and a white insect-repelling daisy, *Euphorbia lathyrus*, said to repel moles and gophers.

Since its publication in 1984, *The Heirloom Gardener* by Carolyn Jabs has spurred an enthusiastic growing group of individuals, motivated by taste preference, curiosity, nostalgia for bygone days, collectors' mania, or unregenerate vegetal Ludditism, to take up Whealy's call for establishing seed-saver and exchange networks. Like Adams, Jabs gives full credit to Whealy's Seed Savers Exchange for stimulating her research and her writing.

At Fort Collins, Whealy told the U.S. government seed representatives that he hoped he could get at least part of his preserved heirlooms into official government collections to provide a "frozen" backup for his own efforts, and obtain badly needed help in getting material out to plant breeders.

He then gave them the full force of his astonishing accomplishments and aspirations: "During these last twelve years, I have put together a network of over one thousand growers, an incredibly diverse group, which includes university breeders and researchers; elderly backyard gardeners and hobbyists; amateur plant explorers maintaining hundreds of varieties; rooftop gardeners in New York City growing one heirloom tomato; traditional peoples—Anabaptist-descended sects in the Northeast and pueblo dwellers in the Southwest—still growing their people's seeds; and farm families of every ethnic background. The main strength of this diversity is that the amassed collections are extremely decentralized, so that any catastrophic loss through fire, flood, or other act of God met by a couple, or even a handful of our members, will be more than offset by the preserved seed banks of all the rest. Seeds are the guarantees of our survival."

CHAPTER 13

Weeds: Guardians
of the Soil

THE IDEAL FOOD to feed the world would be a grass, heavy-headed with seed, which could grow perennially like the wild grasses of the prairie, taking nitrogen straight from the air. Such grasses once grew uninterruptedly across the whole Midwest, from Allegheny to Rocky Mountain crest, some short, some so long they reached the underbellies of the first frontiersmen's horses.

Now, except for isolated islands still being chivalrously preserved by a handful of sensitive farmers like Fred Kirschenmann of North Dakota, the wild prairies have all but disappeared. Pervasive monoculture has led to the wholesale eradication of an extended happy company of prairie plants, which for centuries danced in variegated hues of spring, summer, and autumn flowers, with no more additives than sunshine and a shower of rain, naturally fixing nitrogen into the hungry soil.

In their place, thousands of acres are now plowed, year in and year out, to plant a single crop of wheat or corn. These man-developed grasses have bigger heads of seeds, but in the process have lost the strength to reproduce perennially, obliging farmers to waste weeks of labor, tankfuls of fuel, and tons of fertilizer to reseed them annually.

But what if edible proteins could be grown without the huge energy-wasteful expenditures for gasoline-fueled tractoring for plowing and harvesting, and petroleum-based chemicals for fertilizing and weed killing? What if food grasses, married to hardy weeds, could be made perennial without having to be replanted annually, yet produce as abundant seed as hybridized corn, barley, wheat, or oats? This was the dream that led one son of America's prairie heartland to give up his comfortable position as a tenured full professor of environmental science at a California uni-

161

versity to return to the Smoky Hill River in Kansas to begin a long-term effort at reversing one of the most tediously unvarying and soil-debilitating practices in agriculture: plowing and sowing.

We found Wes Jackson in Amarillo, Texas, giving a lecture on "Star-Wars in Agricultural Choices for a Sustainable Future." When asked how it felt to have his newly coined term—sustainable—universally and unattributively coopted, he demurred, saying it had long been "in the air," that Lady Balfour, founder of England's Soil Association, might well have been its originator.

Wes Jackson.

A rumpled plant geneticist who appears to have been dropped into his clothes from above, Jackson, in his early forties, founded the Land Institute in Salina, Kansas, but left it, he told us, "because I had concluded that the deterioration of the environment was outwardly mirrored in the present inner human condition and, as such, required not endless discussion but practical action. Most of my students were only passively absorbing my lectures, then going out the door to forget them, taken up with the blandishments of citified preoccupations far from the concerns I was trying to instill in them. Admonished by our children to get off our duffs and practice what we'd been preaching, my wife, Dana, and I left our poshly comfortable life in suburbia to do what Thomas Wolfe said couldn't be done: "Go Home Again."

On his hundred acres of native Kansan prairie, of postage-stamp size compared with the expanses of heavily fertilized wheat growing for miles all around it in every direction, eroding soil into river systems, Jackson was struck with how his virginal setting, perennially teeming with life, contrasted with its surroundings, drearily and artificially inseminated. Why couldn't some of his naturally-growing plants be coaxed to match, or outyield, in terms of nutritious grain harvest, the artificially-planted acreage? Was it possible to get members of such a nonhybrid polyculture to produce seeds in copiously sufficient quantities for human needs? Could plants in an ecosystem that had fed millions of bison and other wild animals not sponsor their own fertility by fixing nitrogen from solar energy alone? Could not the germ plant of so-called "weeds," considered so pathogenic to ordinary cereal crops, be enlisted as allies in the grain-growing art?

The old English word *weod*, since become *weed*, had a double meaning as it was spoken in the Kingdom to about 1100. Its pejorative overtone, when it referred to plants tending to overgrow or choke out more desirable ones cultivated for food, was applied not to their being entirely worthless but to their being without value when growing in certain locations. Weed could also mean herb—derived from the Latin *herba*—or any plant that does not develop persistent woody tissue but like grasses dies back at the end of each growing season. To add to the confusion, the term *herb* also applied more restrictedly to a class of plants having savory, aromatic, or medicinal value.

This old Anglo-Saxon dichotomy suggests that weeds, like people, can be foes or friends, depending on how or in what circumstances they are viewed. What may appear unsightly, troublesome, useless, or injurious in one context can, in another, be comely, effectual, convenient, and benign.

In Romance languages there is no word per se for weed. The French call them *mauvaises herbes*—"bad herbs" or, because *herbe* itself specifically means grass, "bad grasses." In Slavic languages, such as Russian, they are collectively known as *sornaya trava*, or "rubbish grass," or more simply, in the singular, as *sornyak*, an epithet equivalent to *nudnik*, which characterizes a human being as a tiresome bore.

Spanish priests, who spelled *herba* with a "y," could refer to a fragrant Californian plant of the mint family as *yerba buena* (good herb); whereas Californian tarbushes, as well as a Mexican oil-yielding plant, were *yerbas santas* (saintly herbs); and various South American vegetal specimens could be *yerbas sacradas* (sacred herbs).

When the conquistadores disembarked, they found native Indians cultivating and caring for all manner of wild plants, using them for both food and medicine, not only for their own bodies, but to nourish and revive the soil, interplanting them with domesticated crops to increase their harvests. The vegetally-unsophisticated priests, seeing beans, corn, squashes, and pumpkins flourishing side by side with what to them appeared to be totally useless companions, dubbed the nonproductive offenders *malezas*—a word vaguely implying a sense of moral depravity. Going by the established standards of Iberian agriculture, they preached that the "bad" weeds should be rooted out and burned, like heretics, to leave the fields as bare as the apses of their grim cathedrals.

Jackson took a different path, surveying the literature to discover wild perennial winter-hardy grass species with good seed-production records. Consulting the Plant Introduction Center in Pullman, Washington, as to which relatives of high-yielding plants from all around the world might be worthy of experimentation, he received, to his astonishment, 4,300 accessions, all of which he planted, each in a three-foot row.

In a several-years-long trial, he settled on two of the most promising. The first was giant wild rye, a native of Siberia and northeast Asia, with seed heads twelve to fourteen inches long, which had been prized by Genghis Khan, whose Mongol cavalry hordes carried it with them on their campaigns of conquest almost all the way to the Baltic Sea and the straits of the Bosphorus.

The second was Eastern gamma grain, with startlingly high 27 percent protein-content seeds that could be "popped," like corn, made into tamales or bread, or used as an especially rich animal feed.

Looking to convert sorghum, a tropical Old World grass similar to Indian corn, from a cold-abhorring plant into one as winter-loving as any Eskimo, he crossed one variety with its sorghum cousin, the noxiously weedy Johnson grass, to produce ten progeny, three of which survived the harsh Kansas winter to be recrossed into a succeeding generation of 1,500 grandchildren, 450 of which made it through the winter. Currently he is working to find out if his newly-created vegetal specimens can be safely propagated or whether their seeds may be so weedily aggressive as to represent a threat.

Still another prairie specimen being researched by the Land, as Jackson's institute is called for short, is a 34 percent protein Illinois Bundleflower nitrogen-fixing legume, which manufactures its own fertilizer and is prized for its seed by quail and wild turkeys, thus suggesting its potential excellence as poultry feed. Jackson's young ag interns who come to work at the Land during one or more forty-three-week growing seasons, had already eaten it in birthday cakes. Preliminary extrapolations suggest that one superior strain may yield as much as 3,000 pounds of seed per acre, whereas 30-bushel-per-acre wheat produces only 1,800 pounds.

Of several alleopathic candidates, which suppress neighboring species by exuding naturally herbicidal toxic substances from their roots, Jackson has selected, as most expectantly successful, Maximilian sunflowers.

"What we're after, in an effort that will take years of dogged work," said Jackson, "are not just perennials per se, but perennials that can thrive among other plants so as to take advantage of the natural integrities within a complete system."

To get at the heart of this problem, John Piper, one of Jackson's young plant experts, has, on a transect of the Land's chastely unsullied prairie, clipped aboveground portions of all its myriad plants to determine the ratio of "grass-ness" to "legume-ness" to "sunflower-family-ness" to other "nesses" in the whole biomass. This study, it is hoped, will reveal best-bet constellations of plants for meeting what Jackson calls "the expectation of the land," the important main goal being to find the appropriate mixes that can grow harmoniously and productively.

To date, this new research has laid bare various never-before-considered anomalies and curiosities. If Bundleflower, for instance, is planted separately from its natural neighbors, it is found to be prone to a splash-borne fungus bombarded from the earth below onto the underside of its leaves, which are attacked and pruned from the bottom up by the fungal infection.

"The usual reaction of plant breeders to such a situation," Jackson ironically explained, "is to launch into a seven-year, highly expensive program to breed resistance to that particular pathogen into the plants, tying up the services of a whole crew of geneticists and plant tenders. But, when my people had a close look at our native prairie to see how Bundleflowers were faring in natural conditions, they were found completely free of the ubiquitous fungus because the entire grassy mat surrounding them absorbed the bombarding raindrop splashes to which its plants were naturally resistant and thus protected the Bundleflowers from attack.

"So what's clear, in the first instance, is that plant specialists can volunteer to be 'heroes' in a needless battle, while in the second they can, with far less bravado, take advantage of nature's wisdom to manage, easily and costlessly, an otherwise ominous threat. This illustrates the elegance of horticultural *simplicity*. It shows up the difference between a natural complexity in native prairies and a *complication* in man's so-called 'scientific' approach."

The saddest side of twentieth-century agriculture is, to Jackson, its having become overoperational and underobservational, with the modern-day farmer losing the ability to "look" that was common to any nineteenth-century naturalist.

"Agriculture has been subjected in its often nonsensical research to a kind of military attitude that pervades research everywhere, an attitude that sells itself with misplaced concepts of freedom. Our traditional idea of going west to find freedom, abandoning worn-out land to cultivate virginal pasture, is part of the same syndrome. But if we had looked at the land as a nurturing mother, as the Russians did, we might have seen that true agricultural freedom is not freedom to *go*, but freedom to *stay*. What Star Wars and chemical eradication of weeds have in common is that they both aim at putting a blow torch to the planet. Weeds do much more to protect the soil than to harm it."

When Wes Jackson chucklingly says of his Land Institute: "We're out to save the world from sin and death," he is only half kidding. The fact that his ideas of supplanting conventional annual grain crops with perennial alternatives runs counter to most of what has been taught for decades in state agricultural colleges is to him a sure sign he is onto something real.

Speaking at the Salina County Farmers Union, Jackson ridiculed his audience for wearing their silly visored caps advertising the products of their corporate chemical overlords, comparing them to so many medieval serfs.

Jackson has no illusions that his program can produce rapid results. "What we're trying to do," he says, "is build a symphony orchestra from scratch. Then we have to persuade an agricultural Lincoln Center or a Covent Garden to schedule its premiere performance. That's going to take at least a couple of decades. We've got a tremendous job ahead."

Garth Youngberg, executive director of Maryland's Institute for Alternative Agriculture, thinks that some 350,000 farmers reporting annual sales between $50,000 and $100,000 apiece—occupying a niche between hobby farms and giant spreads—will be open to new approaches such as Jackson's because of their desperate search to cut costs.

Richard Harwood, deputy director of Winrock International, a development institute with overseas interests, is even more sanguine. He talks of replacing usual annual cereal crops with unusual perennials in a "decade or less," providing the Jackson insights can be institutionalized by rooting them in a nationwide network of private and governmental professional experts.

According to Charles A. Francis, agronomy professor at the University of Nebraska, who is introducing new ideas learned at Robert Rodale's research center in Pennsylvania, this trend is underway. If so, the families of *malezas,* so genocidally decimated since the first Spanish landings in America, may yet be reborn to play their role in a sane and gentle agri-practice that is truly cultural.

Whereas monks and pilgrims, equally misguided as they arrived in the New World, set about destroying weeds wherever they appeared, beginning in the process a debilitation of America's greatest asset, its soil, the richest they had ever seen, the *malezas,* like their Indian tenders, may yet bring back sense to agriculture in the Americas.

An enlightened mind on the healthy role of weeds, Professor Joseph Cocannouer, widely-traveled Oklahoma soil scientist, who learned much from the Indians and has written extensively on the use and functions of weeds, in particular in *Weeds: Guardians of the Soil,* points out that, by foraging far down into the subsoil like prairie grasses, weeds, with deep-diving roots, bring to the surface elements beyond the reach of most cultivated crops. They also pump up moisture, raising it by capillary action along the miles-long surface of their root systems, breaking up hardpan in mistreated soils that can range from an inch or two to several feet deep between the surface of the ground and its lower strata. "It is the weeds' unique ability, developed over millennia-long struggles for existence," says Cocannouer, "to seek food and water under adverse, mostly man-made, conditions, to 'eat' their way through concretelike compaction by virtue of special dissolving substances exuded from their probing roots."

Expanding on this theme, Cocannouer stresses that most domesticated crops, pampered by man, have lost the ability to probe deep into the earth possessed by their wilder ancestors. But, once room has been made in the root tunnels of the weeds, cultivated crops can follow in search of sustenance; many normally shallow-feeding crops will forage deep into soil if conditions are made right for them.

Both Cocannouer and Jackson stress that one of nature's valuable laws is that unrelated root systems do better when growing together than when that of a single plant is grown alone. First to go down are the anchorage roots, to support the plant so it can stand high to reach the powerhouse of sunlight with its endless supply of energy for photosynthesizing food in the leaf laboratories. Anchorage roots have to be rigid, yet flexible enough to stand severe strain. Next comes the great mass of roots known as the food hunters, ranging in size as they plow through the soil from many inches in diameter to slender threads.

Yet, when these rover roots have reached their source of food material, they cannot themselves take it up. This must be done for them by the tiny invisible one-celled root hairs, which protrude mostly from the smallest rover roots. These delicate absorbing rootlets, visible only under a microscope, are very short-lived, developing on the spot, living and dying in quick succession as they absorb food and water through a tenuous cell wall and inner membrane. The water streams up through root and stem in what Cocannouer calls "the greatest watercourse in Nature," climbing to the very top of the tallest trees, depositing nutrients in the leaf factory to be turned into the sugar without which there would be no human life. The remaining water, transpired through the leaves' mouthlike stomata, is returned, along with gases, to the atmosphere in an endless living cycle.

Organic farmers point out the danger of soil saturation with soluble chemicals such as NPK, because the plant, obliged to suck up water to transpire, and normally selective in what it wishes to chelate and absorb, cannot screen out the excess ionized chemicals and thus becomes engorged with an imbalance, which, though it may cause the plant to grow fast and flush, fails to give it, and those who eat it, the *balanced* nutrients they need.

And the rootlets' ability to function efficiently, drawing up substance from the soil, equally depends on the soil's condition. They cannot do so if the earth lacks suitable fiber, causing it to be too compact, or too loose, affording no support. Nor can the rootlets develop in soil that is too wet, too dry, or too cold.

Hence the usefulness of weeds. Live weeds break up the soil with their own roots. Dying, they bequeath to the soil the fiber of their bodies, rendering the hard clay spongy, the loose sands firm. And all weeds—from sunflowers to carpetweeds—help make the world's best fertilizer, either as compost or, when turned back into the land, adding, in biodynamics, cosmic and telluric forces to the soil.

Anna Penderson Kummer, in *The Role of Weeds in Maintaining the Plains Grasslands*, adds an almost metaphysical dimension, pointing out that, were it not for the constructive work of several important pasture weeds, the last wild-grass pasture areas in the United States—where the deer and the buffalo roamed—would today be as barren as desert. Weed growth, she says, is vital to the return of grass to land where the grass has been seriously thinned through overgrazing, sheet erosion, or a long period of drought.

When adverse factors kill the grass, it cannot come back to refiberize the soil with its own roots. It must wait for the weed roots, like yeomen for the squire, or nature spirits for the devas, to unlock the tight soil, and then fill it with the fiber of their bodies to reestablish porosity. The process may take one or several seasons. But thereafter the weeds do not drive out the grass. They only get the soil ready for the grass to come back. Grass has the power to rout weeds when conditions are right for the return of grass. Once the soil is rich in fiber, grasses dominate the weeds. That this could happen solely by chance, rather than with the aid of some overlighting intent, is clearly a notion as limited as are the academies that have chosen to subscribe to it.

Wild meadows all contain weeds, many of which are constantly shifting about, imperceptibly improving weak spots so the grass can come back strong and healthy. Some weeds even produce special seeds that can lie dormant over long periods of time, even though sprouting conditions appear to be excellent, waiting for when they may be needed. In this way "nature" miraculously makes sure no situation can arise where varieties of weeds are not available and ready to go to work as necessity calls. Only when the land has been completely peeled by erosion or totally poisoned by chemicals will there be no weeds. And this has happened by the hand of man until half the West is desert.

Professor Cocannouer is convinced that the same laws apply to tall-grass regions, and that the dust storms of Jackson's Midwest can be prevented in large measure by the correct use of plains weeds, such as milkweeds, thistles, tumbleweeds, and others, without which the grasses could not survive.

For potato gardens, Cocannouer recommends planting pigweeds, spaced far enough apart to permit strong root development without crowding the tubers. A husky pigweed, spaced every two feet in a row of potatoes, can increase production and enhance keeping qualities. The same goes for peppers and eggplants. For a tomato or onion patch he recommends a combination of pigweeds, lamb's quarter, and sow thistle, scattered thinly. In a corn field weeds can become "mothers," making way for the corn roots to produce better stalks and ears. "Mother weeds" will also loosen the soil for root crops such as beets, carrots, turnips, parsnips, and rutabagas, which need a deep, friable root zone where food material is easily available.

A crop growing in a field with the right amount of weeds, says Cocannouer, will survive droughts much better than crops grown on "clean" land. Moisture comes up along the outside of the weed roots, checking evaporation from the surface soil. The same weeds can provide fine shade for the ground when the sun is particularly scorching, or prevent a torrential rain from pounding the dirt into cement. Many varieties of weeds also protect domestic crops from insect pests, even having a place in the flower beds as companions to their ornamental brethren. If weeds are carpeted around a rose plant, their roots work down into the soil to intermingle advantageously with those of their hosts.

As the American Indians long since discovered, and many North and South Americans are rediscovering (together with other aspects of a heritage almost completely blotted out by white teachers, priests, and government bureaucrats), a large percentage of the soil-building weeds are themselves succulently edible in one way or another. Most of them, including poke shoots, sow thistle, lamb's quarter, smooth-leaf pigweed, and even stinging nettles, if young and tender enough, are very nourishingly full of elements lacking in domesticated crops. Dandelion and wild lettuce make salads far more nutritious than the water-filled iceberg lettuce dominating the nation's salad bars; they can be cooked in a large kettle and seasoned to taste with anything from salt to bacon. The tender stems of the common milkweed, when boiled in two or three waters to remove the milky sap, are a fair substitute for the best spinach. Milkweed's small pods, when cooked, resemble okra, and are considerably more savory. Milkweed roots, once the original bitter taste is removed by boiling, can substitute for potatoes. The roots of primrose are highly nutritious, and sheep sorrel is delicious either in salads or as fillings for pies. And the fruit of the ground cherry makes excellent preserves.

Few would believe that, in pre-agricultural days, man's choice of foods, including what are now called weeds, was truly awesome. In southern Africa's Kalahari Desert, one of the most inhospitable environments on earth, the native Kung bushmen, exiled to that wasteland by tribes invading from the north, make, or made, regular meals on no fewer than eighty-five wild vegetables, never once setting eyes on the less than a dozen kinds of plants that today make up a full three-quarters of the global diet.

Plants maligned as *malezas* by the black-soutaned, white-collared advisers of "stout Cortez," looking out over the Pacific from his "peak in Darien," could, had they tongues, inform us of many things we need to know about the conditions of our soils, says Ehrenfried Pfeiffer in his book *Weeds and What They Tell Us*. Were we attentive observers we would see that, above all, they are witnesses to our failure to treat soil properly and that their abundant growth takes place only where man has missed the point in regard to them. As nature's corrector of man's errors, they tell a silent story full of subtleties concerning the finer forces through which nature helps soils by balancing and healing.

The powers of plants such as the fungi that produce molds were long unsuspected. For centuries the common mold *penicillium*, growing on bread, cheese, other foods, and even beverages, gave, at best, a clear warning that comestibles were getting old, rancid, or being stored at temperatures insufficiently cool to preserve freshness and combat decay. But when, just before World War II, its nontoxic acids were discovered to produce powerful antibiotics against microbes such as *cocci*, it graduated in man's eyes from its humbler role of mold to that of lifeguard.

Like molds, weeds are both alerters of decay and survival specialists supreme, triumphantly persisting in circumstances where cultivated plants, softened through centuries of human protection and breeding, fail to stand up against nature's odd caprices. Their peculiarities allow them to be classed into three major groups, the first of which indicates increasing *acidity* in the soil. It embraces such species as sour-juiced sorrels; docks, coarse, long-taprooted members of the buckwheat family; fingerleaf lady's thumb and the horsetails, related to ferns. These are the best sentinels of all, for they provide warnings as to when changes for the bad begin in soil, the acidity being due to lack of sufficient air, water standing in the upper surface layer, insufficient drainage, excess of acid fertilizer, and, most important of all, *lack of humus.*

Even where soils are underlain with natural limestone, as in Kentucky's famous "Bluegrass" country, acid-loving weeds may thrive because the topmost soil has been delimed through unbalanced cultivation, as when grain is too frequently sown without rotation.

To the second major group, which reveal *crust formation* or *hardpan* in soils, belong field mustard, horse nettle, morning glory, quack grass, pine apple weed, as well as cresses and camomiles. The conditions for their redolent growth include wet soil turned up by plowing, or an excess of potash.

The third group, like Good King Wenceslas, treads in man's, not nature's, steps, often spreading out wherever he has disturbed nature, thriving on manure, compost, and other products of his cultivation—extensions, as it were, into man's artificially-created realm. These include plantains, chickweeds, buttercups, dandelions, nettles, mallows, as well as lamb's quarters, prostrate knotweed, carpetweed, prickly lettuce, and common horehound.

Weeds of the rose family are one of the surest signs of lack of horticultural attention to detail and insufficient cultivation. Members of the usually useful legume family, also unfairly termed "weeds," prefer light, sandy, or otherwise poor soil, and those of the pink family choose gravelly earth or strips along hedgerows or the edges of woods, where they establish a "borderline" culture between cultivated and uncultivated nature.

Weeds, says Pfeiffer, are dietary gluttons, which, if offered a full menu of soils, will go for those carefully hoed and manured rather than their natural habitats as fast as any bon vivant will go for a brunch of opulently varied dishes rather than a bowl of oatmeal.

To accurately detect the properties of a given soil it is necessary, he adds, to judge by the prevalence of an entire group of weeds rather than of a single individual. Should many distinct groups of "wild" plants begin an invasion into any area where they have previously been absent—the large, coarse fern called *bracken* comes to mind—this is a sure sign of decline in soils.

In the rest of his book Pfeiffer provides an account of what each and every weed among four hundred varieties can tell the gardener and farmer. But, before entering into this detailed discussion, he says point-edly: "If you learn to listen to the lessons nature provides in producing weeds under extremely different conditions of soils, climates and methods of cultivation, then you will have made your first step in the most important combat against them: to put them where they belong and keep them away from where they do not belong."

For centuries if not millennia the *brujos* of the South American jungles have known, just as do the sophisticated sages of the subcontinent of India with their Ayurvedic wisdom, that for every human ailment there exists a naturally growing plant, as often as not, malefically labeled a weed. As fast as we destroy their natural habitat, so, one by one, their benefits will vanish, perhaps forever, from the planet.

In his envoi, Pfeiffer enjoins us always to remember that whenever weeds grow they speak out to us, and that wherever they flourish particularly abundantly, they indicate not their failure, but man's. "There are many, many dynamic plants called weeds," says Pfeiffer. "Go out in the fields and discover for yourself how benign their properties can be!"

CHAPTER 14

Icicles in
the Greenhouse

TENDING THE PLANET'S soil may now be our prime priority for an even more urgent reason: to save the world from imminent glaciation. All our healthy topsoil, all the microorganisms in it, and all the plants that thrive thereon, from lichens to the great rain forests, have received their nourishment from billions of tons of mountain rock dust, ground up and washed away by melting glaciers from the last great ice age, some twelve thousand years ago, to be globally spread by whirling windstorms. That life-giving dust is now used up, and our precious topsoil wantonly eroded. Unless something is done to replenish the soil with rock dust, and quickly, warns one group of worried experts, another great age of ice will do it for us.

Two radically conflicting theories about an impending climate shift have polarized the country's "experts" into rival camps. One claims that the planet is gradually warming, with no immediate danger to mankind, the other that it is cooling, placing us in imminent peril of another ice age, with all the consequences of such disaster, known and unknown. Both schools blame the situation on what they call the *greenhouse effect*.

The notion was first proposed in 1861 by Ireland's renowned natural philosopher John Tyndall, when he suggested that increased concentrations of carbon dioxide (CO_2) in our atmosphere might someday raise surface-air temperatures enough to cause a problem. When coal, oil, and natural gas are burned, the two principal combustion products are water vapor and carbon dioxide, about half of which remain in the atmosphere: as both are transparent, they allow the sun's rays to pass through to earth, but trap the reflected heat, as in a greenhouse.

Proponents of the warming trend, supported by official government agencies, and widely reported by the media, maintain that rising levels of carbon dioxide in the atmosphere, mostly produced by the burning of fossil fuels, are creating hot air, trapped at the equator, which, unable to dissipate back into space, threatens to cause the ice caps to melt.

New York City, they say, will someday risk being not covered again by a mile of ice, as it was twelve thousand years ago, but being subjected to the climate of Fort Lauderdale, its streets equally canalled by a sea rising gradually in the course of the coming century. So no strong countermeasures are contemplated by a petrochemically controlled establishment, other than vague talk of gradually reducing fossil-fuel consumption.

The cooling advocates, mostly tenured professors of climatology and paleo-climatology, maintain, quite to the contrary, that the government's position is based on inadequately programmed computer models that leave out critical data, such as cloud cover, and are politically motivated to protect the continued burning of fossil fuels.

According to these "cooling" climatologists, the greenhouse effect, while it does increase warming at the equator, has an opposite and much more dangerous effect of sucking up moisture in the tropics. In heavy clouds, this moisture is propelled by prevailing winds toward the poles, where it condenses into snow to cause more ice and, more dangerously, to cool the polar oceans.

In the 1930s, Sir George Simpson, then Director of the British Meteorological Office, described what he called the *general circulation pattern* of the winds, whereby air heated in the tropical and subtropical zones rises to a high altitude, where it is moved by the differential in pressure toward the poles, there to be sucked back to the surface of the earth by the expanse of cold snow, causing a cyclic pattern. Increasing carbon dioxide steps up the cycle, contributing more heat to the warm zone, increasing the amount of water vapor taken from the ocean, and increasing the speed with which it is carried north. This contributes to more snow cover, which manufactures more cold air, which sinks faster, and is carried to the south at higher speeds. At the same time, huge masses of heavy cold air fall off the ice and snow banks into the oceans, where currents distribute the cold around the globe. The phenomenon led Simpson to the odd conclusion "that the last Ice Age was not caused by a decrease of solar radiation but by an increase!"

A heavy, colorless, and odorless gas that does not support combustion, CO_2 was discovered through the death of dogs in caves where humans could walk unaffected because the weight of the gas kept it below knee level. It is one of the most important ingredients of the planet's biosphere. Exchanged between plants and animals, between air and sea, at a rate of hundreds of billions of tons per year, the gas arises from and helps sustain all life on earth—so long as it is kept within limits.

By a variety of methods, including analyzing the bubbles trapped in glacial ice, scientists have estimated that in 1850, in a less-industrialized world, CO_2 made up between 250 and 290 parts per million of the atmosphere. To monitor the clearly increasing amount of the gas in the present atmosphere, a measuring device was placed atop the Mauna Loa volcano in Hawaii, where it has shown a rise from 315 ppm in 1958 to the current extremely dangerous 343 ppm.

Well aware that Tyndall was correct about carbon dioxide causing a greenhouse effect at the equator, "cooling" climatologists are equally aware that the planet as a whole is *not* warming, with a loss of 1.5 degrees Celsius in average Northern Hemisphere temperature since 1938; that the arctic is expanding; that growing seasons are becoming shorter; that millions of the earth's inhabitants are threatened by drought; and that drought is due to cooling, not warming. Fred B. Wood, Jr., of the Office of Technology Assessment of the U.S. Congress reports that between 1960 and 1980, based on data for about 400 to 450 glaciers observed each year, advancing glaciers have increased from 7 to 55 percent.

In fact, say geologists, ever since its origin, the surface of the planet, and thus its climate, has been cooling, owing to the slow decay of the original inventory of radioactive material in its core. Twelve million years ago, according to their calculations, the cooling reached a point where the "age of ice" began, a period which has seen a series of ice ages of increasing intensity. Furthermore, say the climatologists, the overall temperature has been falling for the last six millennia, and especially during the last forty-four years.

This flatly contradicts the data of the warming experts, who, according to the climatologists, fail to take into account the effect of *city* warming, a factor quite distinct from the general trend. Nor has any of the warming evidence so far materialized, except for dubious computer models.

Modern climatology, which only began to flower in the mid-1950s, when a number of researchers became interested in what might actually be happening in the world's climate, was largely sparked by the discoveries of Italian-born Professor Cesare Emiliani, head of the Department of Geology at Miami University, published in 1955, now considered the basic modern contribution to deep climate research.

Studying small crustaceans known as *Foraminifera* in cores of sediment dredged up from the Gulf of Mexico, Emiliani was able to trace the climatic history of the planet going back millions of years, and produce the first reliable paleontological framework. What he found was a

Cesare Emiliani.

succession of recent ice ages—some twenty-five have now been clocked
—each lasting about a hundred millennia, with relatively short intervening periods of deglaciation, lasting ten to twelve thousand years—such
as the one we have been enjoying since the mythical demise of Plato's
Atlantis, about 9,000 B.C. As matters stand, warn the "cooling" climatologists, *only* our intelligent technology can postpone or prevent another
cataclysmic freeze, which would wipe out the better part of the human
race.

The earliest scientific description of ice ages and their origin was
made in the middle of the last century by a Scottish philosopher-scien-
tist, James Croll, who postulated that their cyclical recurrence was controlled by regular changes in the earth's elliptical orbit, the tilt of its axis,
and its anomalous so-called "Chandler" wobble. Quickly discredited, the
theory was only revived in the 1930s by the convincing mathematical
calculations of a Yugoslav geophysicist, Milutin Milankovitch, which
supported Croll's original data.

Milankovitch posited a continuous, changing relationship between
earth and sun, with the earth's orbit changing shape every ninety thousand to one hundred thousand years. From being almost perfectly circular, the orbit slowly becomes slightly elliptical, then slowly moves back

to circular, varying the sun's intensity by as much as 30 percent over the cycle. Milankovitch also found anomalous cycles in the roll, wobble, and shift of the earth's tilt, one cycle shifting solar energy from the Southern to the Northern Hemisphere and back over a period of twenty-one thousand years.

Even so, skeptics remained, and it took Emiliani's oxygen-isotope data from deep-sea cores to confirm the Croll-Milankovitch mechanism and arouse climatologists to take a better look at what might be in store for man. By Emiliani's scenario the planet is heading straight for the lockers of a deep, deep freeze.

In a fascinating portrayal of the earth's climatology over the last fifty million years, Dr. John Imbrie of Brown University adds chilling details showing that the end of an interglacial period can be sharp and dramatic, as indicated by mastodons found in Siberian and North American ice packs, so quickly frozen that wildflowers they were chewing were still fresh between their teeth.

Further evidence was produced in the 1960s when Professor George J. Kukla and his co-workers at the Lamont-Doherty Geological Observatory found that Czechoslovakian deposits of loess* indicated ten distinct ice ages, which fit with Emiliani's ocean-bottom figures to show that the interglacial periods have uniformly been short respites between long glaciations, and that the present interglacial is coming to its end.

Roused to action by the threat of imminent disaster, climatologists looked to see what might be done. But they were hardly off their marks when the warming counter-theory, supported by the government, was brought to the fore by Dr. Roger Revelle, then director of the Scripps Institution of Oceanography, and his German colleague Dr. H. Suess— no relation to the famous Dr. Seuss beloved by children, though his notions are considered every bit as zany.

Revelle and Suess simply dusted off Tyndall's greenhouse effect, and by the 1970s had managed, with the help of government funds indirectly provided by the petrochemical companies, to create a growing bias in "orthodox" American science toward the warming theory.

But, for those who understood, the climatological writing was in the cores. Thomas E. Overcamp of the University of Alaska's Geophysical Institute saw the evidence as pointing to cooling, and his colleague at the same institute, Gunther E. Weller, suggests the greenhouse effect may be counteracted by a cooling trend: "We are due for a change toward cooler climates." Wallace S. Broeker, of Columbia University's Lamont-

---

* A pale calcareous clay or loam, generally of aeolian origin, loess is remarkable for its organic remains, which consist chiefly of land shells and bones of herbivorous and carnivorous mammals.

Doherty Geological Observatory, warns that a climate change could happen so suddenly people wouldn't have time to adapt.

In 1972, the first major conference of climatologists met at Brown University to discuss "The Present Interglacial: When Will It End?" The consensus of those attending was that "global climate change constitutes a first order environmental hazard."

Letters were issued by the university to the governments of the world warning of an impending "global climatic disaster."

Two years later the International Federation of Institutes for Advanced Study (IFIAS) held a conference in Bonn, West Germany, which stated in part: "A new climatic pattern is now emerging. . . . We believe that this . . . poses a threat to the people of the world. The direction of the change indicates major crop failures almost certainly within the decade. . . . We urge the nations, individually and collectively, to plan and act to establish the technical, social, and political means to meet this challenge to peace and well-being. We feel that the need is great and the time is short."

By the fall of 1973 the U.S. Central Intelligence Agency had obtained sufficiently foreboding evidence to sponsor a meeting in San Diego of the principal investigators representing the various research approaches. By the second day a consensus was reached that "a global climatic change is taking place, and that we will not soon return to the climate patterns of the recent past."

In 1975, eighty-four climatologists from ten countries attended the First Miami Conference on Isotope Climatology and Paleoclimatology, chaired by pioneer climatologist Cesare Emiliani and by Nobel laureate Willard F. Libby. In a consensus of the conference the latter wrote: "Ice ages have been the normal condition during the last several million years, with temperate climates enduring only 5 percent of the time. . . . Because the global food supply depends primarily on climate, current understanding of climate must be vastly improved in order to meet the challenge of tomorrow's food supply."

Libby then produced the key phrase of the conference: "We possess the methods and techniques to establish climate history and only a concerned effort is needed to do that." By this Libby meant that concerned climatologists, aware that an ice age was coming on fast, have been studying methods to offset its damage with technological expertise. What is needed, they say, is a man-made way to mitigate the natural cycle of cold brought on by the earth's anomalous orbit, and stave off disaster by artificially providing one-tenth of one percent of the sun's heat.

Of the many suggestions proffered—including some expensive and dubious, such as darkening vast portions of the earth's surface with coal dust or the oceans with strands of black polypropylene—the one preferred by climatologists is the proposal of Space Global of California to launch into orbit in near space a number of sun-synchronous reflector systems called Solettas to reflect more sunlight onto the planet, enough to offset the natural loss. Similar reflectors, called Lunettas, could be used to increase moonlight by as much as one hundred full moons, allowing for night work in agriculture, multicropping, and better street lighting. Though what this might do to vital biological rhythms, not to mention Witches' Sabbaths, remains a perplexing query.

The system would require thousands of rockets to place some 1.5 million square kilometers of reflectors in orbit, costing hundreds of billions of dollars. But its proponents assure that the cost could comfortably be amortized over the sixty to one hundred years the system would last —avoiding the considerably greater discomfort of another glaciation.

That the problem is real and that the world is faced with a developing catastrophe of unprecedented global dimensions became clear from two CIA-solicited reports (declassified in 1976), which summarized the scientific literature and showed that the United States was facing a world of chaos. The first, "A Study of Climatological Research as It Pertains to Intelligence Problems," prepared by the Office of Research and Development for its internal planning purposes, starkly presented the findings and opinions of the climatological community that the world was cooling and that the next ice age was imminent. The second, "Potential Implications of Trends in World Population, Food Production, and Climate," prepared by the Directorate of Intelligence, Office of Political Research, added tamely that "if the cooling continues for several decades there would almost certainly be an absolute shortage of food."

But the language of the second report also made it clear that its drafters realized that the economic and political impact of a major climatic shift, with its promise of famine and starvation in many areas of the world, was almost beyond normal comprehension. Boldly, the analysts warned "there would be increasingly desperate attempts on the part of powerful but hungry nations to get grain any way they could." The reporters envisaged a dire spectacle, including war: "Massive migrations, sometimes backed by force, would become a live issue, and political and economic stability would be widespread."

Cold-bloodedly they concluded that "in the poor and powerless areas, population would have to drop to levels that could be supported." Then came the bottom line, presaging future policy: "The population problem would solve itself in the most unpleasant fashion." By which euphemism they meant genocide, a cynical but handy solution to both starvation and overpopulation.

By 1977 the country was beset by serious drought—a clear sign of cooling. The government's answer was to put sixty million more acres of land into production, further stimulating sales of chemical fertilizers, pesticides, and herbicides. The result was a very large crop of wheat and corn. But the per-acre yield was one of the smallest in the history of agriculture, and the nutritional content the lowest. For 1978—as predicted by the CIA and just about every cooling-climatologist—the country still faced a shortfall of food.

And the drought was expanding deserts all over the world. Since 1976 the dry-land farming area in the U.S., which includes Minnesota, North Dakota, and states farther south, has been all but reannexed to what was romantically labeled on nineteenth-century maps "Great American Desert." To irrigate crops, underground aquifers are being pumped dry, adding to desertification, which, say the climatologists, has been going on with a general cooling of the planet for the last 6,000 years.

About 4,500 years ago a green and succulent Sahara region began to turn into a plantless desert. Some 2,000 years later the North American desert was hard on its heels. Similar histories describe the formation of other great deserts such as the Gobi wastes in Mongolia—all due to cooling.

In 1979 another explanation was given for the curious fact that CO_2 in the atmosphere causes not only increasing heat at the equator, but increasing cold at the poles, and that the latter situation will eventually predominate. Dr. George Kukla and Dr. B. Choudhury published a paper in *Nature* in which they showed that carbon dioxide in the atmosphere *not only* traps the infrared wavelengths rising from the surface of the earth, thus increasing the heat; it also filters out the *near* infrared wavelengths as they come from the sun. These are the wavelengths that melt the ice and snow. Hence: more CO_2 at the poles, more ice and snow. And since the snow and ice reflect the rest of the solar energy spectrum back into space, the growing snow and ice are manufacturing increasing amounts of cold.

Already in 1978 the alarming increase in snow cover during the period 1968–72 was discussed in the world's largest scientific journal: *Science*. And ever since, the figures for snow have been well above the pre-1972 peak. Due to the effect of carbon dioxide, Northern Hemisphere temperatures have been caused to drop by a factor of two degrees Celsius since 1938. In the Southern Hemisphere, where there is less snow cover, the cooling has not been so rapid.

A further correlation has been found between increasing cold and the number of earthquakes. The colder it gets, the more earthquakes we get. Ice and snow accumulating on the poles press down on the planet,

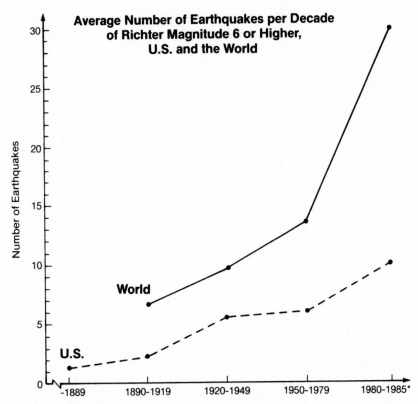

Average Number of Earthquakes per Decade
of Richter Magnitude 6 or Higher,
U.S. and the World

*Prorated for the decade from earthquakes 1980-85.
Source: James M. Gere and Haresh C. Shah, *Terra Non Firma* (New York: W.H. Freeman, 1984).

causing it to bulge at the seams like a balloon. This triggers the pre-stressed earthquake faults into slipping: hence earthquakes. It also causes volcanism—potentially even more dangerous—by squeezing the molten magma and causing eruptions. The colder it gets and the more snow presses down on the poles, the more magma is compressed, and volcanoes act up. †

---

† The 1980 Almanac shows a significant increase in both earthquakes and volcanism; it lists 13 major earthquakes between 1906 and 1942, and 34 for the next 36-year period from 1943–79, an increase of over 250 percent as many. For volcanic eruptions the same source lists 28 between 1906 and 1942 and 144 in the period 1943–79, an increase of over 500 percent for the same-length period. In the past few years the increase in volcanism shows a correlation with increasing snow cover. Eruptions such as that of Mount St. Helens, occurring closer together, are tied to the process of glaciation. Volcanism, predominantly a spring and fall occurrence, is a reflection of the wintertime increase in the weight of the snowfields at the north and south polar regions respectively.

The increase in volcanism gives climatologists much to worry about. Should there be two or three more eruptions of the Tambora or El Chichon types during a five-year period, the veil drawn across the sun by erupting dust, capable of obscuring the sun's energy with fine aerosols for two to six years, would be enough to plunge the planet precipitately into another ice age.

A UN-sponsored Conference of Experts on Climate and Mankind in 1979 warned that the world had entered a cooling period, and that the warming theory was complex and questionable, that cooling posed a threat to life and economic substance. It was pointed out that a drop of one degree Celsius could result in a loss of $19 billion worth of rice, and of $28 billion worth of softwood in the U.S.S.R.

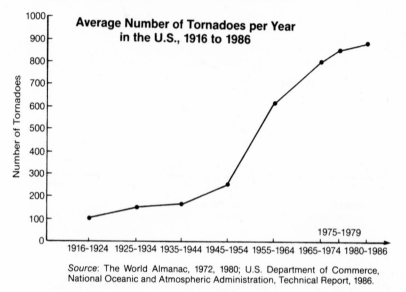

Source: The World Almanac, 1972, 1980; U.S. Department of Commerce, National Oceanic and Atmospheric Administration, Technical Report, 1986.

Larry R. Ephron, author of *The End*, forecasting the coming Ice Age, says there is probably some error factor in this graph due to more complete reporting, but it is probably significantly less than the 900 percent increase shown from 1920 to 1986, and the 350 percent increase shown between 1950 and 1986.

To check on the official Soviet views about cooling, we addressed Dr. Viktor Kovda, head of the U.S.S.R. Academy of Sciences' Scientific Council of Problems of Soil Science and Reclamation of Soils. From Moscow he replied:

As a soil scientist I am much involved in both ecological problems and problems concerning current climatic trends. I am fully informed concerning the existing prognosis of global warming influences by a growing CO_2 concentration in the atmosphere. However, observations by many Soviet scientists in the U.S.S.R. have compelled me to believe that, on the contrary, it is definitely a cooling trend which is taking place over the recent 15–20 years.

The signs of permafrost shifting southward, some shortening of the growing season, and the increased severity of winters, as well as extended freezing of our northern seas, are valid arguments for me.

It may be a question of a periodic fluctuation (like those that have taken place in the past) or it may well be the beginning of the next glaciation. But there are no signs of a forecasted warming, and that is *definite*. A statement of mine, namely that cooling goes hand-in-hand with an increase in arid lands as can now be observed practically everywhere, has been published several times in both Russian and English.

As part of the solution, Kovda added:

I am certain truly modern scientifically-based agriculture *must* be only ecological-biological-organic, with regular application of manure composts, correct plant rotation (including legume grasses), and, depending on crops, remineralization.

In the United States, C. Bertrand Schultz, professor of geology at the University of Nebraska, warns of an imminent ice age in which Canada will be unable to grow grain, and the Soviet Union will be unable to feed itself. Already in Alberta they no longer have the forty-one frost-free days required for a harvest.

Concurrent with Dr. Kovda's report came another from Colorado to add further fuel to the forebodings of the cooling climatologists. Adam Trombley, inventor of a homo-polar motor, and head of Project Earth—an effort to reveal and publicize hidden facts about the planet's climatic and environmental changes—warned that the problem had been further complicated by a phenomenon kept secret by the U.S. government, related to one of the most profound reasons for the cooling trend: dust in the atmosphere.

Strangely, the story starts with the construction of the Aswan High Dam on the Nile in lower Egypt in the early 1960s. This monster barrage completely blocks millions of tons of agriculturally-rich silt that, throughout history, flowed down the Nile to nourish the Delta from which it "breast-fed" the Mediterranean Sea, allowing for the burgeoning of tons of phytoplankton in its temperate waters, plankton that sucked up CO_2 from the air.

When the Nile delta was cut off from the life-giving silt, the Mediterranean plankton, as revealed by French oceanographer Jacques Cousteau, began to die. Today the eastern part of the sea is fundamentally dead and the western part on the brink of death. Even more recently and even more drastically, Cousteau warns that the whole Atlantic Ocean is being so polluted that it too will be dead within ten years if action is not taken immediately to remedy the situation.

Climatically-speaking the consequences of the Mediterranean debacle are almost unimaginably severe. Because the phytoplankton of the inland sea absorbed and metabolized the increasing amounts of carbon dioxide emitted from Europe's industrial plants since the early 1800s, the sea acted as a huge sponge, or sink, for gas drifting southward, capturing it before it could travel on to northern Africa. But these days the heavy, heat-laden gas, unattenuated by the sea, rolls along over the Sahara Desert, which lacks any biomass to absorb it.

Starting about 1969, unusually violent winds were born over the Sahara, blasting the sand into the upper atmosphere, where it hyperabraded: turbulence ground the sand grains, normally too large to float in the atmosphere, into what is scientifically called a *suspended particulate*, equivalent to dust.

By 1970, as revealed in spies-in-the-skies photographs taken from satellites, an enormous chronic cloud of dust began to be permanently suspended above the desert, over which a column of heat, known as the *Sahara Chimney*, normally rose at night. Carbon dioxide increasing in the rising air of the chimney formed a heat sink, dampening its vertical force, putting a cap on it.

Deprived of its "fuel," the huge "weather machine" constituted by this chimney could no longer drive air into the upper troposphere, whence, after cooling, it could descend to meet weather fronts coming from the southern polar region to build beautiful rain-bearing cumulonimbus and strato-cumulus clouds over the Congo River Basin, site of one of the world's largest rain forests. The same rain that created the forests had also come down on a semi-arid sub-Saharan belt known as the Sahel, which has begun to dry up so disastrously that millions of Sahel villagers have died of starvation, forcing nomads to migrate southward in search of new lands to assure their survival.

This peregrination only compounded the problems of Black Africa, in that the newcomers, knowing no better, began to burn forest acreage to clear it for farming and grazing. In the mid-1970s Sky Lab astronauts could observe smoke from such fires blanketing large areas of the sub-Sahel region. One NASA photo, taken in 1985 at such a height above the Sahara that it reveals ten thousand square miles of land surface, gave the impression of a shot taken from an airplane flying low over Los Angeles on a particularly smoggy day.

"That's ten thousand square miles of dust," said Trombley, who had astonishingly learned from Richard Underwood, for twenty years Assistant Head of Photography at NASA, that the Sahara Desert is moving southward at a rate of five to thirty kilometers per year.

To accent the enormity of the problem, Trombley screened more NASA slides, not cleared for release. One of these, taken from the Gemini space capsule in 1965 over Central Africa, showed the beautiful blue waters of Lake Chad, the size of Lake Erie, still 150 meters deep. A second photo, taken in 1982, revealed the whole lake to have lost almost all its water. "The reason I'm using the 1982 photo," Trombley emphasized, "is because by 1985, the outline of what used to be Lake Chad is barely visible at all."

Another vertical shot, equally dramatic, showed hundreds of square miles of forests burning near the Zaire-Angola border. Hundreds of points of light flickered through what, though they appeared to be clouds, were really smoke plumes, providing evidence of man's desperate effort to create farmable land at the expense of trees. "Those forests," said Trombley, "used to absorb CO_2 and metabolize it. Now, by burning, they've become a CO_2 source instead of a sink."

Still another space photo revealed how the slopes all along eighty miles of Juba Bay on the once beautiful tropical island of Madagascar were becoming denuded of forests, cut for their exotic woods for export to Japan. A second photo showed from on high how migrants from teeming Brazilian cities were denuding Amazonia and other regions to open private farmsteads in a rain-forest devastation more than matched by "Big Mac" razing of forests to grow pasture for hamburger-producing cattle.

Severe droughts, widely considered to be episodic, that have racked Africa and other areas over the past several years are due, in large part, to the permanent layer of dust that now affects not only the Sahel but most of the African continent, eastern, central, and southern.

By 1984, the electrostatically-stratified ever-building dust cloud began to move westward over the Atlantic Ocean, so clearly revealed in a NASA photograph as to take its migration out of the realm of theory. When it reached the Gulf of Mexico, it did not stop, but moved inexorably across it like pea-soup fog across London. The result was Gloria, a hurricane with the largest radial extent in history, followed in 1988 by the even more powerful Gilbert and others, their winds partly generated by the dust engendering the largest hurricane season in the history of the Gulf.

And still the dust did not stop its westward movement, proceeding all the way to the Hawaiian Islands. There, a lot more dust was measured than anticipated, with four times the thermal capacitance, or ability to hold in heat. When dust traps heat and prevents it from surging northward, cold air advancing southward can proceed unimpeded further and earlier than normal, causing colder weather in various parts of the United States, slowing precipitation in the normally rainy Pacific Northwest.

CO_2, says Trombley, absorbs heat not only from solar infrared radiation but also from the earth's mantle, which transfers it to the earth's crust. That heat, instead of escaping into space, gets locked up in the CO_2-filled atmosphere, leading to a tremendous amount in the crust, as proved by greatly increased surging in glaciers since 1975, from the Himalayas to Alaska and Siberia, with more calving of Antarctic icebergs, including the greatest ever in late 1987.

As a parting shot, Trombley showed a photo taken from the space shuttle in 1985. "That's the blue edge of the Caribbean," he said, his tone no longer earnest, but appalled. "That yellow haze is a mask of dust. Ten thousand square miles of it. Cousteau's measurements confirmed the greatly decreased sunlight reaching the surface of the ocean. If you add the fact that twenty-eight million acres of rain forest were cut down last year, not including what was *burned* off the face of the earth, to the thirty billion tons of CO_2 emitted into the atmosphere by industry, you get an idea of the *major* problem which only *real, grounded, practical* activity will correct. It's lucky the dust is only in the troposphere, eight to ten miles above the earth's surface. If it were lower, in the stratosphere, we'd be in even bigger trouble. But Mother Earth is sick enough as it is. And when Mom gets sick, what happens to the kiddies?"

CHAPTER 15

Dust for Life

IN 1982 two determined proponents of rapid soil remineralization, John Hamaker and Don Weaver, in a book aptly entitled *The Survival of Civilization*, sounded a clarion warning that an imminent ice age was upon us. They also offered a solution, provided it be acted upon with haste. Hamaker, a seventy-year-old engineer-farmer and climate theoretician, with Weaver, his young collaborator, warned that the catastrophic climate shift would begin within the decade, predicting that trees would die the world over, growing seasons would shorten, winters would become more severe, summers harsher, and unless a world mobilization to counter these effects was initiated in the 1980s, famine would be our common fate by the 1990s. And that is just what's happening.

Hamaker's suggested remedy, quick, easy, and relatively cheap, requires a massive program of world-wide remineralization and reforestation. He wants to grind up glacial gravel (which contains all the required elements, and of which there is an almost inexhaustible supply at an economical price), spread the dust far and wide, then plant trees as if there were no tomorrow. To facilitate the operation he has patented special grinders, and has hopes of eventually enlisting the world's air forces to distribute the dust, a suggestion that might put these deadly forces to the first productive use of their existence.

Already in the late nineteenth century a German chemist, Julius Hensel, taking issue with Liebig and the proponents of chemicals for agriculture, proclaimed that all that was needed to produce luscious, healthy crops was the aboriginal food of plants: ground-up rock dust. His method, once put into practice, would not only free the farmer from heavy yearly expenses for artificial fertilizers, but gradually wean back his exhausted fields to their virgin state of fruitfulness.

Something of a mystic, Hensel liked to quote a Hindu saying that "God sleeps in stone, breathes in plants, dreams in animals, and awakens in man!" His appeal to the world was simple: Turn stones into bread!

By revealing the inexhaustible nutritive forces, hitherto unrecognized, that are stored in rocks, air, and water, he hoped to feed the hungry, prevent epidemics in man and beast, make agriculture once more profitable, and return the unemployed to a healthy life in the country.

Two hundred farmers in the Palatinate, delighted with the first few years of experimentation with rock dust on their fields, supported Hensel's claims. They testified before a court of justice that fertilizing with stone meal showed far better effects than those achieved with chemicals.

But the greater Hensel's success, the greater grew the opposition to him. Though the rock dust was successfully tried out on a large scale on the estates of the Grand Duke of Luxemburg, as verified beyond any doubt by a company of teachers and editors who made minute comparisons with neighboring fields cultivated with different fertilizers, by that time even royalty could no longer stand up to the chemical interests. To quash the interloper, they launched an expensive and rabid campaign to denigrate Hensel, keep his books out of print, and put a stop to his "heretical" notions that NPK might be a toxin in the soil.

By the time the German chemical companies had amalgamated into the vast I. G. Farben conglomerate and brought their Fuehrer to power, the last of Hensel's books was consigned to the flames.

The loss to Germany from not listening to this prophet was to be a devastating plague of dying trees. By 1987, 50 percent of the forests in West Germany were dead, including 90 percent of the famed Black Forest, a disaster attributed by Hamaker and Weaver not so much to their being doused with acid rain as to the fact that the soil in which the trees struggle to live has been deprived, through constant demineralization, of its essential nutrients. Throughout the past ten thousand years of our interglacial respite we have taken minerals from the soil into the plants we eat, or leached them away down river and stream, as Steiner clairvoyantly warned, without replenishing the stock, a disaster greatly increased by the use of chemical fertilizers, which exacerbates erosion. In 1984 the world lost an estimated 22.7 billion tons of topsoil to erosion, and another 25.4 billion in 1985. In the United States alone, four million acres of cropland (the size of Connecticut) are lost each year to soil erosion. As trees weaken from starvation, they become vulnerable to pests, industrial pollution, and increasing forest fires. When they start to die—as clearly

John D. Hamaker, from photo taken by
Steven Johnson, January 1986.

indicated by recent decades of narrowing tree rings—they fail to pull excess carbon dioxide from the atmosphere. Whole forests weaken; stray lightning bolts start huge forest fires; more CO_2 is spewed into the air.

Trees on hilltops, says Hamaker, go first because they catch more wind than those down in the hollows, plus the fact that the soil is thinner on the top of hills, an assertion recently proved all over the deforested Alps. Acid rain puts on the finishing touch, destroying the few remaining minerals. In acid soil more and more microorganisms are eliminated. What the acid rain accomplishes, according to Hamaker, is acceleration of the death of trees already dying of starvation. Efforts to replant seedlings on such demineralized land are equally doomed: life in the soil is already too far gone. There is little for the plants to grow on.

Kenneth E. F. Watt of the Department of Zoology, University of California at Davis, adds that the available evidence strongly suggests that the mass tree mortality in Europe and North America, solely explained in terms of acid rain, may in fact equally be accounted for in terms of the downward trend in summer temperatures over the last four decades, and downward deviations from that trend in particular groups of years. The unnatural cold renders the trees susceptible to acid rain.

According to a UN-FAO estimate, half the world's tropical forests have disappeared since 1950: 37 percent in Latin America, 66 percent in Central America, 38 percent in Southeast Asia, 52 percent in Africa.

And reforestation is slipping far behind.

Hamaker warns that man's gross deforestation of the tropical rain forests could trigger the next ice age as the amount of carbon dioxide in the atmosphere approaches 350 parts per million. Nicholas Shackleton, and other scientists in Great Britain, writing in *Nature* in 1983, showed that the last glacial period had begun when the concentration of carbon dioxide in the atmosphere reached a mere 290 ppm.

According to Shackleton, the last interglacial behaved much like ours, with carbon dioxide rising over the last five thousand years to a peak that ushered in the ice age, after which the cold waters of the oceans quickly absorbed the CO_2. Shackleton also implies that the growth of CO_2 was caused by the destruction of vegetation through developing drought and cold as an interglacial came to an end. Geneviève Woillod, a Belgian palynologist, shows that France went from deciduous forest to treeless tundra in 150 years, plus or minus 75, with a 20-year change-over to glacial conditions.*

Like the cooling-climatologists, Hamaker and Weaver predict worldwide drought, followed by increasing numbers of forest fires, earthquakes, volcanic eruptions, high winds and tornados, the latter caused by the increasing expanse of arctic snow cover interacting with increased heat from the tropical zones. The scenario for the coming ice age includes, as a permanent feature in the middle latitudes, 100–200-mile-an-hour northerlies. In the spring of 1988, 170-mile-an-hour winds were clocked on a mountaintop in North Carolina.

Other Hamaker-Weaver predictions, like so many vultures, are already winging in. Nineteen eighty-three produced more than a 25 percent increase in major earthquakes around the world. Japan had the biggest in a quarter century in May 1984. The 1983–84 winter broke all records for cold, storms, and earthquakes in the Northern Hemisphere, especially in Canada, Iceland, and Russia, followed by a similar pattern six months later in the Southern Hemisphere. The November 1983 winter storm in Soviet Europe was one of the earliest of the century. December 1983 was the coldest in American history. Oklahoma, Nebraska, and Texas had record-breaking winters. Blockbuster snowstorms, quite unseasonal, pounded Colorado and Wyoming. Utah recorded 65 below in January of 1984. Snow blanketed Italy as far south as Florence, while in the Rhône delta of the sunny French Riviera thousands of flamingos died because they could not get at their normal prey in brackish water covered with a layer of ice. For the first time in history it snowed in the Persian Gulf near Abu Dhabi.

---

* In the winter of early 1988, the highest winds in centuries virtually deforested most of the timber stands in southeast England.

Meanwhile, south of the equator drought ravaged Brazil, Australia, Africa, as millions died—over a million in Ethiopia's Sahel region alone. Millions more were threatened by a repetition of the drought in 1988. Vast drought-induced tropical forest fires—a critical factor, according to Hamaker, in speeding the course of glaciation—blazed up to release more smoke and CO_2, as documented by the semi-secret NASA photos taken from U.S. satellites. While relatively minor fires raged through Alabama, Tennessee, and South Carolina, destroying over 500,000 acres, a massive fire in Indonesia in February of 1984 burned for five months, devastating an area equivalent to Massachusetts and Connecticut combined. Considered perhaps the most severe environmental disaster the earth has suffered in centuries, it wiped out plant and animal life, including hundreds of thousands of giant mahogany trees, countless birds, bears, deer, pigs, civets, forest cattle, and rodents, leading to the extinction of many species, a disaster that didn't even make the *New York Times*.

Ecological "experts," who had considered the region—formerly Borneo, now East Kalimantan—one of the dampest parts of the world and ecologically stable, began reevaluating existing ecological assumptions.†

Yet, despite this gruesome accumulation of evidence in support of the "cooling" school, the warming theory continues to be the basis for "non-official" government policy, though some of its supporters are beginning to waffle.

Boldly Hamaker accused the administration of lying or of using evasive language for political and economic reasons. He claimed the warming theory to be no more than a conspiracy of the financial complex aimed at preserving massive investments in the carbon-dioxide-producing fossil fuels—oil, coal, and natural gas—along with their related industries. "Such a massive conspiracy of silence," wrote Hamaker, "is understandable only when one realizes that official announcements of our situation would plunge the world into a financial debacle. . . . The bankers have been wringing their hands over this prospect for half a dozen years; meanwhile our chance of survival gets weaker every minute."

Ringing in the bankers was not likely to widen Hamaker's popularity or credibility in Washington, or elsewhere among the multinational conglomerates. Yet opposition to the established position was beginning to

† Although the immediate cause of the fire was unknown, forestry inspector-general Hendri Santoso did not rule out the possibility that logging concessionaires, who controlled some 5.2 million acres, ignited the already-dry forest, hoping it would cover up their failure to carry out obligations to plant a seedling for every tree they felled. A West German estimate put the loss of salable timber at $5.6 billion.

be voiced in widening circles. The Massachusetts Audubon Society's magazine *Sanctuary* published an article accusing the Reagan administration of McCarthy-like scientific coverups.

In the Ozark highlands of southeastern Kansas, its limestone sculptured by fast-flowing streams, John D. Hamaker and his wife, Anita, had only recently moved into a still-unfinished bungalow with a long, gently sloping metal roof.

Already a martyr to the cause of eco-agriculture, John's stooped frame is riddled with Agent Orange, accidentally sprayed onto him by a passing truck while it was still legal in this country. Randomly the poison breaks out on his tortured body, causing great discomfort and a much-weakened constitution.

Yet his strong Midwestern voice, like gravel rolling in a river bed, resoundingly warns that if we do not hurry it will all be over by 1995. By his calculations we are already in the crucial stage, with only a few years in which to stop the onslaught of the ice. "We must cut down on fossil fuels, immediately! And we must plant fast-growing trees to absorb as much CO_2 as possible. Hopefully they'll mature in time to be harvested for turning into alcohol, a fuel which does not contaminate the air. And we must develop solar heat, and other healthy alternatives for energy, many already invented but censored by our rulers, yours and mine."

Basic to John's thesis is the urgent need to remineralize the planet's soil not only to save the trees and agriculture, but to provide the vital nourishment man needs to retain his health and rational wit. Only with remineralization, says John, can the soil's microorganisms obtain the nutrients they need to reproduce, lay down their bodies, and make the stable colloidal humus vital for plants, animals, and humans to thrive on as they once did before we demineralized the earth. Like most great discoveries about nature's secrets, John's was made by a chance observation: rain water running over gravel from a concrete parking lot, a sort of milky fluid that disappeared into a pile of rocks to produce a dandelion of rare proportions.

"I took home enough of it that night to have a mess of greens," said John, "just from that one plant. And there was enough left over for three more servings. I mean, that was some dandelion, and it had excellent taste. The next day I looked around where that gravel was, and I could see that the feeder roots, way down, had ends that were white. But back on the starter root I saw something which I recognized as humus, attached to the roots, and that gave me something to really think about."

John's wife ushered us to the kitchen table, where she had prepared a supper of organic vegetables, and because the Agent Orange has sapped his strength, a meat loaf for her husband.

"It's not what kind of food you eat," said John, "vegetarian or meat. The Eskimos lived healthy lives on fish and blubber. What matters is that the foods eaten carry forward the protoplasm of the microorganisms grown with a natural balance of the elements. Man's intestinal tract is a root turned inside out. The purpose of eating food is to recreate a population of soil organisms in the intestinal tract. Protoplasm from the microorganisms can then be absorbed right into the blood."

It was a novel way to approach one's supper. But, as it developed, John Hamaker's thesis made more and more sense, though not of the sort provided in Biology 1 at Harvard. Between mouthfuls of his own organic home-grown food, John spelled out the substance of his spiritually and physically nourishing discovery, one with which he hoped to encourage a healthy agriculture in a healthy world—providing we manage to stave off the coming ice age. All plants, animals, and humans, says John, live on protoplasm, and microorganisms are the only living thing that can make organic protoplasm from inorganic elements.

"The cycle of life is really the story of the travels of protoplasm as it goes from microorganisms to all the life above the soil, then back into the soil. That dandelion started me into all kinds of new experiments." ‡

"Saucers all over the place like Petri dishes in a lab," said Anita, smiling benevolently. "All kinds of things in 'em."

John, unperturbed, picked up where he'd left off. "I arrived at the inevitable conclusion, never taught in Aggie colleges, that the plant was sucking protoplasm directly out of the microorganisms, leaving just the skins behind. Fresh organisms, not consumed by the plant roots, dehydrate and join the bank of fertility with almost no loss to leaching or erosion. Ultimately, I realized, what's left over is the makings of real good healthy humus."

The facts fit precisely with Podolinsky's description of the genesis of humus: but the notion that the roots were carnivorously devouring the insides of the microbes was arresting. In Podolinsky's case, his roots devoured the entire contents of the humus jar, the living with the dead.

‡ Never yet successfully analyzed, protoplasm is described as a living substance that fills the cells of microorganisms, plants, and animals. It contains all the compounds of life in a mixture of such minute components that it has defied analysis with microscope.

Surprisingly, but not unnaturally, Hamaker's idea of protoplasm dovetails with that of the theosophists. According to Annie Besant, the prana of the Hindus, an energy from the sun that parallels but is different from electromagnetism, builds up minerals. Functioning as the controlling agent in the chemico-physiological changes in protoplasm, prana leads to differentiation, and to the buildup of the various tissues of the bodies of plants, animals, and men.

In the veiled language of *The Secret Doctrine*, Helena Blavatsky speaks of prana as invisible and fiery "lives," which supply the microbes with "vital constructive energy," enabling them to build the physical cells, "the size of the smallest bacterium relative to that of a 'fiery life' being as that of an elephant to the tiniest infusoria."

"The ag colleges," said Hamaker, "funded by the chemical companies, have all along insisted that roots cannot absorb anything larger than an ion in solution, meaning their NPK, ruling out the ingestion of whole molecules of humus, and therefore any advantage to placing organic additives in the soil."

He waved his fork for emphasis: "The chemical food faddists' concept that minerals are taken in by the roots only in the form of ions, and in some unproven way are built into proteins by photosynthesis, is false. The protoplasm of the organisms is simply transferred from the microorganism into the plant cell to perform the functions required by its nucleic acids. Each higher form of life uses the protoplasm transmitted up the ladder of life to make compounds specific to its needs. Protoplasm in water is slightly milky and slightly yellow. Both effects are probably from lipids (fats) in a state of colloidal emulsion. The result is sticky to the touch."

Reaching for the sideboard, John placed on the table a small jam jar half full of what looked like nice brown earth. "Put your finger in that jar and you'll feel it's sticky. That's the protoplasm. It's in British Columbia glacial gravel dust, instantly activated by a pinch of good local soil with its normal complement of microorganisms. Put into the ground, it's ready to go! I spread ten tons of it on one-fifth-of-an-acre garden. If it had been finer I'd have only needed about two tons. The finer it is the bigger the area it covers and the more it is available to be chewed up by the microorganisms.

"They feed on the minerals of all the mixed rock in the top layers of the earth's crust, plus carbon, water, and the gases and sea salts from the air, all of which they turn to protoplasm. The rootlets find the microorganisms, and the invisible root hairs suck up this protoplasm. But they can only do so with fresh microorganisms, many of which are provided by the castings of earthworms or by other larger organisms, which feed on the dehydrated carcasses of smaller ones."

At last the sequence of underground events was falling into order: Hensel's fine rock dust, in Flanagan's colloidal suspension, via Schatz's chelation, is the elemental food of microorganisms, which in turn become the food of other microorganisms. They, and they alone, according to Hamaker, are capable of transmuting the inorganic rock dust into the mysterious living protoplasm; plant roots, sucking in this basic ingredient, pass it up the ladder of life. An original conceit, it waits for confirmation in the halls of science, but leaves no doubt that plants must ingest more succulent and nourishing sustenance than the synthetic chemicals dished up in NPK.

Since no microscopic analysis of the underground process is as yet possible, anyone's description is as plausible as the next. Dr. Hans Jenny,

Professor Emeritus of Soil Science at Berkeley, the world's authority on the subject, whom we consulted on the phone, admits that, although the orthodox term "contact exchange" is restricted to the exchange of ions between soil and root, he does not exclude using the term "diffusion" to cover the intake of whole molecules, providing they are small enough; small, in this context, being a matter of opinion.

Dr. Patricia Jackson, of the USDA in Beltsville, Maryland, maintains that the size of ingestible molecules is limited by the size of the pores of the cell membrane to ten angstroms, an ingestion witnessed in cell cultures in the laboratory, but never in living plants. Yet ever since the 1940s careful researchers have maintained that plant roots can absorb much larger molecules. Recent research shows that, thanks to chelating components in humus, plants can capture and ingest enzymes, hormones, and colloidal particles by means of the gelatinous mucigel which they exude.

Dr. Bargyla Rateaver, a petite but fiery supporter of the organic method—who single-handedly and successfully fought for the introduction into California's university system of regular courses on organic agriculture for degree credit and for transfer credit, against constant and obdurate opposition from university deans, supported by the chemical companies—is categorical in her assertion that root hairs ingest large molecules, and even whole microorganisms. Quite recently, Dr. Rateaver was given an award by the Committee for Sustainable Agriculture, not, as the speaker put it, for her brilliant work in support of organic agriculture, but "for the amount of hell she raised in trying to set the record straight." In various publications she has shown how the tiny single-celled root hairs of such nitrogen-fixing legumes as clovers, alfalfa, and vetches "invaginate" their tips to allow the entry of chains of Rhizobium microbes. Inside the root, the Rhizobium congregate and multiply by the million to fill nodules where they fix nitrogen from the air for their own benefit, for that of their leguminous hosts, and for the soil to which it is eventually bequeathed.

An explanation of this remarkable process of cell ingestion was provided in the December 1987 issue of *Scientific American* by Mark S. Bretscher. In an article on how animal cells locomote, he describes what is known as the "endocytic cycle," in which a cell's plasma membrane indents into a pit coated on the inside with protein. This pit balloons inward, taking with it the material to be ingested, is pinched off to form a vesicle, moves into the cell and, in plants, releases its load of nutrients in sizes thousands of times larger than the ions of the NPK solution. Comparatively speaking, the difference is between ingesting a mouse and an elephant.

A typical bacterium is about ten times the size of a colloidal particle;

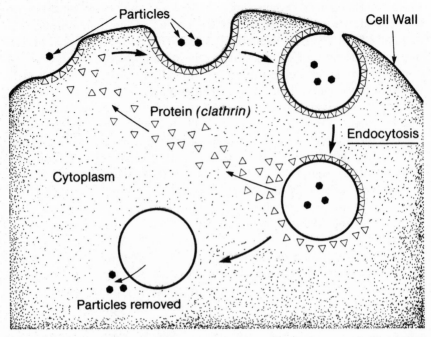

How cells form pits in their cytoplasm to ingest particles. From an article by Mark S. Bretscher in *Scientific American*, December 1987.

the particle, in turn, may be two or three hundred times as large as a small protein, which, in turn, is made of amino acids whose molecules are larger by far than ions. The scale is not peas to watermelons, but peas to dinosaurs.

And yet, as Dr. Rateaver points out, somewhat scathingly, the chemical companies, while insisting that plants can only take in ions in solution, maintain, speaking out of the other side of their mouths, that large molecules of their systemic pesticides can somehow be ingested by the plant so that anywhere a bug bites the plant it is bound to die. No mention is made of what such a bite might do to humans.

John paused to butter some bread. "In nature, that is the way that plant roots feed, ingesting through the membrane of thin cells. After a while, the older parts of the root become coated with the proteinaceous substance of the humus, which gradually turns to a natural brown. These cells become sealed, so all the intake comes from the tips, which in the soil grow toward new supplies of microorganisms. J. J. Dittmar, of Iowa State University, found that the total length of roots and root hairs of a single rye plant was seven thousand miles, and its total surface area seven thousand square feet. The roots alone grew more than three miles per day in search of microorganisms. And only the finest particles of mixed rock dust can cover the phenomenal surface area required to grow significant quantities of something so small as a microorganism. And such particles are only a tiny fraction of the topsoil; if they aren't replaced, the soil wears out, and then it's dead."

John took a mouthful, chewed it deliberately, then pointed to his mouth. "A lot of the solution's in the grinding. The smart farmer will include very fine sand-sized particles in the silt he spreads, so the earthworm will have grit for his gizzard with which to grind up silt along with any organic matter. Professor Emanual Epstein at the University of California, Davis, has estimated that plants suck up five times the billion metric tons of minerals mined each year by man. And the earthworm grinds up even more than that. He's a big operator, and he works cheap. While the worm cleans the roots, he is grinding silt and old roots to make fresh organisms, which come out of his tail to feed the dandelion root ends. If we fail to keep free carbon throughout the topsoil, which gives it that black color, the earthworms disappear."

He wiped his mouth: "It is a reasonable conclusion that without the carbon for his energy requirements he just can't do the work of grinding. Without grinding, new fine silt is not there to replace what's been removed in the crops. When the stock of protoplasm in the soil is gone, growth of crops comes to an end. With pitch, which is mostly carbon, and with rock dust, I did countless experiments."

Anita rearranged the dishes, asking if we cared for custard. "It's made with honey. There's never any sugar in this house."

John smiled, then took a more positive tone. "We can build enormous per-acre tonnages of protoplasm in the soil in a very short period of time, enough so that the sun energy reaching the plant becomes the only limit to its growth. Photosynthesis way up in the leaves is still just as essential as is protoplasm down in the roots, and that's the limiting factor. Whereas record crops have usually been produced by the use of large amounts of composted manure to produce the organisms, such record crops can now be grown on every acre of land by using rock dust with some carbon."

John estimated that the national average yields in fifteen years could run about three times what they are at present. Protein content of grains, the indicator of health-giving potential, could run two to three times the present protein figures.

Despite his warnings of impending doom, the creative farmer in him was still sanguine about the wonders we could perform with his solution.

"When we have rebuilt our soil fertility to eighteen inches in depth," he mused, looking eastward at the bare wintry landscape, "the Mississippi River will have an even flow of silt-free water all year round. Annual flooding in the lower Mississippi Valley will cease. Underground aquifers will again load up with water."

The bottleneck to enacting Hamaker's vision remained a lack of grinders. His solution awaits a wider manufacture and distribution of these patented devices, easier access to the bottomless pit of glacial gravel, and the eventual cooperation of the world's air forces to help spread the dust in a do-it-or-die program to save mankind. If enough trees are planted and fed with glacial rock dust, and a stop is put to the decimation of the great rain forests, there may still be a chance, says Hamaker, for man's best friends, the trees, to suck up excess CO_2 from the atmosphere and postpone the advent of an ice age long enough for scientific minds and enlightened legislators to come up with a means for more permanently avoiding such disaster.

"I don't know of any commercial grinders that are suitable for direct installation on the farm," said John. "The motors are too big for farm wiring; they are too costly; and the power and maintenance costs are too high. I would say that a ten-horsepower motor, or even less, would take care of most requirements. If only there were recognition of the need, there would be no problem. Industry would produce the capital to manufacture grinders. But all this is going to take government-sponsored programs on a crash basis. In World War II we learned to build Liberty ships as fast as we built cars. Now, if we don't act soon, we're going to have hunger all over the world, on the broadest scale, together with a crisis in medical care, catastrophic flooding, food riots, and revolt at the polls, or in the streets, against a system which, for profit, substitutes lies for truth and generates national leadership you can't distinguish from the Mafia."

He sighed, his last mouthful of meat loaf as chilled as his prognosis for the future. "Congress could take action," he said wearily. "But they will only do so when they realize they're threatened with impending death. By then it may be just too late. To get them off their duffs we must, each of us, prevail upon our congressman to act, and act immediately. It is still in our hands; and the job can still be done."

CHAPTER 16

Life and Death in the Forest

AT THE BEGINNING of the 1980s, the leading German weekly news magazine, *Der Spiegel*, rang the first public alarm bell warning that all over Europe forest trees were dying, shedding their deciduous leaves or drooping their sharp needle clusters like soggy rope ends.

Only a few ecologically-aware specialists, such as Hubert Weinzierl, soon to become president of the Bund Fuer Umwelt und Naturschutz (League for the Protection of the Environment and Nature) had been trying, for over a decade, to arouse public consciousness to an issue which, by 1984, Munich's forestry professor Dr. Peter Schuett could characterize without hyperbole as the "greatest ecological catastrophe in the history of humanity." A few alert owners of large expanses of forest, whose own survival depends on the health of their trees, called for action. Among the first was Hermann Graf Hatzfeldt-Wildenburg, a dashing and handsome owner of extensive forest lands on the banks of the Sieg River in Germany's Rheinpfalz province.

An economist trained in the universities of Princeton in the United States, Ibadan in Nigeria, and Basel in Switzerland, Hatzfeldt spent time in the late 1960s working as a Ford Foundation staffer in Thailand. But, as most of his revenue comes from his forest in Germany, he returned there in 1970 to take care of his estate.

To Hatzfeldt's amazement, there seemed to be no one in Germany responsible for the administration of forests at the federal level of the Republic, nor academically for study of them in universities and research institutes, nor commercially for management of them as a prime source of income. No one seemed to have mustered sufficient concern to ana-

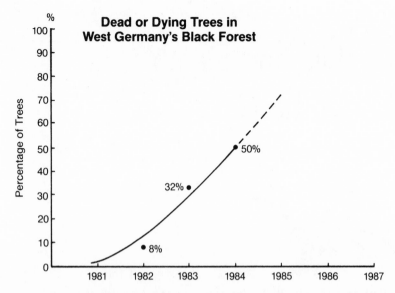

Source: Embassy of the Federal Republic of Germany, Washington, D.C., 1985.

Dead or dying trees in West Germany's Black Forest.

lyze the many dimensions of the mysterious new affliction that was taking on the aspect of a sylvan holocaust.

In 1981, as the young owner of Schloss Schoenstein and its surrounding woodlands, Hatzfeldt took it upon himself to call a meeting in the city of Kaiserslautern, to which he invited a score of experts concerned with forestry to discuss the *Waldsterben* crisis, or "death of the forest."

Hatzfeldt and his colleagues were trying to establish a clear connection between the now baldly evident *Waldsterben* and its supposed underlying causes, the pollution of the atmosphere with carbon dioxide and acid rain, the direct result of the ever-increasing industrial use of fossil fuels for power that was creating the same havoc in European forests as it was in the troposphere over the Sahara Desert.

In 1982, in the German Federal Republic alone, well over a million acres, or a Rhode Island-size tract of fifteen hundred square miles of forested lands were in various stages of illness. European-born Americans returning to visit their birthplaces were most easily shocked by the disease-caused depredation. When Lehigh University physiologist and historian of psychology Dr. Josef Brožek visited his boyhood playground, Zaichny Rokle (Hare's Gorge), near the little Bohemian town of Police in the Steny (stone) Mountains along the Polish border, he found that every one of the trees in that three-quarter-mile-long almost canyonlike valley were as stone dead as if they had been burned in a forest fire. Yet in his native country Brožek could find no official willing to talk about the impending doom.

The nightmare of formerly forested hills and mountains becoming devoid of trees, wild animals, and even singing birds was so overwhelming that the data had to be kept from the Czech people by their leaders as it has been in Poland, other East European countries, and, until the new turn toward "candidness" initiated by Soviet Party Chairman Gorbachev, in the Soviet Union.

Statistical facts on the Czech sylvan disaster, revealed by Dr. Wolf Ochslies of the West German Institute for East European and International Affairs after careful digging, illustrated how dire conditions have become. Of all taxonomic groups of fauna, 60 percent of amphibians, 35 percent of mammals, and 30 percent of reptiles, birds, and fish are currently threatened with extinction. Almost all the partridges in the country have vanished, along with 80 percent of the once-abundant hares and nearly half the pheasants.

Ochslies quotes an article from the German newspaper *Die Welt* (July 20, 1984) dealing with a fact-finding visit to Czechoslovakia by Austrian government functionaries, one of whom remarked in astonishment: "This forest is no longer dying, it is already dead!" Of other woodlands, the functionary continued: "The impression on all of us was devastating,

One of thousands of posters in West Germany: "Here dies the forest." (Credit: Reinhard Janke)

with forests looking like sylvan cemeteries, not green, but gray, brown, or reddish in color, and completely lifeless, all deer and wild boar non-existent and not a single bird singing! No one would believe it unless they could see it for themselves: how, for instance, mile upon mile of stands of seventy- to a hundred-year-old trees now resemble forests of telephone poles!"

Seeking to maintain the momentum started by the first Kaiserslautern conference, Hatzfeldt called a second one in the spring of 1983. Hatz-feldt introduced the seminar by questioning whether the woods could be saved with forestry practices alone, concluding that foresters could only treat symptoms but not the prime causes. "We are between Scylla and Charybdis," said Hatzfeldt, "and we have little hope, but we cannot give up the struggle, even with little scope in which to find a solution."

And indeed he could see little scope for action in what amounted to a web of conflicting explanations for the disaster, or allegations that it did not exist.

Because of his personal preoccupation with public activities on behalf of his country's welfare, and his constant peripatetic travels, Hermann Graf Haztfeldt was a hard man to meet. When finally caught up with in the lobby of Frankfurt's Hessischer Hof, he walked up dressed as if he had just come from a day-long tramp on some wild moor, in farmer's brogans, rough whitish-green cord pants and tattered woolly sweater over a white shirt from whose collar billowed a colorful silk foulard. Thick glasses gave him a scholarly mien. A slight limp, due to a back injury, gave the impression of a cross between an ancient German *Ritter* returned from a campaign and an anti-business-suit environmental activist.

At an Italian restaurant around the corner Hatzfeldt outlined the campaign he had mounted over six years to arouse German public opinion to the devastation it faced with the demise of its woods. He had just returned from a trip to Poland to discuss with foresters what was happening in their land.

"I went to Poland," said Hatzfeldt, "to have a personal look at a country which, due to its avid urge for more and more industry, has ignored the effects, caused by the factories it spawned, on surrounding nature. Even our almost hopeless situation does not yet match what is taking place in parts of Poland and neighboring Czechoslovakia. Poland's ever-optimistic national anthem, '*Jeszcze Polska Nie Zginela*' ('Poland Has Not Yet Perished'), may still prove false, not in the sense of human politics, but because of its callous attitude toward the effects of the so-called benefits of industry on its forests. The trees in Polska no longer take a decade to die, they die in a couple of years!"

On the way to visit Hatzfeldt's estates in the Sieg River Valley, as our train pulled into the station at Wissen in the valley of the Siegland right-bank region of the Rhine, filled with smoking industry that had piled up over the centuries enormous mountains of slag, the problem of pollution was evident enough. But Hatzfeldt's Schloss Schoenstein (Beautiful Rock) towered sublimely over the banks of the Sieg River, surrounded by still beautiful woods with centenarian oaks. There we were met by Hatzfeldt's associate, Dieter Deumling, who explained in perfect English that he had been living in Oregon for almost a decade until he had received an urgent call from his friend Hatzfeldt, to come back to Germany to help save the woods, a request with which he had immediately complied.

In a four-wheel-drive Land-Rover he chauffeured us up into a vast tract of forest several miles above Schloss Schoenstein. Snaking through a deep wood of dying beeches, he indicated the ones on the left were mere adolescents of 30 to 40 years, the ones on the right just maturing at 150 to 180. "At this relatively low altitude," said Deumling, dejectedly shaking his head, "deciduous trees, like these beeches, are already as badly affected by *Waldsterben* as the evergreens. It is the same all over Germany. In the northern and middle reaches of the Federal Republic, a third of our beeches, maples, oak, and ash have the disease. It's not quite as bad for our 60- to 80-year-old spruces, all of which are affected, but it is a ghastly enough outlook, nonetheless. And it disposes of a crowd of optimists who are vociferously claiming that a perfect solution would be to replace the dying conifers with hardwoods."

Jumping out of the Land-Rover, Deumling walked over to the edge of the road to cut a spruce bough from a sick tree and another from a tree as yet unaffected with *Waldsterben*, or at least not appearing to be affected.

"See, here," he said, pointing his gloved finger. "On this healthy bough the growth is symmetrical and there are needles growing out all along the central branch; you can clearly make out the annual growth by counting the segments. But on this sick one there are no needles on the central branch. You will note little abnormally emerging shoots, called *adventitious budding*, where they shouldn't be emerging. In biology *adventitious* means 'appearing in an unusual place or in an irregular or sporadic manner.' In German it's more heartrendingly called *Angstriebe*, or 'anxiety growth,' a term that concisely illustrates that a tree is doing something abnormal in an attempt to save its own life. Firs even put out such growth from their trunks, which doesn't help because most of the firs in Austria and Germany are already dead. The white fir is extinct."

He threw away the healthy branch and concentrated on the sick one: "For a tree this old, the needles should be much longer. Look how measly they are! They should be at least twice as long. And there are many other symptoms you could look for. You could find some on every branch of this tree, even on apparently healthy branches if you look close enough. It's as frightening as if you were to find pimples or rashes breaking out all over your body."

As we drove back down to Schloss Schoenstein Deumling said wryly: "The whole problem is as thorny as any spruce needle; it's enmeshed not only in seemingly endless scientific debate, but in a political maze worthy of Machiavelli. To offer an eventual solution, whether lime, or rock dust, or something else, will be to attract salesmen in droves. The chemical companies won't be left at the starting gate, and when *they* stage their sales act they'll surely recommend that forest owners cover their bets by mixing a variety of products together; and in terms of profit, the more the merrier. I'm surprised they haven't already got to it in your country, where the trees are dying, not quite as fast as they are here, but almost. In the Smokies your balsam are going, and in New England and Canada your precious sugar maples will soon be gone."

He sighed and brought the car to a stop in the courtyard of the castle. "It's just like the situation with American farmers. They're constantly being frightened by chemical salesmen with the threat that if they abandon the use of what amounts to a whole medicine cabinet of chemical products, and turn to a more wholesome way of treating their land, they risk going broke, which they are doing anyhow! It's a cruel system designed to keep them permanently on tenterhooks. There seems to be no solution."

But a solution there was, and not far away, near the little Austrian village of Grimsing on the Danube's *rive gauche*, just down from the riverine city of Melk, dominated by its huge Benedictine monastery, now a boarding school for boys. It was a solution to vindicate both Hamaker and Hensel. There, in the summer of 1980, Rudolf Schindele, a manufacturer of fine veneers, knew, like hundreds of other European foresters, that something sinister was happening in the nearly one square mile of hilly forest he owns outside Grimsing.

As perplexed and worried as Deumling, and little realizing he would make a discovery that could go a long way toward the solution that was evading the academic fraternity, Schindele decided to build three kilometers of logging roads through some woods he had bought that were severely afflicted with *Waldsterben*. As he began to excavate material for the roadbeds right out of the side of a small forested mountain, the derived substance turned out to be a crumbly, multicolored metamorphic rock, geologically known as *paragneiss*, in deposits which Schindele figured might total three million tons.

In existence for over two hundred years, the area had once been covered by an ocean so old that no living organisms, large or microscopically small, inhabited it to leave fossil remains in the sediment, which ultimately turned into rock. During the road construction, a certain portion of this paragneiss was reduced to powder by heavy equipment and blown by high summer winds to parts of the forest adjacent to the roads. Just four weeks later, the spruces in these areas, whose needles had been growing increasingly yellow, a sure sign of *Waldsterben*, were turning back to a radiant dark green. The total area of recovering trees extended over some 13 acres. During the next four years, new growth on the accidentally treated trees looked better and better.

Because of the particular susceptibility to the *Waldsterben* syndrome of firs and spruces—now decimated by the millions—attention in Germany, like Schindele's in Austria, has focused on these evergreens.

To see for ourselves what Schindele had accomplished—and could demonstrate—we took a train to Melk in mid-November 1985, and there were met at the station by Maria Felsenreich, Ph.D., a dynamic Austrian environmental activist, who owns a large organic herb garden twenty miles from Vienna and has organized a campaign to save forests throughout Europe's German-speaking region. She introduced us to our host, dressed appropriately in the forest green of a Styrian woodsman. Over lunch at the *Gasthof* in the town's central square he told us the saga of his success in bringing trees back to life with rock dust, adding, with conviction, that he believes that by ingesting such *Gesteinsmehl* animals and even humans can regain their health. Stroking his salt-and-pepper hair, he claimed that by imbibing two teaspoons of the dust every day he had managed to turn his snow-white locks back to gray, a claim attested to by various newspaper clippings, which showed him before and after.

To see what the dust had done for his trees, we set forth in Schindele's 380 SE Mercedes to climb high up into his forest property, over ground covered with a foot of fresh snow, stopping to put on chains when the narrow road grew so steep the wheels spun ominously close to a vertical abyss.

Progressing through trees hung with dazzling whiteness, as if in a huge, endless Christmas card, Schindele pointed toward some handsome beeches with slate-gray trunks and yellowing copper-colored leaves still clinging to their branches. "Deciduous trees, like beeches, are also being affected all over by *Waldsterben*. But it's intriguing that in my forest many of them are beginning to keep their leaves much longer into the

Robert Schindele, described
by the German magazine *Horzu*
as sitting on millions of marks'
worth of rocks for rock dust.

season than before, now that they have had the benefit of rock dust." At
a curve in the road we stopped to look down a slope to where baby
spruces were growing healthily, eight to ten feet high, two to three times
taller, Schindele explained, than had they not profited from the rock
dust that had fallen on them "as from the hand of God!" Further down
the slope, their shorter, sicklier counterparts had received no dust.

"How long," we asked, "can the European forests survive, with no
real help, either from rock dust or some other agent?"

His answer shot back like an arrow: "About five years. Ten at the
most."

Proceeding "downhill," Schindele provided more gruesome details:
"Evergreens normally hold their needles seven years before dropping
them to make way for new growth. Trees that have benefitted from
Gesteinsmehl now hold their leaves for almost five years; but those affected by *Waldsterben* are dropping them within two to four years."

To see more evidence of the efficiency of *Gesteinsmehl* in other parts
of Austria we visited a Tyrolean agronomist-engineer, George Aber-

mann. Independently of Schindele, he had made some experiments with *Gesteinsmehl*, this time ground from rock taken from a quarry in the Tyrolean ski-resort village of Kitzbühel. In Innsbruck's Grauer Baer hotel, Abermann, a lean, graying, fit-looking man of forty-two, with pale-blue piercing eyes and the self-assurance that comes with high intelligence, motivation, and the joy of cause-oriented work, offered to show us unequivocal proof of the dust's effectiveness.

Trudging through deep snow into the *Matzen Naturschutz*, or Nature Reserve, halfway between Innsbruck and Kitzbühel, Abermann explained that one of the reasons ground-up rock from quarries had proved disappointing in the past was that the dust had not been ground fine enough to be easily available to the plants—or, in Hamaker's terms, to the microorganisms. "This," said Abermann, "permitted the chemical industries to proclaim that rock-derived materials are useless in agriculture."

But in 1980 Abermann met the owner of a Kitzbühel stone quarry and gravel works who had been turning out gravel from a basaltic stone known as *diabase* since World War II.

The crushed rock, extremely resistant to crumbling, and thus to deterioration, was used mainly for railway track beds. In the process of crushing, a great deal of apparently useless dust, or powder, effloresced as a by-product, until the quarry owner noticed that farmers would come to his works to truck it away for free and apply it to their lands in quantities of about one thousand tons a year.

The quarry owner donated Abermann twenty-five tons of this stone flour, a single kilo of which could cover an estimated 2,600 square meters. "On agricultural crop land," said Abermann, "the trick is to mix the dust with dried cow manure from which the dust appears to remove the odor, providing the soil with organic as well as mineral fertilizer. In the forests it seems to do the trick by itself."

About half a mile up into the snowy woods, Abermann caught his breath, and said with a sly grin: "I thought we'd have no results whatsoever to show off before at least two years had gone by. But when we dutifully broadcast the dust by hand not just around the trees themselves, but over this whole area, to treat every square meter of soil, to my amazement, five months later I could see the little trees burgeoning with new healthy explosive growth!"

Pointing to a stand of spruce over one hundred feet tall, whose branches indicated they were already half dead, Abermann added: "It's a bit difficult to photo-document the changes because the trunks are bare almost halfway up, and the greenery is way up in the air, inaccessible, except maybe by helicopter. But you can easily spot changes for the good

Agronomist-engineer George Abermann in an Austrian Nature Reserve near Innsbruck pointing to a tree recovering with the use of rock dust.

if you look in this clump of young spruces."

Standing next to one of them, about as tall as himself, Abermann went on enthusiastically: "Before I started experimenting with this little tree it had no needles at all. The whole aboveground portion seemed stone dead. Then it was treated with *Gesteinsmehl* and it didn't die. Its needles regrew, very copiously, as you can see! What we've demonstrated on a few trees can easily be repeated on millions of them, even hundreds of millions, whereas all the young trees which were planted here in reforestation efforts died before we began our experimentation. You should also know that the pollutants in this area are loaded with copper and cadmium, highly toxic to the soil, and that the trees recovered despite the fact that there was no abatement in the cadmium emission."

Bidding us follow him over to another batch of spruces next to a towering larch, Abermann pointed to one of the trunks: "By looking at the spacing on the trunk between the vertically-spaced series of branches you can easily see where these spruces stopped growing, then, after treatment with rock dust, put out three to four feet of new upward growth each year for three years. It's fantastic," he exclaimed, raising his hands as if in disbelief, "millions upon millions of *Schillings* have been spent by our federal and provincial authorities to reforest areas dying of *Waldsterben*, yet most of the newly planted trees, so carefully put into the ground at such expense, have all died. Now even the local forester is beginning to say that by using my rock dust there is no more need to plant young trees; thanks to its effects plenty of new ones are growing spontaneously from seed."

Asked what the effective difference would be between dust ground from Schindele's paragneiss and the diabase quarried in Kitzbühel, Abermann replied: "Not much! If, and when, the idea of using massive amounts of *Gesteinsmehl* is adopted in official circles, it will be of no real importance whether it is ground from paragneiss, diabase, basalt, porphyry, or certain other rocks, because all of them produce dust that works in similar ways. The only problem is to get the *Gesteinsmehl* adopted in official circles, and for that one needs a great push, most likely from on high."

When we came to parting, this man of heart tightened a steely grip, and said with a smile: "*Jawohl*, my friends, it will take an evangelist's fervor to bring about the required change in thinking."

By the end of 1987, Schindele had built what he calls the world's largest mill for grinding *Gesteinsmehl*, which he exports all over the world, for treatment of both forest and agricultural land, and for addition to human diets. Pointing to his own darkening hair, Schindele recommended a daily intake of two spoonfuls of finely ground rock dust, explaining that its high content of silica, aluminum, potassium, iron, magnesium, and other trace elements are essential to health, that vitamins taken in the form of supplements are without effect unless trace elements are provided with them as co-factors.

News of his remarkable rock dust was soon spread world-wide by radio, television, and the press, with the result that so many customers turned up in front of Schindele's plant that there were lines of cars several kilometers long. Schindele even claimed that, as a result of his sales of rock dust to the general public as a health-giving dietary additive, pharmacists reported that in parts of Germany drug sales were off as much as 40 percent. Reaction from the pharmaceutical industry, as might have been expected, was rapid and deadly.

According to Schindele's unconfirmable estimate, millions of dollars were spent on a media campaign to claim there were deleterious amounts

Rock dust being sprayed from a tank to revivify dying forest in Austria. (Credit: *Soil Remineralization*)

of chrome and cobalt in Schindele's product. The results were a cutback in his sales of rock dust and a prohibition by the Federal Republic against distributing it in Germany for human consumption. And so artificially incited were the good people of Grimsing against the rock dust they petitioned the authorities to prevent Schindele from parading his dusty, noisy trucks through their clean and quiet town. This cost Schindele the trouble and expense of a new road to bypass the town. But he was lucky enough to obtain registration for his dust as a "mineral dietary supplement" in another European Common Market country, and this enabled him to sell it in all participating countries.

When the University of Vienna found that Schindele's product worked against radioactivity—a claim confirmed by a Soviet institute for atomic physics in the Ukraine—the Soviets sent a truck to pick up two thousand kilograms of his *Gesteinsmehl*. Analysis under a micropolariscope revealed an alteration in the molecular and atomic lattice, which had an effect on ionized radioactive particles taken into the body.

This led plant scientist Dr. Gernot Graefe, in Austria's Burgenland province near the Hungarian border, to add Schindele's rock dust to an organically processed product he has developed over the past ten years

Root grown in soil treated with rock dust and organic Biovin (extracted from the residue of grape pressing) compared with the untreated smaller root on the left. (Credit: From Larry Ephron's documentary on *Stopping the Coming Ice Age*)

from tons of residue that follow grape harvests. With it he was able to bring large sterile acreages back to fertility. He therefore developed a homeopathically dosed spray—a kind of etheric humus as he calls it—which, distributed onto the surface of polluted ponds and lakes, has been found to bring water back to its formerly pristine condition. Even more remarkably, he claims it can be injected into rolling morning fogs to be carried for kilometers—even hundreds of kilometers—through forests where, coming into contact with tree leaves and needles and, via the soil, with roots, it holds promise of resuscitating vast *Waldsterben*-afflicted tracts of woods.

The still invisible and intangible positive effects of the spray have thus far been objectified only through measurements made by dowsing with a pendulum to supersensibly reveal data unattainable by normal human senses. Still considered by orthodox science as mumbo-jumbo, if not outright charlatanry, the art of dowsing has nevertheless been very successfully employed as a diagnostic tool. Aubrey Westlake, M.D., an English physician who used it extensively, stated a year before his death at ninety-two: "I believe that the rediscovery of the dowsing faculty is not fortuitous but has been vouchsafed to us by providence to enable us to cope with the difficult and dangerous stage of human development that lies immediately ahead. For it gives indirect access to a supersensible world, thus extending our awareness and knowledge. The faculty should be regarded as a special and peculiar sense, halfway between our ordinary physical senses, which apprehend the material world, and our to-be-developed future occult senses, which in due course will apprehend the supersensible world directly."

All of which is no more than what Rudolf Steiner was up to almost a century ago.

CHAPTER 17

Savory Soil

WHAT IF BILLIONS of tons of Schindele's rock dust were readily available in America, as effective on crops, trees, and even humans as his *Gesteinsmehl?* In a narrow valley south of Salt Lake City, blessed with a profusion of pink hills, cobalt lakes, and azure skies, a geological prospector, Rollin Anderson, has discovered just such a treasure.

In a hundred-year-old adobe farmhouse on a hill, surrounded by centenary Lombardy poplars, we found Rollin, though already in his nineties, acting like a "crusty young fellow." Like Schindele, he has been swallowing down a spoonful of his native Utah soil with every meal—not just ordinary soil, but a special montmorillonite clay.

"Some scientists," said Rollin, "think my rock stores up energies of sun, earth, and water, only releasing them as needed for the growth of plants." He spread his hands as if accepting bounty. "And Robert Ripley claimed that Sun, Earth, and Water are represented by the Hindu sound AUM; so I thought of calling my ore Anderson's *Utah Mining;* but I refrained. Instead I called it Azomite: or A to Z Of Minerals, Including Trace Elements. And they're the secret to its great success."

One sunny day in August, as we breakfasted on scrambled eggs and Azomite, Rollin told us how he had come to discover his precious substance, and how he had come to eat it. Half a century ago, as a contracting engineer in his forties, he had become fed up with city life in San Francisco, convinced that what was wrong with America was its food, and therefore the soil from which it derived. Sick soil, said Anderson, means sick people. And somewhere there had to be a remedy.

Told that gypsum might help neutralize alkaline soils, and that if mixed with fertilizers it could help grow better crops, Rollin packed up and moved to his native Utah to exploit a gypsum mine owned by his father. But, before he could obtain the necessary equipment, World War II broke out, to scotch his every effort.

Roving the river district of Sanpete County, he came upon a range of terraced hills with a pink sheen, twenty-one of them to be exact, rising two hundred to five hundred feet from the arid desert terrain, all with a pinkish ore. Intrigued, he took samples to Salt Lake City to his friend Dr. Charles Head, ranking scientific expert and chief microscopist at the U.S. Bureau of Mines. Head placed a piece of ore beneath the lens of his microscope and let out a long, low whistle. "How much of this stuff do you think is out there?" he asked, in no way attempting to disguise his excitement.

"Several billion tons," replied Anderson. "That's what I reckon."

Head's excitement, it developed, was not because the sample contained nitrates, considered valuable as fertilizers, which it didn't, but because it was a colloidal clay containing quantities of minerals very similar to the caliche rock of Chile and Peru from which the world's nitrates have long been mined. Between 1919 and 1925 Head had been seconded by the U.S. government to study Chilean and Peruvian nitrates in South America. There he had developed the conviction that the benefit plants were deriving from South American nitrates was not from the nitrates themselves but from minute quantities of trace elements, which served as catalysts—a word coined by the great Swedish chemist Berzelius to describe substances that speed up chemical reactions, but come through these reactions without themselves changing.

In the "gay twenties" few men in the scientific field, especially in agriculture, knew much about trace minerals; and ever since, because of the obscurantism of official bureaucracy, Head was obliged to be careful what he said, lest he lose his job. The prevailing opinion considered trace elements impurities that would contaminate food. The notion gave birth to so-called refined foods, from which these "contaminants" were deliberately removed for a supposed improvement in nutritive qualities.

Now at last Head had a chance to check his own theory. Would Anderson please grind up some of his montmorillonite ore, put it on his plants, and see what happened. Anderson, like everyone else in those days, had a wartime Victory Garden, and was happy to oblige by pouring powdered montmorillonite onto the ground around his vegetables, leaving several rows as controls to see what difference might develop.

Jutting his bulldog chin with evident pride, Rollin told of his early successes. "The first tomatoes we planted with the dust came up fine and healthy, whereas the controls were attacked by hideous long green worms. We picked off the worms wherever we found them, but they ate a lot of leaves. On the Azomite plot, not a worm. The plants were stronger, held fruit well, and had great flavor. Once you've tasted a vegetable grown with Azomite you're spoiled for life. The beets in the control plot were juiceless and woody. The ones with Azomite dripped with juice and were tender at all ages of their growth. By fall, one measured seven inches across, just as tasty as the young ones. The same with tomatoes, cabbages, and peppers: and everything kept better when canned or frozen. We couldn't help feeling that Dr. Head's theory about minerals, trace elements, and catalysts was definitely proved. Here was a substance that gave results that you could see without the aid of any microscope."

Rollin Anderson's mine in Utah containing billions of tons of Azomite, a natural colloidal silicate with twenty-five or more mineral and trace elements. Colloids are the pantry in which plant food is kept and gradually released as needed.

Ushering us into a living room lined with some eighteen hundred volumes, covering subjects from Agriculture to Zionism, many of them dealing with the occult and the esoteric, Rollin seemed happy to have someone with whom to share the story of his early discovery, and the remarkable results that had ensued.

"None of the local geologists or mineralogists seemed to know what we'd found in those hills: some called it *brecciated rhyolite*, a glassy volcanic rock similar in composition to granite. Others called it *diatomite*, a mineral made from the calciferous bodies of tiny marine algae. Others called it diatomaceous earth. But to Head it was good old montmorillonite, an aluminum silicate clay admixed with various minerals, rare in the United States, and even in the world, but greatly prized by medicine men of Indian tribes. On Head's instructions, I obtained samples of rhyolite from most of the known deposits in Utah, as well as from the surrounding states; but none was similar to our ore, though all were similar to each other. Ours was definitely a first-class montmorillonite clay. Now geologists consider it to be an ancient oceanic deposit brought to the surface by volcanic action, a form of heavy sedimentation on the sea floor, a mixture of mineral elements and marine life such as seaweed, shrimp, and algae. The clay contains all the essential mineral trace elements in a balanced ratio, as laid down by nature. In this form the minerals are naturally chelated, as in plants and animals, in an organic, easily assimilable form."

Rollin poured a teaspoon of the pinkish-gray Azomite into the palm of his hand to show that it was as fine as a lady's face powder.

"The problem," he explained, "was how to get the stuff out of the ground and refine it in a wartime emergency, which preempted obtaining machinery of any sort. Only when the colonel in charge of a U.S. Ordnance Depot in Tooele, Utah, had the good sense to order several tons for an experiment were we able to acquire a small hammermill and an ancient Fordson tractor. With this rudimentary equipment we set about mining the ore from the pink hills, grinding it to various sizes."

He looked up with satisfaction. "We now know that Azomite aids the soil in fortifying the natural mineral balance. It helps satisfy the 'hidden hunger' in soil caused by mineral depletion or deficiency from continued use over long periods of time. Soil without humus is half alive, and without bacterial action humus is dead. The reason the bacteria in the soil fail to function properly is because of the lack of natural trace elements and catalysts."

A gust of wind drove down the valley, turning poplar leaves from green to pewter, bending large red poppy blossoms almost to the ground.

"I learned the power of Azomite," said Rollin, "experimenting with earthworms. I didn't know, until I tried it out, that earthworms can be kept alive in a metal container, filled with just the soil the worms are found in, from early spring until late fall, and all year round where winter is not a problem, by the simple addition of a small amount of Azomite, and, of course, moisture. The worms will be lively and healthy, with firm body tissues thanks to all the elements. They not only grow but multiply. Any boy who has been fishing knows that after a few hours worms in a container will bunch up, become slimy, thin, and sort of transparent, then die, unless you keep replenishing the container with fresh soil or mulch and moisture. Well, I have kept as many as two hundred earthworms in a twelve-quart pail filled with the soil they were dug in, from early June until the middle of November, with nothing added but a heaping tablespoon of Azomite mixed with the soil at the time the worms were dug up. The worms were as fresh and active at the end as when I first took them from the garden. Just try it!" Rollin looked to his wife, Elsie, as if for confirmation, then hurried on, as if his time were running out.

"We further learned that by applying Azomite directly in contact with the seed or root structure one could get much quicker action. We tried it on lawns, but people complained they had to cut the grass too often. On pasture and perennial crops the best results were obtained by applying about fifteen hundred pounds to the acre. Results were even more noticeable after the second or third year."

He waved toward the valley, where fruit trees grew in an orchard. "Trees seem to respond to Azomite about as readily as any vegetation, especially fruit trees. In one orchard where leaf curl, sluggish growth, poor-quality fruit and many pests were the problem, Azomite corrected the conditions within a year. By the end of the third year, none of the conditions existed."

Azomite, Rollin explained, should be applied to trees in the fall, just after harvest, starting about eighteen inches from the trunk and spreading as far as the drip line, then disked in, anything from two hundred to three hundred pounds.

"But the real payoff," he said, smiling broadly, "came when we fed it to cows through silage. Animals showed a definite preference for pasture grown with Azomite. Cows, horses, sheep, goats, rabbits, turkeys, all preferred Azomite-treated hay. I've had animals walk right through belly-deep lush-looking pasture not treated with Azomite to get to that which was, and then eat off it until you'd swear there was nothing left to chew on. Failing to get an adequate supply of any one trace element, animals have difficulty breeding, calves are small, litters of pigs are weak. Beef cattle fail to make the best use of their feed. Dairy cows produce less milk; sheep have thinner fleece."

Elsie, tall and slim, got up and headed for the kitchen. "Tell them about the chickens," she said. "It was amazing."

Rollin drew breath, his satisfaction evident. "We got started with poultry quite by accident. It was difficult to get all the Azomite ground to a fine powder. There were a lot of pea-sized nodules left over. So I had the bright idea of feeding the chunk-sized Azomite to poultry as a grinding agent. When a neighbor placed some Azomite in the pen where culled hens were housed, by morning it was gone. None of the hens died; all started laying again. Baby chicks would take Azomite from the very

Corn grown on soil in which inorganic nutrients have become "locked up" or have been depleted by the use of commercial fertilizers. Compare with corn grown with Azomite *(below)*, which brought 100 percent increase in crops.

Corn grown with Azomite.

first day, if it was ground fine enough; it seemed to stimulate their appetite. They developed more evenly, feathered out sooner, and later gave a greater percentage of fertilized eggs. Pullets were laying a week before they were supposed to, and their shells, which had been fragile, were now much harder. Did you know that it costs the U.S. poultry industry $60 to $70 million annually for broken eggshells?"

Turkey growers suffer heavy losses when their birds become infected with "weak leg condition" (staphylococcus infection) which makes them unable to stand.

Rollin paused for us to appreciate the importance of the remark, then hurried on. "With turkeys we had even greater success. Azomite gave them earlier maturity, greater weight, stronger legs, and a greater number of prime-grade quality. Then we found that it was just as good for cattle. A farmer's cow got loose in the barn, where she found a bucket of Azomite and licked it up as if it were lush feed. So we spread the word and cattle ranchers starting mixing it in with feed. One rancher wrote that since he'd included Azomite the average gain per head per day was more than four pounds. Prior to feeding Azomite the cost per head in the feedlot for three months had been $140 a head. Since Azomite it was down to $95, and the quality of the beef was greatly improved. Another

Turkeys fed on Anderson's Azomite show strong legs, good feather bloom, and no debeaking.

farmer wrote that seven Holsteins which had been bred four times artificially failed to settle until 5 percent Azomite was mixed into their daily feed. On the fifth breeding, all the cows settled. So we fed it to hogs, and by market time the runts had caught up to the others. With goats we managed to breed culled ewes past lambing with a ram that was supposed to be infertile; and we got plenty of kids, plus 50 to 60 percent more wool from the sheep."

To make his point, Rollin waved a small booklet: *The Story of Trace Minerals* by Dr. Melchior Dikkers. Already in 1931, Dr. Dikkers, as Professor of Biochemistry and Organic Chemistry at Loyola University, was so struck by the properties of montmorillonite clay—claiming it to be one of the most amazing and unusual materials he had ever been fortunate enough to come in contact with—he launched an extended research program. Years of intensive study convinced him that trace elements were the key to all living organisms, essential to the structure of certain complex chemical compounds that influence the course of metabolism, a vital factor in the health of every living being.

Metabolism—the sum total of all chemical reactions that proceed in every single cell of the body twenty-four hours of each day—is what keeps us all alive. Some thirty trillion cells are at work, constantly, in each and every human body, twenty million in the human brain alone. In each cell, the process by which foodstuffs are synthesized into complex elements is carried out by enzymes, large proteins which are themselves synthesized by the cells. And it became clear to Dr. Dikkers that trace elements were essential to the creation of these enzymes, to act as catalysts to bring about chemical changes by their mere presence, without themselves undergoing change. It is a phenomenon for which science has no real explanation, but which clearly cannot occur without both the enzymes and the elements taking in and radiating energy to achieve specific effects.

Combinations of trace elements have been found, under certain conditions, to acquire entirely new properties, very different from those of individual elements acting singly. There is a noted interaction among trace elements, such as iron and copper, both of which are concerned with blood formation. In plants, iron and magnesium are associated in chlorophyll formation.

Without chlorophyll there would be no life on earth, the very first green plants being the understood link between the energy of the sun and life on the planet. Only green plants and certain microorganisms are able to absorb the sun's energy, store it, transform it, and then transfer it to man in the form of wheat, corn, vegetables, and fruit. Uncooked and unprocessed food will supply enzymes directly to the blood. Some two thousand different enzymes, every one a protein, are synthesized by every cell from amino acids furnished by the blood, obtained from ingested food, best eaten raw.

Any heat over 119 degrees Fahrenheit destroys enzymes, as does pasteurizing. Many chemical substances—fluorine, chlorine, lead, barbiturates, Benzedrine, amphetamines, nicotine, carbon monoxide, nitrates, sulfur dioxide, DDT, and most other pesticides, herbicides, and chemical fertilizers—inhibit enzyme activities, as do water and air pollutants.

The activities of enzymes are extremely susceptible to foods. The mere presence of chemical additives in food may cause some trace elements to become unavailable. The same applies to chemical fertilizers in the soil. They can cause trace elements to become unavailable to plants. Enzyme reactions are influenced by a deficiency of any functional nutrient.

Dr. Rudolph Abderhalden, Director of the Laboratory for Endocrinological and Enzymatic Diagnosis in Basel, Switzerland, and Professor of Biochemistry at Halle University in Germany, believes the majority of all diseases may be enzymatic in origin. He asserts that metabolism is synonymous with enzyme activity, and that disease is a disturbance in the harmonious pattern of enzyme activity, an activity dependent on the presence of trace elements. Breakdown of the enzyme system results in disease or death of the cell. Many nutritionists and physicians now agree that there is really only one disease: malnutrition; that all the other ills derive from it.

"We now know," said Rollin, "that the synthesis of all known natural mineral elements is the secret of the harmonious synergetic function that forms the basis of healthy living matter. Azomite is a complex compound of natural colloidal silicate minerals and trace elements. Some thirty-two trace elements—iron, cobalt, magnesium, zinc, copper, etc. —occur in such minute quantities they must be measured in parts per million, yet they appear to be basic in the complex chemical and electrical mechanism that makes up the human body. The form in which the major part of the natural inorganic nutrients are assimilated by animal and vegetable consists of material in the colloidal state.

"In plants, rootlets and root hairs are generally in intimate contact with the colloidal sources of the soil nutrients they feed from. Plant nutrients are thought to pass from the soil solids to the plant without leaving the sphere of colloidal influence."

Rollin laid down Dikkers's book, to make the basic point about his precious Azomite. "Trace elements need to be ingested in a balanced manner, because they interact. A little too much of one can produce a critical deficiency of another that is present in barely sufficient amounts. Trace elements function as activators, as catalysts, within the living cell, be it plant, animal, or human; and they are the root of all living processes, with an influence out of all proportion to their size. While the quantity of any one element may be small and effective, compared to another, no element functions alone, but only in conjunction with others, equally important." *

We were back to the colloidal glacial sediment in Hunza water, full of the same trace elements, electrically charged, which proved to be the source of its vitality. Could Flanagan, we wondered, put an extra charge into Rollin's Azomite by subjecting it to his vortical method?

Rollin was immediately receptive.

"Colloidal," he emphasized, "is a condition, not a mineral. Fine dust-like mineral particles pass into the colloidal state of fineness upon reaching a critical size when their activity prevents them from settling out as

---

* "All organic activities and processes of growth in any living organism, or any metabolic function associated with sustaining life of any kind is an electric phenomenon and requires elements of ion exchange. *Cations* (+) carrying a positive electrical charge, and *anions* (−) having a negative potential are constituents of acids, bases, and salts, which become active as electrolytes (or conductors of electricity) in aqueous solutions. Life in its broadest sense is electrical—derived from the interplay of chemical elements. And the entire electrolyte system must be kept in constant flux, moving blood, digestive fluids, and body fluids." (*The Body Electric*, by Robert O. Becker, M.D.)

molecules of their particular inorganic element. Particles larger than one micron are generally in an available condition ready for immediate use by plant, animal, or human." †

To physicists, a piece of material can be subdivided into smaller pieces only so far before these cannot be seen with the most powerful microscope. At a further stage a limit is postulated beyond which particles cannot undergo subdivision without losing their chemical character: this they call the molecule. The smallest particle visible in the microscope is still about one thousand times larger than the largest molecule. In this twilight zone of matter are found the peculiar forms first called "colloidal" by Thomas Graham in 1862.

"By colloidal," said Rollin, "Graham meant those materials which readily crystallize and have the vital function of diffusing readily through animal membranes, as opposed to amorphous masses, which do not diffuse readily or at all through animal membranes, and cannot therefore be assimilated."

And here, we realized, may lie the explanation for the extraordinary vitality of colloids, as well as for the surprising facts of homeopathy, in which the smaller the dose, the more powerful the effect. Copper is said to be effective in plant life when present in a concentration as low as one part per ten million (dry matter); molybdenum is effective when present in one part per two hundred million, and cobalt is effective when present in one part per billion.

By the laws of physics, the smaller an element is divided, the larger is the area of surface exposed by all the pieces. A one-inch cube has a surface area of six square inches; the same cube divided into eight cubelets will together have exactly twice that surface area. By the time the cubelets or particles become microscopic, their cumulative surface is enormously increased. And the larger the surface exposed, the larger the particle's potential to be charged with energy.

In colloids the ratio of exposed surface area to volume of material becomes extremely large. As electrical charge tends to repel particles from each other, colloidal particles are kept separate, in suspension, retaining their vitality. But if the charge decreases (reduced by light, heat, electric fields, etc.) the particles tend to snap together and coagulate. With coagulation the system loses its colloidal behavior and becomes "dead," in both organic and inorganic systems.

All of life is found in the colloidal form and has many characteristics also in inorganic colloids, which led German Nobel laureate in physics Wolfgang Pauli to conclude that colloids provide the most important

† A micron is a millionth of a meter. There are ten thousand angstroms in a micron.

known link between the inorganic and the organic, a clue to the very source of life.

One of the keenest supporters of Rollin's Azomite is a veterinary doctor, C. S. Hansen, who attributes the extraordinary powers of trace elements to the microwaves they radiate. He maintains that insects have an innate intelligence that respects a vigorous growing plant, capable of producing seed for reproduction, and will somehow have the sense to avoid it. He said that when the natural trace-element material, such as Azomite, with the proper microwaves present, is supplied to a growing plant he has failed to find any insects present. Insects avoid such treated plants. But when a plant is not of vigorous growth, and capable of carrying on as a perfect species, nature gives insects the job of clearing it up.

"Anything that becomes inferior in quality," says Dr. Hansen, "becomes food for insects, so that only the healthy plants capable of developing seed for reproduction are left to mature. Imperfection in life has a way of being destroyed if left to the devices of nature. Food products from a deficient soil should never be used for human or animal consumption, and they should never be used for reproduction again as feed."

To demonstrate the effectiveness of the microwaves radiated by trace elements, Dr. Hansen took a bag of Azomite and spread it on the ground around an orange tree with mature fruit ready to be plucked. "The tree," he explained, "was full of heavy metals: zinc, lead, mercury, and insecticides. Within four minutes after the Azomite was spread, there was not an orange or leaf on that tree that wasn't free of the harmful effects of the heavy metals, DDT, and other chemicals."

So amazed was he by these results he repeated the experiment several times. His explanation is as simple as it is amazing: "Microwaves from the trace elements in the Azomite catalyze the heavy metals into harmless compounds, which the plant or tree can then use or automatically return to the soil."

Hansen says the effects of different forms of radiant energy on colloids and protoplasm are being extensively studied, and that it is known that different wavelengths and frequencies may produce structural effects on colloids and organisms: ultraviolet rays can slow down or stop the streaming of protoplasm, causing increased viscosity or coagulation.

Rollin sat back and sighed, partly pleased about getting his points across, and partly in despair about the world. "We have ganged up on nature by taking the attitude that insects are invading our fields and destroying our crops. So we kill the bugs, thinking it correct. Instead we are killing ourselves. But the bugs are only destroying our crops because we are not feeding the crops their proper food. We are not giving the plants the natural trace elements which give them access to the benefit of the microwaves of creation."

What he meant by the microwaves of creation was not to become clear to us until we met later with a brilliant ornithologist turned entomologist. But first we had to deal with an eminent biologist.

CHAPTER 18

Biomass Can Do It

THE ROCK DUST is there, by the billions of tons; the organic material is there in billions of tons of garbage and sludge; the U.S. population alone produces twelve thousand pounds of excrement per second, while, in that same second, U.S. livestock are producing another quarter million pounds; and there are 2 billion acres of unused or marginal land in the world, 62.5 million in the United States alone. What would it take to stop using poisonous chemicals, stop burning fossil fuels, and instead create biomass with all this dust, sludge, and acreage, to thus reduce the danger of CO_2, and feed a world population increasing by the billion?

The answer is not the fantasy of some crackpot dreamer, but hard data spelled out by the U.S. Department of Agriculture's vast Beltsville Research Facility, a multi-million-dollar outfit spread across miles of the Maryland landscape just north of Washington, D.C., designed, at taxpayer's expense, to improve the conditions of agriculture for the farmer.

Not that this particular approach hasn't been put forward before, time and again, over the past quarter century, by a series of experts writing in Charles Walters's *Acres U.S.A.* It is only that now the proposal comes from an official government agency, in serious form, through the lucid writing of one of its professionals, a tall, jovial Doctor of Botany and Taxonomy, James A. Duke, expert in the study of hallucinogenic plants, whose office looks out across the greensward at the enormous USDA library—an institution that fails to carry a single book by either Steiner or Kolisko!

With just the acreage that now lies fallow, says Dr. Duke, we could be self-sufficient in energy and not have to burn another pound of fossil

226

fuel. At the same time we could have a large surplus of proteins from legumes and grains; and we could remedy the nation's appalling balance of payments by $60 billion. All of this simply by planting our marginal soil, all 62.5 million acres of it, and imitating the American Indian method of intercropping—legumes such as alfalfa with cereals such as corn—to create "energy farms" on soil not presently exploited. Such farms could not only feed the nation, with a surplus, but produce abundant fuel from crops, eliminating the need to import crude oil from abroad. And all this without taking into account the 125 million acres presently devoted to hay and corn, 90 percent of which is grown for livestock.

James A. Duke, Phi Beta Kappa, Ph.D. in Botany, has been for many years with the USDA at Beltsville, Maryland, and is responsible for over a hundred scientific publications and several books, including his recent *Handbook of Medicinal Herbs*, and a videotape *Edible Wild Plants*, a guide to a hundred useful wild herbs. He is holding an Australian chestnut that is being tested by the National Cancer Institute for its chemical–castanospermine–as a therapeutic hope for AIDS. (Credit: USDA)

To make auto fuel from fresh plant tissue is just as easy as making it from the fossilized remains of plants and microorganisms. But plants have an enormous advantage: they are renewable, yearly and indefinitely. From fresh plants low-pollution fuel is economically available to replace both gasoline and diesel fuel. This would greatly reduce the greenhouse effect by cutting down on the industrial proliferation of CO_2. At the same time it would create a great mass of vegetation to absorb the present surplus, further offsetting the greenhouse effect. Organic wastes from all this bonanza would help rebuild a degraded soil.

Alfalfa, says Duke, grows well in the cool months, producing enough vegetation per acre to yield the energy equivalent of two to seven barrels of oil. Basing estimates on average alfalfa yields, Duke concludes that we could get nearly a ton of edible leaf protein per acre of alfalfa (and that's only one-seventh of what Harold Aungst was able to get using Sonic Bloom).

"The trick," said Duke, "is to intercrop a legume with a cereal. True, if you grow grain alone, you'll get more grain; and if you grow legume alone you'll get more legume. But if you grow the two together you'll get a greater biomass, and that is what you're after."

Corn is one of the more productive plants, in a category known as C-4, which photosynthesize best in the heat of summer. With the aid of sunshine, its stalks and leaves alone produce the energy equivalent of twenty barrels of oil per acre, plus another six barrels from the grains if these are used for energy. To achieve this output requires only two barrels of oil per acre, nearly one barrel of which goes for nitrogen fertilizer. But alfalfa, like most legumes, takes nitrogen from the atmosphere and puts it into the soil at the rate of about two hundred pounds per acre, comfortably compensating for the one required barrel of oil. Other highly fuel-productive C-4 plants include rice, sorghum, and the taller grasses, such as those that Wes Jackson is improving at "the Land."

The 55 million tons of protein derivable from the 62.5 million acres now lying fallow would be about ten times what Americans need for their diet. The residues remaining after protein extraction would yield the yearly equivalent of 250 million barrels of oil in the form of alcohol from the cellulose broken down to sugar. This alone could significantly cut oil imports from Persian Gulf countries, and eliminate the need for patrolling dangerous waters.

Revitalizing Pfeiffer's dream, Duke suggests that, if we were to fertilize with sewage sludge, our 62.5 million acres of corn and alfalfa could probably reduce imports by *one million* barrels of oil *a day*, from our present daily import of nearly seven million barrels.

Already in 1979 Dan Carlson had submitted a proposal to the Department of Energy offering with his Sonic Bloom to increase the annual production of fuel derivable from an acre of corn grain alone (without the leaves or stalks) from a normal 250 gallons to a bumper 650, and possibly achieve two crops in a year, which would comfortably raise the flow to over 1,000 gallons, renewable annually. But DOE, in the throes of nearly being aborted by the newly elected Ronald Reagan, never made an official assessment of this sanguine proposal.

Yet there still remains the patrimony of 125 million acres to work with, presently used, or misused, to grow hay and corn for livestock. Were these green acres to be made into "energy farms" of appropriate combinations of legumes and cereals, we could, according to Duke, after harvesting for local consumption and for export of 100 million tons of legume protein, produce more corn cereal than we have ever harvested before, and generate 3.5 billion barrels of oil from the residue.

This would take care of the country's entire energy requirements. And just as appealing are all the other benefits accrued. In becoming self-sufficient via organic energy farms, we could, says Duke, generate employment for the depressed farming, housing, and automotive industries. More hands would be needed to plant, cultivate, harvest, and process energy crops.

Small factories would be needed near the energy farms to convert energy crops into renewable fuels like ethanol (grain alcohol), methanol (wood alcohol), and methane gas, all of which generate less pollution than gasoline.

Detroit, says Duke, could reverse its slump by manufacturing converters needed to run our cars on renewable fuels. Decentralizing the fuel-production process, eliminating the transport of fuel halfway around the world, would stimulate depressed local economies while conserving energy in fuel transport, to say nothing, adds Duke with a smile, of removing the oil producers' fingers from our economic throats.

By converting to organic renewable fuels, we would generate research and jobs for America rather than for OPEC. Price shifts following such a conversion might make it possible to fulfill the long-held dream of U.S. farmers of trading a bushel of corn for a barrel of oil.

Air in Los Angeles and Denver might once more be fit to breathe. And then there is the problem of water, becoming increasingly scarce in such states as Arizona. Current efforts to deal with our energy problems call for massive use of western waters—water that will be needed by farmers to grow their crops. More than half the water used in the United States now goes to uneconomic livestock production, 2,500 gallons being needed to produce one pound of meat. This, says John Robbins in his eye-opening *Diet for a New America*, would make the cost of common hamburger meat $35 per pound were the water used by the meat industry not subsidized by the U.S. taxpayer.

The organic energy farm, says Duke, will alleviate both water and energy problems. Water removed during processing of crops for energy and protein can be piped back to the fields. In addition, the buildup of humus in dry western fields would help to hold the scanty rain. Another creative way of turning a disaster area to advantage is Duke's suggestion that strip miners convert the torn-up land into energy farms, interconnected by canals dug with their earth-gouging machinery. Stripped coal could then be barged out, and sewage sludge barged in to fertilize and rehabilitate the land. Before long, says Duke, barges would be hauling renewable fuels to urban centers and sludge back to the energy farms, bringing a knowing smile to Pfeiffer's ghost.

As early as the 1960s, Donald Despain, a maverick economist and industrial-relations counsel, had proposed a fundamentally new industry to transform agriculture from only a source of food supply to a supplier of industrial products, which would create a degree of agrarian prosperity never before experienced in America.

"With agriculture entering a long depression," he told audiences in 1972, "and farmers getting the same prices they got nearly twenty years ago—while paying prices three times higher—their growing crops for *power alcohol* could pull them out of a slump and into prosperity."

Despain quoted a Dow Chemical Company executive, William S. Hale, as telling the U.S. Senate's Subcommittee on Agriculture: "Alcohol, which can be manufactured from any farm product containing sugar crystals, is the only outlet in mass form we have for excess agricultural products."

So, why is it, if the fact has been known and proved for over half a century that internal-combustion engines can run on alcohol as a sole fuel or on gasoline with an alcohol additive, either substituting for gasoline or stretching it by 100 percent, that this bonanza is not available to one and all?

In the 1930s Dr. Leo M. Christensen in a pamphlet, *Power Alcohol and Farm Relief*, dug deeply into the extensive scientific literature on the use of ethyl alcohol as a cheap fuel for all combustion engines. All the investigators agreed that from the standpoint of national economic welfare alcohol was the best fuel because of its many established advantages, plus the fact that it could be produced within each country, whereas petroleum usually has to be imported.

Most attractive to farmers was the chance to distill their own fuel on their own farms, or make larger amounts—as much as ten thousand gallons a day—in a community distiller, which could process any crop containing sugar or starch.

Opposition of the oil companies, says Christensen, was organized and brutal: they went about distributing to filling-station operators across the nation cost-free mimeographed material to scare the public into believing that alcohol was inefficient or dangerous. The Petroleum Institute, with branches in every state, went into action, and in the nation's capital money gushed like oil to lobby senators and congressmen.

Intrepidly fighting for the "Farmer's Alcohol," Charles Walters, Jr., carefully documented in a series of articles in *Acres U.S.A.* the cynical attitude of the "Big Oil" cartel members before, during, and since World War II, as they fought bitterly against the distillation of grain, even grain so spoiled as to prevent its consumption by humans, and even when survival of America and its Allies was at stake in World War II. Only on orders from FDR's "no red tape" Baruch Committee, did B-29 bombers eventually fly on a mixture of high-octane 100-proof alcohol. But right after the war the government closed down its alcohol refineries, even though Dr. G. E. Hilbert, Chief of the USDA's Bureau of Agricultural and Industrial Chemistry, reported, on the basis of extensive testing, that "farm alcohol makes low-octane fuel equal to regular gasoline," emphasizing that it could economically provide a vast market for surplus grains. To distill a billion bushels of surplus grain, he said, cost only $30 million —a small amount compared to the $200 million required to build increased grain storage facilities, "which in no way solve the problem."

While grain was being stored in bins, elevators, vacant lots, tents, ships, and even on the main streets of towns, each bushel of grain containing better than 2½ gallons of ethyl alcohol, superior to premium gasoline, the farmer was being subsidized to retire land.

The move to alcohol was even supported by Truman's Secretary of Agriculture, Charles F. Brannan; and a USDA expert told one U.S. Senator there was no reason why all damaged grain could not be used for producing industrial alcohol. But, when Dwight D. Eisenhower was elected in 1952—as doyen of the military-industrial establishment, the machinations of which he warned against just before leaving office—a special commission was formed to look at American postwar agriculture. Blatantly ignoring seventy-five years of alcohol experience in Russia, Poland, Italy, France, and England, the commission concluded that it had found "no encouragement for believing that, in the present state of knowledge and under present economic conditions, the use of industrial alcohol for motor fuel can be justified." It was a specious statement predicated on oil's being available at 4.6 cents a barrel at Ras Tanura, and other Saudi Arabian oil refineries, *ad infinitum.*

Whereas the commission used the excuse that alcohol was not efficient as a fuel, Walters pointed out that it is equivalent to gasoline in power, burns cleaner, produces lower emissions, and causes no unusual engine wear. Cars in the Indianapolis 500 use 40 to 75 percent alcohol in their engines, and the world's speedboat record was made with 100 percent alcohol fuel.

That the petrochemical companies, now faced with imminent depletion of their oil reserves, may be looking to control the source of biomass is evident from the manipulation of farmers into debt and expropriation by foreclosure. A straw in the windy politics of the nation's capital was seized by Vice-President George Bush as he girded himself to carry the chemical banner into the presidential race of 1988. In *Chemical Marketing Reporter* of August 10, 1987, he was quoted as touting ethanol and methanol as a step toward "energy independence, less smog in cities, and more American jobs."

Already in 1983 Duke was commissioned by the Northern Agricultural Energy Center of the USDA in Peoria, Illinois, to prepare an unbiased comparison of two hundred of the more promising renewable energy species of plants. Included, along with such common energy grasses as sugar cane and the all-too-familiar oilseed, the peanut, were more exotic specimens such as the "petroleum plant," euphorbia, and the "gopher" plant, whose milk, according to Nobel Laureate Melvin Calvin, can produce fifty barrels of oil per acre per year; and diesel trees like the huge copaifera, which bleeds like a rubber tree to give fifty barrels of diesel per acre per year; and the kerosene tree, sindora, another large tropical tree, which is bled for its resin; and petroleum nuts like pittosporum, a fast-growing tropical legume tree grown for firewood to burn for electricity; and the fast-growing fuel-wood species like leucaena, the Philippine tree from whose fruit kerosene is readily derived.

In Hawaii, says Duke, it is economically feasible to produce electricity from leucaena. In the Philippines *Pittosporum resineferum* bushes could satisfy the kerosene needs of every Philippine family. Four percent of Panama planted to leucaena could satisfy Panama's energy requirements.

And Duke points out that *all* of U.S. petroleum requirements could be satisfied with the hydrocarbons derived from planting acreage the size of Arizona with the "petroleum" plant, euphorbia, commonly known as "spurge," a shrubby plant with a bitter milky juice that survives in arid areas.

Then there is the family of oil palms, considered the third among the plant families important to man after legumes and grasses. They produce quantities of oil, and can grow very well on marginal, or even desertified, land. According to the Office of Technological Assessment (1984), about two billion hectares of tropical lands are in various stages of degradation, a wasted potential asset. Technologically improving such degraded lands with sewage sludge and planting energy trees would offer an organic solution that would lead to higher productivity of energy sources while temporarily but vitally tying up CO_2.

Tropical countries, especially humid countries, with few or no fossil fuels, bankrupt by high energy costs and hungry for energy alternatives, must, says Duke, look to what natural resources they have at hand. For the Third World, he suggests a variety of palm oils that could make many of those countries self-sufficient in fuel. Much degraded land requires expensive irrigation and desalinization, but the nypa, a palm of southeast Asian mangrove swamps, grows even where it is inundated once or twice a day with saline tides. The nypa can give two to three times as much alcohol per hectare as sugar cane; and the Philippines alone have 400,000 hectares suitable for nypa production.

Upgrading the OTA's two billion degraded hectares to give twenty-five barrels of oil per hectare per year, could, said Duke, reading from one of his serious papers, facetiously entitled "Reading Palms into the Future," produce the required fuel to run the world. "OPEC," he added with a smile, "might become an acronym for Oil Palm Exporting Countries. Oil palm trees, representing a standing biomass of about ten to one hundred metric tons per hectare, would meanwhile tie up a lot of CO_2 in previously unproductive land that tied up very little."

The babassu tree (*Orbignya barbosiana Burret*) is reported by OTA to yield more than a ton of fruit per year. During World War II liquid fuels were derived from the babassu; they burned easily and cleanly in diesel engines. Residues were converted to coke and charcoal. In Brazil nearly 100,000 people are employed on 15 million swampy hectares described as "probably the largest vegetable oil industry in the world." It is wholly dependent on wild plants, developed from an indigenous cottage industry, capable of further expansion. Of the fruit, 10 percent is kernel, 50 percent of which is oil, indicating a yield of about forty kilograms of oil per tree, or a barrel for every four trees.

Ironically, Brazil, a leader in developing alcohol from energy crops, producing a billion gallons of alcohol a year, mostly from sugar cane, is obliged to import diesel fuel. Yet it is admirably suited to producing diesel from palms that have twice the energy content of sugar cane, and are easier to grow.

Duke is convinced that the oil palm (*Elaeis guineensis*) can outyield other varieties, such as those in the *Aleurites* and *Sapium* genera to produce ten to sixty barrels of oil per hectare per year, *renewable yearly*. Transesterified palm oil is an excellent substitute for diesel fuel, with a far lower polluting effect.*

* A chemical term for converting an organic ester into another ester of that same acid.

Clement and Mora Urpi (1984) suggest that *Bactris gasipaes* may yield four times as much fruit as the date palm, or 11 to 30 metric tons per hectare, with up to 55 possible. Its oil yield might be as high as that of oil palms, with a more nutritious residue. They speak of a yield of from 35 to 105 barrels of oil per hectare per year, *renewable*. Malaysia, presently at the forefront of palm-oil production, has twenty-four-hour pipelines relaying palm oil from the interior to the coast.

To supply the whole world's requirements in fuel oil would take two billion hectares of palm oil. But if it were possible to double the yield through biotechnology such as developed by Steiner, Carlson, and others, that acreage could be halved. By OTA figures there are 4.8 billion hectares of land in the tropics, of which only 1.8 are in forest, leaving 3 billion to develop for energy plantations. To increase their potential, Duke suggests screening clone tissue cultures for increased tolerances to aluminum, cold, drought, salt, and salt-water irrigation.

Opting for a green world instead of a greenhouse, Duke points out that anywhere on the planet we can increase the rate of photosynthesis to sop up CO_2 to make simple and complex sugars we can decrease the magnitude of the greenhouse effect, a solution simpler, cheaper, and more practical than some of the farfetched and expensive suggestions of worried climatologists.

A hectare of leucaena can fix 25 metric tons of CO_2 per year, or 2,500 metric tons per square kilometer, up to maturity when the trees slow down almost to a stop. Balick and Gershoff (1981) mention another palm species also found in Latin American swamplands, and in upland forests: *Jessenia baua*. It is a rich source of both food and oil, and could tie up plenty of CO_2 in marginal swampy land. It should take only a million square kilometers or 100 million hectares of leucaena to soak up the 2.5 billion tons of CO_2 we put into the air each year. And if all the leucaena was harvested to give energy (instead of our burning fossil fuel) the effect would be doubled. CO_2 could be stopped in its tracks.

But Duke warns that palms are presently an endangered species, their fragile family disappearing about as fast as the energy resources they could help replace. He urges concerted effort to analyze *all* palms for their economic potential, while they are still around.

"Then there's conservation. A good half of the energy this nation uses—more fuel than is consumed by two-thirds of the world's population—could be saved through conservation, which alone could decrease the U.S. contribution to the greenhouse effect by 50 percent, denting the world total by one-third."

Each North American consumes about 2,900 gallons of fuel oil equivalent per year, or nearly 70 barrels per capita, 17 percent of which goes for producing food, whereas the world mean is only about 11 barrels.

Americans use much more energy to produce, process, retail, and pre-pare food than there is energy in the food produced. And each year the average American consumes about as much wood in the form of paper as people in the Third World use to cook their food.

It takes as much as 300 gallons of oil per acre to cultivate land in America. Ninety percent of all grains, including corn, grown in the U.S. goes to livestock to provide the animal protein that Americans crave, or have been maneuvered into craving. A meat-centered dish is the most resource-expensive of all diets. An American steer eats twenty-one pounds of plant protein to produce only one pound of protein in steak. A stunning twenty-five thousand calories of energy are expended for every thousand calories of beef protein produced, which only goes to putrefaction in the human gut. If America were to take the presently unpalatable step of going vegetarian, says Duke, all that grain could be saved for energy production, resolving the energy crisis and greatly im-proving the health and energy of humans. As John Robbins points out, 1.3 billion human beings could be fed on the grain and soybeans eaten by U.S. livestock. And if the U.S. population were to reduce its intake of meat by a mere 10 percent, they could adequately feed the 60 million people who starve to death in a year worldwide.

Energy conservation, says Duke, does not require the curtailment of vital services; it merely requires the curtailment of waste. Often a dollar invested in energy conservation makes more net energy available than a dollar invested in developing new energy resources. Thirty to 50 percent of the operating energy in most existing buildings could be conserved, and 50 to 80 percent could be saved in new buildings.

Duke makes several encouraging suggestions for improving the coun-try's future, many already popular with organic gardeners: planting gar-dens on rooftops; developing two- and three-tiered forest ecosystems in lieu of monocultured orchards; developing desirable vines to climb over houses during summer to function as natural air-conditioners, conserv-ing energy and cutting down on CO_2; filling every window with culinary herbs or ornamental plants. But, like Hamaker, his prime suggestion is to keep planting trees; he suggests the addition of fast-growing species to the existing slow-growing ones. At the present rate, an acre of trees disappears from the United States every five seconds.

A firewood farm that generates fifty metric tons per hectare per year instead of twenty-five will tie up twice as much CO_2. Fallen firewood in the forest, harvested and burned instead of fossil fuel, frees up more space for green plants. And Duke recommends using living fence posts —as practiced throughout Latin America—instead of energy-consuming metal-electric fences.

Richard Saint Barbe Baker, the English forester who pioneered the movement to save California's redwoods, and whose bioethic philosophy led to the planting of an estimated 26 billion trees around the world, proclaimed that man's existence depends as much on trees as it does on plants, and that trees are as essential to agriculture as to breathing. The minimum for safety, he insisted, is tree cover encompassing a third of the total land area of the planet, a ratio we have imperiled to the point that we are losing an acre of rain forest every second.

Saint Barbe, as he was familiarly called, discovered that in an agricultural area if he devoted 22 percent of the surface continuously to trees he could double the crop output of the contiguous cleared area. Trees create microclimates in which crops flourish; they reduce the speed of wind, lift the water table, feed an increased population of worms. A single eucalyptus tree, forty-five feet tall, will transpire over eighty gallons of water a day, and a willow can transpire five thousand.

Only 2.8 percent of the world's land is fertile enough to grow wheat indefinitely without the assistance of trees. In England, one field that has raised wheat continuously for a hundred years is surrounded by oaks whose roots go deep, tapping minerals to feed its leaves. When the leaves have served their function, they fall to earth and rot. Surfacing worms carry down their residue overnight to replenish the soil with essential trace elements. And yet, for decades, the trend has been to fell trees and plant pasture, which rapidly erodes, further imperiling life, as nature's long-term wisdom is sacrificed to man's short-term gain.

Saint Barbe and others have cajoled thousands of men and women into planting millions of trees, and have pleaded that millions more can, and must be, planted. One of the adherents of his world-wide "Men of the Trees" movement, Charles Peaty, after a lifetime spent creating, managing, and harvesting forests in Europe, decided to do something about reforesting man-made wastes of western Australia, which less than a century ago were covered with hardwood forests of timber and shrubs that each year sprang into a blaze of colored flowers across the entire countryside.

When farmers cleared this land, a thin layer of topsoil was swept away in a single generation by cyclonic winds and downpouring rain. The fertilizer-laden runoff accumulated in creeks and rivers to turn them into salt bogs, while the trees that had flourished along their banks turned into matchwood. To create new tree stands and shelter belts Peaty invented a special method of planting trees in desert country, with minimal, even no, watering. Over the past six years he has planted on treeless farm acreage millions of specimens of false mahogany, Bald Island marlock, blue mallet, flat-topped yate, wandoo, cypress, two kinds of acacia, eleven varieties of eucalyptus and casuarina trees, one of the world's oldest varieties. To date the results of his efforts are a thousand farms with shelter belts and with millions of dollars' worth of pine plantations 170 miles south of Perth.

Peaty told many audiences of Australian farmers: "If each of you plants trees to slow the wind, your ground, your water, your whole environment will be brought back into balance. Flocks of birds, long since departed, will return."

That Peaty's vision is not utopian is proved in a moving tribute written by Jean Giono, one of southern France's most lyrical writers, to a simple French peasant, Elzeard Bouffier. From pail after pail of collected acorns and seeds, Bouffier is credited with having single-handedly planted a forest of a million trees, covering a vast expanse of previously unparalleled desolation, now a thriving countryside within a splendid French national preserve. If only a million people in this or any country were each to plant a single tree, the feat could be duplicated, and the number would increase exponentially as more people planted more trees. Only thus, and by bringing new life to a remineralized soil, can we hope to save what passes for civilization, and recover the bounty of life on this planet, the secret to which, as is patent, lives in its soil.

Our problem is with time. If Hamaker is right, we have let slip by the chance we had to plant the trees in time to save the planet from disaster. But one last hope remains: microscopic in size but gargantuan in power, believed to be the oldest, hardiest plant form on the planet, a survivor, through billions of years, of all the hazards, imaginable and unimaginable, dished up by an indifferent fate: the one-celled blue-green algae, known as Aphanizomenon (actually not a plant but what is now recognized as a bacterium). Proliferating at great speed, the algae could, according to Daryl J. Kollman, scientist, author, and educator, dispose of surplus CO_2 and feed the world. Grown in man-made ponds all across the world, especially in such vast spaces as the Sahara Desert, the metabolizing biomass could suck up vast quantities of CO_2. No plant, says Kollman, grows fast enough to create the biomass to get us out of trouble.

To propagate blue-green algae all one needs is a pond, a pond liner, more water, and rock dust as a nutrient—all infinitely easier, cheaper, and more effective than any of the climatologists' far-out suggestions. The water does not have to flow; it needs merely to be stirred so that the infusion of algae all get exposed to the sun. Proliferating, the algae draw in CO_2 from the air. Harvested, says Kollman, the algae are the world's best nutrient, sufficient with their protein to save the lives of millions of starving Africans and Third World peoples. And if, for any reason, a batch goes bad, it makes first-rate organic fertilizer.

Daryl J. Kollman.

Ancient organisms, algae look like bacteria but have cell walls and a far greater capacity to photosynthesize, making them the most efficient chlorophyll-producing organism in existence. Mono-cellular, each individual is self-sufficient. Having no circulatory systems as do plants, they are mostly microscopic, though some grow into giant seaweeds, hundreds of feet long.

To Kollman, the blue-green, standing as it does at the very bottom of the food chain, is more basic to biological life than even the regular bacterium. For billions of years it has dwelt in every drop of water and every inch of fertile soil, transforming minerals, gases, and sunlight into viable foods for bacterial, plant, and animal life, responsible directly for about 80 percent of the world's supply of food.

A few years ago, Kollman came across a supply—almost inexhaustible—of blue-green algae in Klamath Lake in southern Oregon, near the quiet lumber town of Klamath Falls. The lake is 130 square miles in area, the only known accessible and unpolluted source of such algae growing wild and in abundance. Geologists estimate that for the past ten thousand years the lake has had an annual procreation rate of 200 million pounds of algae, a rate that can persist indefinitely without disturbing the lake's pristine and healthy ecology.

Entirely surrounded by the beautiful Cascade Mountains, with

Mount Shasta in full view some fifty miles to the south, the secret to the lake's bonanza lies in its location, a natural trap for the nutrient makings of life. Rain and snow that fall on four thousand square miles of rich volcanic soil of the Oregon Cascades wash into Klamath Lake millions of tons of nutrient topsoil. All the required minerals are carried down from the glaciers, available for hungry algae to turn, with their strong supply of chlorophyll and with the power of the sun, into chelated organic molecules of super food—food to feed a starving planet.

To add to this nutrient supply, much of the algae have accumulated through millennia into rich sediment, which now covers the bottom of the lake to a depth of thirty-five feet. The top one inch of this sediment alone, according to Kollman, could support a massive algae bloom for sixty years to come without any new nutrients entering the lake. And its waters are unpolluted. In an area devoid of industry, town sewage, or the chemical toxins of agriculture, the rivers and streams that enter the lake are pure, clean, and potable, with little recreational boating where algae carpet the surface.

Kollman came to his discovery by a circuitous route. A teacher and administrator with a master's in Science Education from Harvard, he was trained in Italy in the Montessori method. Over a twelve-year span of teaching young children he noticed a steady increase in problems normally classified as "learning disabilities," problems he soon recognized as being associated with the diminishing quality of the children's diet. Undernourished or poorly nourished children were not capable of absorbing information, and therefore, of learning.

We came upon Kollman in a Los Angeles suburb in the house of one of his supporters, just as he was beginning his campaign to run for President of the United States on an ecological platform. Tall, balding, quiet-spoken, in his late forties, with a deeply lined face that easily breaks into a pleasing smile, Kollman explained: "I knew that concentration was the first requirement for learning, and that it was getting more and more difficult for children to do so. If I wanted to be successful in the classroom, something had to be done to help the children's diets. An extensive computer search through existing literature revealed that micro-algae were being used in Japan and other Far Eastern countries for the remediation of poor educational performance in schoolchildren."

In 1976 Kollman and an associate became the first researchers in the United States to grow systematically and experiment with spirulina and chlorella, forms of green algae now widely marketed. But Kollman wasn't satisfied: he didn't like the idea of having to grow the algae artificially in man-made ponds; and the cellulose cell wall of chlorella made it difficult to assimilate. The discovery of algae growing wild in one of the world's

Daryl Kollman's double pyramids, built in Oregon, to the scale of the Great Pyramid, the minor half buried in the ground.

richest natural "nutrient traps," completely free of artificial influences, answered for him both problems: the Aphanizomenon's cell wall was found to be composed of a substance nearly identical to glycogen, making the algae 95 percent assimilable by humans. And the algae contain all the trace elements essential to animal and man.

To get the blue-green distributed even more widely, a system was devised for harvesting the crop during the summer, then freeze-drying it to protect the beneficial enzymes and heat-sensitive vitamins, guaranteeing both the algae's nutritional value and the lake's ecological integrity. The result is a 100 percent food substance, 69 percent protein, with all the trace elements in a colloidal state, readily assimilable, the highest source, according to Kollman, of natural vegetable protein and chlorophyll in the world, containing all the essential amino acids in perfect balance, almost exactly as in the human body.

As a food—in powder or capsule—he says it has no peer: one gram of blue-green algae has tested out as containing about 1,400 micrograms of beta carotene. To obtain that much beta carotene one would have to consume 14 grams of liver, 70 grams of carrots, 14 eggs, or 5 quarts of milk. It is also a rich source of neuro-peptides, quickly absorbed to nourish both the nervous system and the brain.

Kollman is said by his supporters to have been driven to his algae and to his running for the presidency by an unusual mixture of humanitarian spirit, scientific ingenuity, and social vision, a combination with which he seeks to launch a revolution in American health and in American nutrition.

In Fallbrook, California, between Orange County and San Diego, Daryl Kollman has been building, at a cost of between half a million and a million dollars, a huge double pyramid, a hundred feet on four sides, buried fifty-five feet in the ground and rising fifty-five feet above it.

Divided into ten stories, the pyramid's purpose is to grow enough food for a large community to feed itself and have a surplus to distribute. Almost any plant can be grown in trays, fertilized with rainwater, rock dust, and blue-green algae, one gram of which contains upward of five million microorganisms.

The upper pyramid, glazed with Plexiglas, allows for the passage of solar and cosmic forces; the lower pyramid absorbs geomagnetic energy from the earth. Some other mysterious energy inhibits the growth of mold or yeast, despite the great quantity of water that passes through the structure.

Kollman also plans to produce small pine trees that people can buy and transplant. Other sites for double pyramids are Sedona, Santa Fe, and Albuquerque. To heat his pyramids in colder climates, Kollman envisages a machine invented by a friend in New Mexico that burns wastepaper and wood, leaving nothing but a trace of water vapor and carbon dioxide, the latter, in the enclosed pyramid, serving as extra food for the plants.

A second pyramid in Fallbrook is planned as a restaurant, general store, and bakery. It will also serve as a communication center to spread the word. "We have to train hundreds of thousands of people to live in such communities," says Kollman, "and we have very little time in which to do so."

At dinner with Eddie Albert, of *Green Acres* fame, also a devoted supporter of organic farming and of authentic ecological revival, Kollman expanded on his program: "The first order of business is to get the people of this country healthy. That means cleaning up our agriculture, and restoring health to the soil. If we don't act now to clean up the environment, we may lose forever the opportunity to do so. We are close to the upper limits of reversibility. And the United States is the only country with the power and the influence to lead the world into a massive environmental and economic cleanup. So far, we've left it to the 'experts,' and you can see the mess they've made. They've left us 375,000 toxic waste sites to be cleaned up in the U.S.A. alone.

"To FDR's Four Freedoms we must add the freedom to have clean air, clean water, vital food, and the right to pass on to our children a world that's fit to live in. This earth can support a lot more people. With a little bit of intelligence, we can support them all in a way we've never seen before. There's glacial till that's thirty to fifty feet thick in North and South Dakota; we can grind it up and move it into Iowa, Illinois, Kansas, Nebraska, Colorado, Texas, and Wyoming for vital soil and vital crops. Meanwhile, until we grow more healthy food, the algae can bring us all the elements we lack for a revitalized, enduring health. The blue-green algae is a gift from heaven. But it may be the last we get if we don't shape up. We can talk about the future, and visualize the future, but if we want that future we must act."

CHAPTER 19

Purified with Fire

IT TOOK THE Chernobyl disaster to arouse the
Soviets to some action, belated, defensive, and far from what was
needed. In the rest of Europe, alarm at the fallout consequences momentarily took people's minds, confused and helpless, off the problem of
their dying trees.

Not so in America, where the plague was spreading. Satellite photos
—taken five hundred miles above the earth—revealed panoramic shots
of mountains dotted with dead and dying trees. These were supple-
mented by earth-based closeups of yellowing needles and lifeless
branches. To struggle with the problem, an international five-day conference was convened at the end of October 1987 on the shores of Lake
Champlain, in the threatened Green Mountain State's city of Burlington. Formally entitled "The Effects of Atmospheric Pollution on Spruce
and Fir Trees in the Eastern United States and the Federal Republic of
Germany," the conference brought together a large number of forestry
experts and other scientists from both countries who were studying the
health of trees from space, or the condition of their roots in the soil, and
everything else in between.

The general consensus admitted, as it had for nearly a decade, that
American and German trees were rapidly dying for a series of reasons as
complex as they were mystifying, apparently from both man-made and
natural causes. But *that*, as the *New York Times* put it in a summary
article, was about the extent of the agreement.

It took Dr. Viktor Kovda, director of the U.S.S.R. Academy of Sciences' Institute of Soil Science, to turn attention away from what he
considered more than an ample discussion of the industrial pollution of

243

Damage to trees at Camel's Hump, Vermont, as of 1984, recorded by Hubert W. Vogelman, professor of botany and chairman of the Department of Botany at the University of Vermont in Burlington. Twenty years ago, says Professor Vogelman, the evergreen forests on the slopes of Camel's Hump, a high peak in the northern Green Mountains of Vermont, were deep green and dense. The trees were luxuriant, the forest fragrant, like a primeval paradise. Red spruce, more than 300 years old, rose 150 feet into the sky. Today the red spruce are dead or dying. Since 1965, 50 percent have died. Scientists look to this devastation as the first signal of an approaching environmental disaster.

the atmosphere to a potentially more intractable problem: the introduction of health-threatening heavy metals, such as lead, mercury, cadmium, aluminum, arsenic, and selenium into the soil, and thence into the food chain, a process he feared might be irreversible.

But a surprise was in the air. As if in answer to some universal prayer, there came a message from the subcontinent of India, brought by a Hindu Brahmin, that the process was *not* irreversible, that something could be done to fight the planet's polluting plague. Arriving in Baltimore, all the way from a small railroad town south of Calcutta, he claimed to have been specially sent by higher authority with a surefire method for cleaning up the atmosphere, restoring health to forests, and bringing back to their verdant fronds flocks of happily singing birds.

The message, he said, was from a venerable sage responsible for reintroducing to this planet the ancient purifying wisdom of the Vedas, Parma Sadguru Shree Gajanan Maharaj, said to be a Kalki Avatar—or "ascended master," who needs not return for another life to this wen of pollution, but does so for some higher ideal—come to preside over the destruction of pollution on our planet, Kalki being pollution.

His messenger, a younger Brahmin named Vasant V. Paranje, born in the same small town of Kakrapur in India, having rid himself, in the manner of the Gautama, of all his possessions, arrived in New York in 1972 without a penny. In the great city Vasant walked the streets until people spontaneously offered him a visa, a green card, money, whatever he might need to set himself up. Declining all favors, he insisted he had a mission to perform and that destiny would guide him. It did, first to Johns Hopkins, then to Washington, D.C., where the rector of St. Stephen's Episcopal Church, Father William A. Wendt, the first priest to promote the ordination of women, took Vasant under his protection and helped him get started with his mission. He was to spread the ancient science of Agnihotra with methods every bit as astonishing as those of Steiner's "spiritual science."

Again the key ingredient turns out to be cow dung, raising the question of whether the Hindus may not all along have had some cogent and highly beneficial reason for tending their cattle as they do, allowing them to wander unmolested through country lane and city street.

This time the dried dung is placed in an inverted copper pyramid, the size of a monk's begging bowl, stepped like a ziggurat, along with a spoonful of ghee, a handful of rice, and a pinch of redolent sandalwood. This strange assortment is set ablaze—to the accompaniment of a mantra chanted in Hindi—as curling pearl-gray smoke rises from lapping red-and-blue flames to purify, or so its devotees claim, the surrounding atmosphere, miraculously increasing the quantity and quality of fruits and vegetables grown in the area. *Agni* in Sanskrit means "fire," and *hotra* "the act of purification."

In a building on a farm on the outskirts of Baltimore the first devotees in this country of the exotic practice have kept an Agnihotra fire alive for the last ten years. During all that time a dedicated score of individuals have taken turns reciting an uninterrupted mantra, around the clock, day and night, summer and winter. To satisfy our own curiosity we traveled to Baltimore, and at dawn of what was to become a bright sunny day in May of 1987, sat cross-legged in a small building on a hillside just beyond the beltway, barely fifty feet from where the uninterrupted mantra was being dutifully but cheerfully recited by its chain-firing devotees. We were to witness a private Agnihotra ceremony performed by a man with Middle Eastern features and a thick black mustache who sat in a yoga position facing an inverted copper pyramid. Beside him stood a gallon jar that must have once held mayonnaise, now half full of translucent ghee, a round biscuit tin filled with dried cow dung, and a tall tea box containing special basmati rice. With these unlikely ingredients, the yogi prepared his fire in the pyramid, breaking up the cow dung, pouring on the ghee, and scattering in the rice. As blue-red flames danced up from the opalescent copper, and pale gray smoke rose toward the blackened ceiling, he intoned a Hindi mantra by means of which he hoped to inject into the atmosphere sufficient nutrients and fragrance to stimulate plants and neighbors to grow happily together.

In the trees outside, a flock of birds was already warbling a cheerful morning chorus. According to our guide, Noni Ford, a young acolyte with prettily beaded hair, the fire must be lit precisely at sunrise and sunset when certain energies rise out of the earth with a quintessential sound audible to sensitive ears. To establish the exact moment of both dawn and dusk, anywhere in this country, the devotees have computerized the location of every tiny hamlet in the United States, and can produce a computer printout at a moment's notice.

Later, in the main house, the owner of the farm, John T. Brown, a jovial fellow of forty, father of two teenagers just on their way to school, explained how he had gotten into Agnihotra after years as a member of a group that follows the teachings of the famous Indian yogin, Paramahansa Yogananda, whose autobiography is a classic in the genre.

Our main objective, said Brown, was not to create any formal organization. "People simply began to show up, out of the blue, and the message spread by word of mouth. There was no publicity, no proselytizing. Vasant's message was simple enough: Be happy! Be happy here and now. And practice Agnihotra to clean up the planet."

Brown smiled as if indeed he were a happy man, convinced that a solution to the toxins had been found. Then he elaborated: "Gradually the practice spread throughout the Baltimore area until now there are six hundred families involved in the regular burning of dung, rice, and ghee in copper pots. To spread the message, Vasant has traveled during

Vasant V. Paranje, the Hindu Brahmin messenger from the Hindu sage, Parma Sad-gurn Shree Sajanan Maharaj, performing the Agnihotra cleansing ceremony in our West Virginia kitchen. He is burning cow dung, ghee, rice, and balsam wood in an inverted copper ziggurat. (Credit: Doris Presley)

the last few years all over the country, and all through South America, Asia, Europe, and Africa. Practitioners of Agnihotra are everywhere, especially in such unpleasant and far-flung dictatorships as Chile and Poland. In one place in the Andes, Cochiguas, in the Elqui Valley of Chile, an Agnihotra fire was started eight years ago by a little old woman under a thatched roof beside an ancient pre-Inca wall. Now hundreds of people congregate there daily, thanks to the enormous success local farmers have had raising their crops with Agnihotra ash, or 'miracle dust' as they've come to call it."

Noni Ford showed us out into the garden. "You should see our corn and potatoes. Their color and taste are remarkable. Last year we had raspberries right up into the frost. And, despite the drought and the freezes, we had wonderful pears, apples, and peaches."

"Just smell that air!" said Brown. "Last year we tried to get government people to come out here to test it. But they said: 'If your air is bad we will come. If your air is good, why should we bother?' "

Brown explained that up until the previous year he had had a business with about a hundred employees, whom he could feed from the three acres of land they cultivate. He pointed to some trays of what looked like cow dung drying in the sun. "We soak our seeds in cow's urine for two hours, then coat them with cow dung and lay them out to dry. Then we plant them in rows in the ground with Agnihotra ash and say a mantra. The ash seems to stabilize the amount of nitrogen and potassium in the soil. And a chemist in Colorado who works for the U.S. government has found from repeated soil tests that the ash greatly increases the solubility of phosphorus."

Brown described the Vedic technique of Vanjya, a process of purification of the atmosphere through the agency of fire. Vanjya is tuned to rhythms of nature, to radiational effects, and to astronomical combinations. "Vanjya," said Brown, "injects nutrients into the atmosphere. The smoke gathers particles of harmful radiation in the atmosphere and neutralizes their effect at a subtle level. Nothing is destroyed, only changed. A powerful change takes place in universal *prana*—the life energy that pulsates through us and connects us with the cosmos—creating a healing effect on body and mind. It also leads to a better absorption of the sun's rays by the water resources of the planet. Fire produces out of a normal state of matter an ideal state, which allows energy transformation at a different level. Bursts of energy emanate from the Agnihotra copper pyramid, depending on the phase of the moon and the position of the earth in relation to the sun."

As we looked in through the door of the small unprepossessing cinderblock building in which the uninterrupted mantra was being sounded, Brown elaborated: "Clairvoyants say a strange phenomenon occurs when plants are grown in an Agnihotra atmosphere. An aura-type field of energy is generated from the plant and persists around it while the ceremony is being performed in its proximity. This enables the plant to maintain maximum growth and yield levels. If you breathe in that smoke, it quickly goes into the bloodstream via the lungs and has an excellent effect on the circulatory system."

The main object of the exercise, according to Brown, is to heal and improve the land rather than pollute and destroy it with chemical poisons and thoughtless farming practices. "We grow superior crops without the use of chemical fertilizers, pesticides, or herbicides. By spreading Agnihotra ash over the soil before tilling we make the seeds more disease and pest resistant. Under polluted conditions, elements of the earth begin to change. Agnihotra is a process whereby the molecular components are reconstructed. Without it, unknown elements begin moving into the intersphere, setting off a chain reaction of disasters. Soil in many places becomes unable to support plants. Trees die for unknown reasons. Forests disappear, not because they are cut; they are choked to death due to lack of nutritional content in air and soil. Clouds in many parts of the world are seeded with hydrocarbons and other toxins. Barium and cadmium become catalysts and cause a change in the genealogical structure of man. Fissures in the atmosphere cause seepage of radiation. A change in the nuclear structure in plants due to pollution makes it impossible for humans to extract nutrients necessary to survive, unless plants are grown in an Agnihotra atmosphere. Plants are starving, and the nutritional value of edible plants decreases, as does the medicinal potency of plants. The amount of chlorophyll the plant is able to produce is impeded by the concentration of pollution in industrial areas. Diseased soil transmits cancer to domestic animals, especially pigs. Cattle carry poisons into the systems of those who slaughter them for food. The meat humans consume will be dangerous to eat. All forms of red-blood animals used for food are beginning to cause cancer on a very large scale."

At which point the "messenger" himself made his appearance, a quiet, peaceful man, with powerful dark eyes, the gray beard of a prophet, and the mellifluous tones of the subcontinent. He said he had just returned from the Erzgebirge mountains in Saxony at the Czech–East German border, where he had recommended building platforms in the devastated forest on which to practice Agnihotra to encourage newly planted trees. We had caught him, he said with a smile, on his way to Chile and then Bolivia, where he was invited to start some Homa farms. Homa, he explained, was Sanskrit for the ancient science of growing plants based on the rhythms of nature—sunrise, sunset, full moon, no moon, equinox, solstice, etc., all with a pyramid of fire. In Germany, he added, several thousand families were practicing Agnihotra, but the federal authorities had taken issue with the ashes being used medicinally; so he had moved on to Poland, where the ash was greeted, as in Chile, as a miracle performer.

Vasant said the Soviet government was putting pressure on scientists and the medical profession to find some solution to the problem of radiation. But although more people near Chernobyl were getting sores on their bodies, the Soviets would not yet use the Agnihotra dust, waiting for its effects to be proven by science. "A pity," said Vasant. "It can do no harm. After all it is nothing but cow manure, ghee, and rice. Why not give the poor people a chance?"

With a grave expression he warned that children in Europe will start to get cancer, the first symptoms being diarrhea, actually a dehydration of the bowels. And he was afraid that people in the United States living near radioactive sources would be developing sores. He warned that pollution leads to hybridization of insects, which become immune to insecticides, and that in a short time we may be hearing of giant man-eating ants being added to the problems of South Africa.

That his prophecies might not be as farfetched as they sound was indicated by his foretelling, several months before dead and dying dolphins surfaced along the Atlantic Coast and seals were found mysteriously dying in the North Sea off Britain, that large shoals of fish would be dying as a result of pollution in the oceans.

Vasant explained that in the language of the Vedas when pollution goes beyond a certain limit there comes a change in the nuclear structure of plants, and they become unfit for human digestion. Agnihotra is the only solution, and it is his hope that chains of Agnihotra centers will spread around the United States and then around the globe. He claims it will even help solve the problem of the ozone hole developing over Antarctica, a potentially serious danger to the planet.

"The United States," said Vasant, "is a blessed country, and it is the divine will that from this country the whole planet should be saved. But we need in each state a place where we can show how Agnihotra works, and where clean produce can be grown abundantly in a small area, which becomes a place of healing with just the use of a single simple mantra. When we hear of anyone interested in practicing Agnihotra we consider it our duty to go to them, at our own expense, and teach them how to do it. Their only investment, apart from a little cow manure, some ghee (easily made from fresh butter), and some basmati rice, is a copper pyramid, which costs a mere $10—a small investment to help clean up the planet, and at the same time grow luscious, healthy crops. It has been easier for us to start with organic farmers; they are more open, and they have already taken the first step by getting rid of pollutants on their land. Even if you get only the same yield, we tell the others, you don't have to spend money on fertilizers, pesticides, and herbicides. Your land and water are safe. And the taste, texture, and quality of your produce will be superb."

Asked what he considered to be the formative force in Agnihotra, Savant replied without a moment's hesitation: "Sound. If you test Agnihotra with an oscilloscope, you will hear a special sound coming from the fire. It is a sound that heals. All the other physical things are there, nutrients, vitamins, minerals; but the key is the sound. If you are subtle enough, you can detect it. Fire produces sound, but it also reacts to sound. If you sing special vibrations while the fire burns in the pyramid there is a resonance effect. Ancient science states that it invigorates the cells of plants and helps the reproductive cycle. Resonance plays a vital part in nature. We have to consider a healing molecular spectrum far beyond the infrared, indeed beyond the whole electromagnetic spectrum."

As a parting shot, Vasant recommended to us the use of Agnihotra as a means of ridding one's property of pests. Presenting us with a couple of complimentary copper pyramids, he enjoined us to help clear the pollution in our respective areas, and at the same time rid ourselves of unwanted mice or roaches in the cellar.

"Just sprinkle Agnihotra dust wherever you see an infestation," he said with a smile as benign as it was clearly heartfelt, "and you'll be surprised how quickly it clears up. It doesn't kill the pests. It simply drives them off."

As we drove off, discussing these strange events, we were struck by the number of similarities between Steiner's "Spiritual Science" and the practices of the devotees of Agnihotra, including their common use of cow dung, and their similar remedy for insects.

From Europe we received reports of a group of scientists in Rovinj, Yugoslavia, experimenting to establish just what Agnihotra does, and how. Their interest had been aroused by the discovery that after they had burned the required ingredients in the copper pyramid their instruments failed to pick up radioactivity in the immediate area, an anomaly since the Chernobyl disaster, which irradiated, along with large parts of Europe, even their small Adriatic seaport on the Istrian peninsula in the province of Croatia. The Yugoslavs also learned that groups of subcontinent Indians living within the borderlands of the Soviet Union who used dried cow dung to seal their huts were unaffected by the radioactive contamination. Intrigued by these mysterious developments, the Soviets have invited one of the Yugoslav scientists, Mato Modrić, a biologist, to visit the Soviet Union to demonstrate the method of Agnihotra in the hope that it may be of value to its citizens.

To check out this curious data, and the amazing parallel between Homa farming and Steiner's biodynamic farming, we traveled to Rovinj, the former Italian town of Rovigno, a charming Adriatic seaside community with whitewashed, tile-roofed houses bunched together on a tiny oval peninsula jutting from the Istrian coast, surmounted by a tall white steeple, its cypresses and oleanders reminiscent of the Monterey peninsula of California.

Mato Modrić, a stocky welterweight in his late fifties, with a broad, strong face and piercing eyes, an expert in electromagnetic fields, dowsing, and geopathogenic zones, lives with his wife, Maria, a petite dark woman weighing only ninety pounds, in a small duplex overlooking the harbor.

Over a supper of gnocchi and roast chicken thighs, Modrić, who speaks English remarkably well, and German and Italian fluently, says he became involved in the Agnihotra phenomenon through his interest in pyramid energy, or "waves from shapes," along with such allied subjects as what the French call *radiesthesie*. Trained in physics, he was particularly curious about the role of the special vessel made of copper (or gold) and its specific ziggurat shape, a form related to the horn antennas used in high-frequency transmissions. What high frequency, he wondered, might be being amplified and broadcast by such an antenna to affect the human aura, its nadis, chakras, or its kundalini? That the ash could produce disinfectant, anticoagulant, and tissue-contracting effects on living matter he said was well established. And he said he believed Vasant when he claimed the ash had pesticidal and fungicidal properties and that it might ultimately solve the problem of mineral deficiencies. It remained to be established what trace elements were in the ash, research into which was going on in such disparate places as Yugoslavia, Germany, and New Mexico.

Modrić explained that he believed he was dealing with a complex that could potentially affect the whole environment, countering the toxins of modern technology developed over the last century by the industrial revolution, and that the process might have enormous implications for our very existence.

He added that he believes that Agnihotra ceremonies performed at various specifically spaced points on earth, if done exactly at sunrise and sunset, could affect an energy associated with the earth, one such as described by both Steiner and Reich, the enhancement of which would have a healing effect on the environment, difficult as it might be to understand or prove in terms of modern science.

He explained that in his conception the Agnihotra ceremony was energetically quite complex, involving at least three energetic aspects, or field phenomena, having to do with the fire and the ash, with radiation of an undefined nature, and with ESP, or psychism. He said that a lot of research was needed to lift the subject from the purely speculative, there being at play biophysical and biochemical interactions, and that much time and money for research might be required. "We believe we can establish the fact of an electromagnetic radiation during the ceremony," he elaborated. "But we are in an area beyond what conventional science considers rational, into an area of informational transfer through intermolecular and interatomic processes mediated by ultraviolet photons. It is logical to conclude that some kind of energetic mechanism is being activated which can be translated into physical meaning linked to concrete information systems that are as yet unknown, but connected to systems of resonance. We are in an area where it is not easy to prove anything. A lot of work will be necessary, and it will depend upon the cooperation of very many people."

His remarks reminded us of Lily Kolisko, working for thirty years, almost alone, to reveal some of nature's secrets in *The Agriculture of Tomorrow*, most of which were ignored for years.

But Communist Yugoslavia, in which nowhere can a Xerox copy be made of anything unless it is first inspected and approved with an official stamp, was not conducive to the required effort. Though the Serbs are remarkable people, with no false *politesse*, tough guerrilla fighters who held down thirty-five to forty Nazi divisions throughout the war—thereby contributing to the safety of Moscow from Hitler's Barbarossa attack—they now do not even have enough money for the upkeep of their existing buildings. Belgrade is dilapidated, with a foul-smelling yellow-gray smog hanging like a shroud over the city, the result of burning soft lignite coals and other pollutants. As Modrić lamented, the country cries out for an Agnihotra cleansing. Potentially one of the richest countries in Eastern Europe, it is being ruined, so we were told, by the dogmatism of Marx's heritage, with what amounts to a strategy of terror practiced by the "leaders" of its own people.

Nor was Poland, where we pursued our quest for further Agnihotra data, very much better. There, we visited Lech Stefański, playwright and TV producer, author of a Polish best-seller *From Magic to Psychotronics*, in his small three-room apartment in the center of Warsaw. In the dark, damp atmosphere of a North European winter, we found it to be a dismal city reconstructed after its ghastly World War II destruction into endless rows of monotonous shoebox apartment buildings, where a room in one of the few remaining hostels must be booked a month in advance.

Again the people were marvelous, incredibly cheerful and hospitable despite their gloomy political and economic prospects, with food and gasoline prices doubling as we arrived, the native's every move restricted by the inane regulations of a deliberately tyrannical big brother. Agnihotra, they said, laughing, echoing the Yugoslavs, might be their only hope for cleansing not only the environmental atmosphere, but the political. Vasant, they said, had seen in Poland a "different nation," prophesying for it a renaissance, with a significant future.

"You can't imagine," said Lech's wife, Helena, "the effect of the Agnihotra ceremony. Even in a tightly shut room, with no windows open, or air circulating, the air during and after the ceremony becomes fresh and pure."

Officially, Agnihotra was brought to Poland by Vasant in 1971 when he attended a symposium on psychotronics organized by Lech. But to the Stefańskis Agnihotra has come mostly through their twenty-four-year-old daughter, Bogna, who studied the practice in India, along with Sanskrit, Hindi, and Bengali, living in the Agnihotra center in Shivapur, and taking long hikes in the Himalayas nearly all the way to the Chinese border.

Bogna said it was hard to say how many people in Poland were practicing the ceremony, several thousand for sure, but that mostly it is done in the family, privately. She said there was no conflict with the Catholic Church because Agnihotra is not a religion, and anyone of any religion can perform it. A lot was being practiced in the Gdansk area, near the Baltic Sea, so "dead" from pollution it is no longer possible to swim in its once sparkling waters. The ash, she said, was excellent as a remedy for cuts and bruises, and a natural medicine when taken internally, especially for clearing up addiction to this or that or the other drug.

But it was clear to the Stefańskis, as it was to Modrić and to Vasant, that the future of Agnihotra depends on how it is validated in America, where they count on serious research. Pat Flanagan suggested a physical explanation for the depolluting effect of the Agnihotra smoke, pointing out that its colloidal molecules of ghee and cow manure could chelatingly attract and grab pollutants in the air, the way water is purified by being flocculated. The seized molecules, he added, as they settle on the ground would alkalize the soil; and if they came into contact with a plant they would stick to the leaves and act as a time-release foliar nutrient. Physically, because of the ghee and the manure, the smoke would be electrically charged. But when asked what the metaphysical properties of Agnihotra smoke might be, though he admitted they must be there, he merely raised an eyebrow, promising to pursue the matter further.

CHAPTER 20

Tuning in to Nature

RUDOLF STEINER'S recipe for getting rid of mice was every bit as exotic as that of the Agnihotra devotees. Catch a fairly young mouse, said Steiner, at a time when Venus is in the sign of Scorpio, skin it, and burn the skin. The mouse must be a field mouse if you wish to affect a field, and the moon's influence must be supported by that of Venus, because the animal kingdom, according to anthroposophical science, conserves the moon influence even when it is not full moon. In Steiner's words: "The animal carries the force of the full moon within it, conserves it, and so emancipates itself from the limitations of time."

His instructions are to carefully collect the ash and other constituents that remain from the burning. "Take this pepper and sprinkle it over your fields. Thereafter the mice will avoid the fields because: in what is destroyed by fire, the corresponding negative force is pitted against the reproductive power of the field mouse." *

* Kolisko tells a strange story about her first experiment with getting rid of mice. "We began by breeding a large number of white mice in order to carry out the necessary experiment during the constellation of Venus. The mice were kept in glass cages covered with wire mesh in a separate room. Each cage contained a male and a female mouse. The exact hour for the experiment was fixed for four o'clock in the afternoon of the correct day of the constellation. We examined the mice every other hour, and found everything in complete order. At two o'clock we examined them for the last time as we fed them. Some minutes before four o'clock we entered the room and received a real shock. In each of the cages one mouse was dead. The female had killed her mate. In all the cages there was the same ghastly spectacle. The female had bitten through the throat of the male, then opened the skull and begun to eat the brain. Some mice must have started earlier or worked more quickly, because we found different stages of this terrible process. In a few cages the female mouse was sitting quietly beside the victim, looking innocent, as if everything were quite all right. Some mice had apparently first eaten the brain, and then started to eat the other

To be rid of nematodes (by which the translator meant any insect pests), Steiner recommended burning the whole insect, not just the skin. This, he said, had to be done when the sun is in the sign of Taurus, or precisely opposite the constellation in which Venus has to be when one prepares the mouse-skin pepper. In effect, says Steiner, the insect world is connected with the forces that evolve when the sun is passing through Aquarius, Pisces, Aries, and Gemini, and then on to Cancer. "In Cancer the force appears quite feebly, and it is feeble again when you come to Aquarius. It is while passing through these regions that the sun rays out the forces which relate to the insect world. If you thus prepare your insect pepper, once again you can spread it over the beet fields and the nematodes will by and by grow faint—a faintness you will certainly find very effective after the fourth year. By that time the nematode can no longer live. It shuns life if it has to live in an earth thus peppered."

Kolisko explains that the ash of the burnt insect radiates into the surrounding soil, and that the insect does not like to live in an area "whence there streams out the counter force to its own life force, its power of reproduction."

That insects attack only weak and dying plants has, by now, become a truism, thanks to the efforts of Howard, Albrecht, Walters, and a whole train of agricultural experts. One would like to know, however, just how the creatures gain the information, how they know just where to go and when, which plant is sick and which is well.

Now, thanks to the efforts of an entomologist with a special understanding of the mysteries of radio antennas and of infrared radiation, what once appeared as necromancy can be reduced to science.

In the course of a lifetime of investigating the habits and habitats of insects, Dr. Philip S. Callahan, professor of entomology at the University of Florida in Gainesville, and a senior entomologist with the USDA, has discovered that insects are well aware of what goes on around them because they communicate on the infrared band of the electromagnetic spectrum precisely as we communicate with radar, microwave, or radio, using a variety of antennas, as sophisticated as any designed by man. With these delicate and highly sensitive instruments, and the use of

---

inner organs. . . . They had bitten off four paws and placed them symmetrically in a square in the sawdust."

Kolisko's explanation for this amazing scene is that the female mice had acted under the influence of the planet Venus, which had come to its highest effectiveness earlier than they had expected. "No other explanation was possible," says Kolisko. "The constellation of Venus had driven the female mice to kill their mates in this extraordinary way. . . . That we were interfering with the forces of reproduction is quite obvious from the fact that the female killed the male."

Recovered from the shock, Kolisko went on to find the proper rhythm with the planetary forces and had no further problem getting rid of field mice.

L. Kolisko. (1889–1976)

infrared light, they can electronically smell out—at quite a distance—meal or mate. Obversely, by the same infrared, they can be lured to an unexpected death.

Most mysterious of the known electromagnetic wavelengths, the infrared, only discovered in 1865 by the English astronomer Sir William Herschel, have remained unfathomable until very recently: there was no instrument with which to tune to them. A mite longer than the longest rays of light visible to humans, they cover all of seventeen octaves, sixteen more than visible light, a *terrra* of some two million frequencies, largely *incognita*.

Herschel happened on them when he placed the mercury-filled bulb of a thermometer against a colored prism, and was astonished to find that, although yellow was brightest, red was hottest. His astonishment increased when he moved the thermometer past the visible red to an invisible area just beyond it, which produced nothing his eyes could see. There, the thermometer recorded an even higher temperature; and the emitted rays, called by Herschel "invisible light," came, after more than a quarter century of controversy, to be called *infrared* radiation.

Many of the properties in this band are still so inaccessible—as Pat Flanagan was to find in the case of Paul Dobler's experiments in which radiation was emitted from agitated water—that it came to be called the

X-band. Yet this all-pervasive band of frequencies is the one most linked to life. Within its ambit, life bubbles up as if from a spring: life does not require visible light, only infrared, which accounts for healthy creatures in caves and the sea's abyss that have never seen the light of day.

On the outskirts of the town of Wichita, we came across Phil Callahan in a setting as exotic as it was unlikely. From the flat, geometrically sectioned prairie land of Kansas there rose before us the Great Pyramid of Giza, in replica, as sparkling white as the original must have appeared several thousand years ago. Beside it, like an excrescence on a lunar landscape, eight geodesic domes of the Buckminster Fuller design burgeoned from a man-made knoll.

This strange complex, a center for holistic medicine and research, funded by Olive W. Garvey, widow of oil and wheat millionaire R. H. Garvey, was the brainchild of psychiatrist Dr. Hugh D. Riordan. The pyramid, as we inspected it, sixty by sixty feet at the base, and thirty-nine feet high, devoid of electric wiring, plumbing, or mechanical equipment above the floor level, is used, so we were informed, for low-energy research, such as the body's emission of magnetic pulses.

In an adjoining geodesic dome, Phil Callahan, also an expert on pyramids and obelisks, had set up a laboratory to monitor the infrared spectrum by means of a highly sophisticated instrument known as a *Fourier transform infrared spectrometer.* Invented in Cambridge for spying from satellites, the instrument is capable of picking up and identifying the exhaust of missiles, or, if used in a bomber, to identify infrared radiations from buildings, as was done in the sneak raid to target Qaddafi's Libyan quarters.

More peacefully, Callahan is using the machine to analyze the infrared wavelengths broadcast by molecules of different substances. "The night sky," says Callahan, "is filled with waves of electromagnetic radiation. It is also filled with vivid colors: red, blue, orange, and green, from thousands of stars that irradiate our atmosphere. Infrared colors, in varying wavelengths, and ultraviolet colors emitted by constellations, are reflected from our own sun to the surface of the moon and down to earth. All these frequencies from cosmos, stars, sun, planets, moon manipulate molecules on earth, depending on the incoming frequency and the size and shape of the receiving antennas on the molecules."

With these strictly scientific remarks, it was possible, at last, to see how iron on earth could be affected by the planet Mars, or tin by Jupiter.

Callahan pointed through the picture window at an expanse of grassland bounded by some leafless cottonwood trees, muttering, half question, half statement: "Are you aware that in such an acre of land there can be as many insects as there are humans in North and South America combined? An insect is nothing but a satellite covered with antennas, like a cruise missile, winging through a sea of electromagnetic wavelengths. At various times during the night, the gaseous molecules that compose our many atmospheric layers are stimulated to glow at very low intensities in beautiful hues of red, green, near-infrared, and ultraviolet. We cannot see these low-intensity colors with our eyes. The cones of our retina, which work so well in bright daylight, cut out at low intensities of light. But the insects see perfectly well with ultraviolet light, and they communicate with infrared."

And communication on the infrared and other electromagnetic wavelengths, we learned, does not stop with insects. Cleve Backster, whose discoveries about plant communication launched *The Secret Life of Plants,* now shows that microcosmic bacteria communicate with each other, and at quite a distance. Dr. Fritz Alfred Popp, of West Germany's Kaiserslautern University, has shown that individual cells do likewise, by modulated electromagnetic radiation. Electron microscopic studies of bacteria reveal arrays of long rod-shaped elements, closest in form to the sensilla antennas of insects. Antibodies are known to recognize and bind invading microorganisms, while enzymes search out and collect raw materials to convert into biologically useful products. Even molecules, says Julius Rebek, Jr., of the University of Pittsburgh, lure and trap each other. Callahan goes further, positing that chemical elements radiate electromagnetic signals to find, recognize, and join each other.

Substantiation for this "flight of fancy" is adduced by the remarkable drawings of individual elements produced by Leadbeater and Besant with their "siddhi" powers in their *Occult Chemistry* (see Appendix C). As analyzed by Callahan, the horns, spikes, and antlers depicted by the two theosophists are replicas, if not originals, of sophisticated man-made antennas used to communicate in a whole gamut of very high frequencies. By Callahan's calculations, the wavelengths to fit such submicroscopic atomic structures would be in the ultraviolet or X-ray bands. And so, at last, Steiner's strange dicta about the elements being sentient, such as nitrogen sensing where there's water, begin to make some sense. And at even higher frequencies, thought, and what motivates that thought, may well manipulate the very smallest matter. Steiner's and Koenig's notion that stag antlers are antennas for picking up cosmic radiations falls from the stars into the realm of physics. Thanks to Callahan—who considers himself a natural philosopher, in the mode of Goethe—much of Steiner's mysterious talk about "the spiritual forces of the cosmos" becomes reducible, with the help of a machine designed to spy from satellites, to academic respectability.

All of which had its start in Callahan's boyhood, growing up in New

York State in the woods and fields around the small Hudson River village of Menands, close to Albany. There all his free time was spent in the wild, exploring nature, or poring over books about insects, birds, and especially hawks.

Philip S. Callahan with his favorite bird, a peregrine hawk. (Credit: *Acres U.S.A.*)

His other great boyhood fascination, only apparently incongruous with hawking, was an attachment to the spreading technology of radio. Aware that if you twang the string of a piano it can cause a violin string across the room to vibrate in *resonance*—providing both strings are tuned to the same frequency—and aware that his crystal set worked on the same principle, vibrating a transmitting antenna that could vibrate a receiving antenna cut to a matching frequency, he was alerted to a prime function of all life: *resonant communication.* In due course it was to lead him to the momentous discovery that a whole world of subtle communication pullulates with life in the darkness of night or cave, and all through the sinews of the "lightless" earth.

As a boy it caused him to wonder if the bond with his hawks might not also be based on some similar phenomenon. Was he somehow tuned to his hawks? Was it possible that living things—insects, animals, humans—could communicate with one another by as yet unknown electromagnetic signals similar to radio waves?

When Callahan joined the army in the Second World War, at the age of twenty-two, his boyhood experience got him assigned to radio school, then to one of the most unusual assignments given any enlisted man during that conflict. As hordes of American soldiers were being convoyed from training grounds in Northern Ireland to ships headed for the Normandy Beachhead, Callahan, a sort of "Wrong-way Corrigan," found himself proceeding by jeep in the opposite direction. Destination: the Magheramena Castle Radio Range near the tiny village of Belleek between the British province of Ulster and the Irish Free State. There, until the end of the war, he was to be engaged in what Winston Churchill called "the Battle of the Beams."

The Magheramena Range put out a series of highly classified low-frequency radio beams—as yet undiscovered by the enemy—extending northward across a wild moorland known as the Pullan, eastward over Lough Erne, and out over Donegal Bay to the western Atlantic. Using these beams, RAF flying boats loaded with depth charges could soar out over the ocean, no matter what the weather, to sink German submarines pinpointed for them by the RAF Coastal Command, then safely return to their bases. Far out to sea, five hundred miles or more, the planes would use a radio compass to home in on a signal from Farrancassidy Cross, near Belleek. Within a hundred miles of the coast they would switch to a beam from Belleek station, then, nearly home, pick up a radio "Z" signal that told them they were in the air space directly above the radio station.

This devotion to duty, used in conjunction with the intelligence received by cracking Hitler's super code "Enigma," enabled Allied planes to locate German U-boats off the West Coast of Ireland and win the Battle of the Atlantic.

As Callahan watched the great Catalina flying boats snake their way home along his beam, he was reminded of the way male moths weave their way toward a waiting female or a sickening plant. On leave one afternoon by a crooked bridge at the foot of Pullan Brae Hill, Callahan spotted a beautiful moth engaged in odd behavior. With its wings of pure white and dark buff it was hovering over a spot on the grass, fluttering and dancing as if tied by an invisible thread to a skyhook. As Callahan peered at it more closely, he noticed that in one respect at least the moth was different from any other he had ever observed: *Its antennas were extremely short.*

Puzzled by this anomaly, Callahan found in an old bookstore in Londonderry a copy of *The Butterflies and Moths of the Countryside* by F. Edwards Hulme, illustrated with a beautiful picture of the moth, known locally as the "ghost swift." The book explained that this "swift" performs its seemingly aimless hovering flight over spots where dull-colored females of the same species hide in the grass waiting to lay their eggs so that the larvae which hatch from them into caterpillars can feast on the roots of dandelions and stinging nettles.

But how, the young soldier wondered, did the male ghost moth *know* that a female was hidden below? Was it because of some signal from the female analogous to the "Z" signal from Magheramena Castle? Were the moth's antennas really instruments to receive and transmit signals in some frequency of the electromagnetic spectrum as yet undiscovered by science?

The Belleek station sent out its beam of electromagnetic energy as pure radio frequency, known as a *carrier wave*, tuned to by a receiver with an antenna; but the beam of itself said nothing. To carry a message it had to be modulated by voice or code. In the case of Belleek, the Morse code signal varied either side of the beam, causing the plane to fly back and forth across the beam to keep on course.

What, Callahan wondered, caused the moth to follow such a pattern? To find out was to take most of his maturing life.

Visiting Hiroshima, only months after its destruction by the atomic bomb, Callahan was aroused by the interaction of radiation with living things. In the wake of that nuclear holocaust, science was concentrating, one-sidedly it seemed to him, on biologically destructive radiation and the dangers of *high*-frequency emissions at the upper end of the electromagnetic spectrum: ultraviolet light, X-rays, beta and gamma radiation. Nobody seemed to be paying attention to its other end: the lower frequencies on the other side of the tinily narrow band of light visible to humans, descending through the infrared all the way to the very long waves he had used to guide planes back to Ireland, and beyond to the mysterious extra-low frequencies, known as ELF, later discovered to be emitted from the brain and to produce on living organisms, at only one to one hundred cycles per second, all kinds of effects, both salubrious and lethal.

Ever ready for adventure, Callahan traveled on foot across China, Thailand, and Burma into India, thence via a seagoing tugboat, on which he worked a hellish furnace-stoking job, to the Persian Gulf and Iraq's port of Basra, little suspecting that in the desert he would find another clue to the enigma of insects and the infrared.

Hiking past the Basra Petroleum Company, he found himself almost ankle-deep in cindered insect corpses as he watched flames flaring gas from chimneys at the edge of the oil field. It reminded him of a passage in W. J. Holland's *Moth Book* about the disappearance of the rosy maple moth "due no doubt to the combined influence of the electric lights, which actually destroy millions of insects that are attracted to them, and

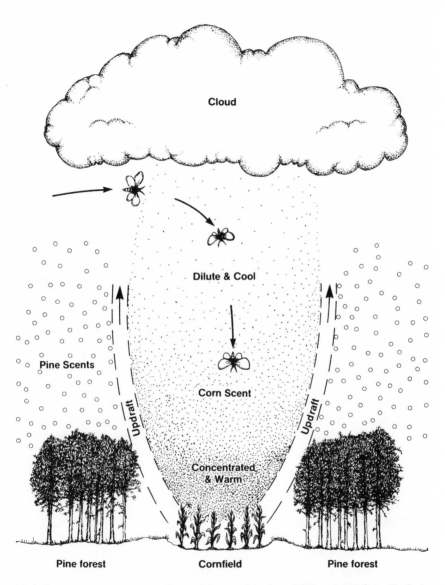

High-flying corn earworm moths navigate at altitudes of 500 to 2,000 feet. To find a cornfield they use their antennae to collect infrared radiation from the molecules of scent rising from the field. Since the scent is warmer and more concentrated near the corn, the infrared wavelengths are longer and fit the long sensillae of the moth's antennae. Higher up, the scent is diluted and cooler; so the frequencies have shifted to shorter wavelengths that fit the shorter sensillae. The insect thus "knows" not only direction but distance to the corn. (Credit: Philip S. Callahan)

the gas wells and furnaces, which lick up in their constantly burning flames other millions of insects." Scooping up a handful of desiccated corpses, Callahan sorted them out into large and little moths, beetles, flies, wasps, and bees, wondering whether it was really the luminosity that had caused their death, or whether it was some other mysterious lure. Many of the insects, he noted, were not flying toward the flames, but swirling around in the night air at the edge of the light. If visible light was the real explanation for such suicidal madness, why, he wondered, didn't insects fly toward the sun or the moon? And why did the dance of the Irish ghost swift so resemble the frantic dances of the doomed moths at Basra Field No. 1?

From the eastern shores of the Mediterranean he took a boat to the southern Italian city of Bari, hiking northward across the Murgie Hills in Apulia to camp near an ancient hunting lodge, the Castel del Monte, where Emperor Frederick II, a scientist of stature, had housed and trained his hunting falcons. As Callahan was writing in his diary, to the light of a thick candle, a small brown moth flew from the grass to make darting passes over the flickering light. First it singed its wings, then dove straight into the flame in a Brunhildian act of self-immolation.

The moth's irrational behavior recalled his reading of the English philosopher and poet Thomas Carlyle's poem "Tragedy of the Night-Moth," which assigned the moths' suicidal actions to passionate love, whereas the French entomologist Jean Henri Fabre, in his essay on the great peacock moth, had, less romantically, seen in its flameward kamikaze rush a desire to "kill" the flame. Neither solution seemed satisfactory to a still curious Callahan.

His next revelation about animal and insect behavior came aboard a commercial freighter, the *Maria C*, bound from Naples, via Algeria, to Philadelphia, with Callahan this time working above decks to enjoy the breeze. Halfway across the Mediterranean, a strange bird with beautiful patterned plumage and an extended crest appeared from nowhere. From its downward-curving bill Callahan recognized a hoopoe, so-called because of its musical "Hoo-po-po" song.

At twilight, when lights aboard the ship were switched on by the bulkhead doors, Callahan was amazed to see swarms of insects around light emitted from under a bulkhead door leading to the galley, but only around this one.

From its perch on a yardarm, the hoopoe darted through the semi-darkness to grab a moth, continuing to dive with great agility, snapping up insects from around the lighted doorway as fast as it could swallow them.

Since he had been part of the crew that had painted that same door the day before, Callahan realized the only difference between the "hoopoe" door and others ignored by the insects was its *fresh paint*, of no special color, just the usual dull gray. Beetles and moths flew in front of it in capricious circles, eventually settling on the rough painted metal, *vibrating their antennas* on its surface.

What, wondered Callahan, could be attracting the insects to that particular door? Noting that its paint had a faint smell of bananas, reminiscent of his mother's fingernail polish, he remembered that Holland had described how night-flying moths were attracted to fermenting bananas and how he had consequently used a concoction of beer and bananas to ensnare them. Strangely, most of the insects were crowding not over the whole door itself, but only on that part of it well-exposed to light. If it was the light that attracted them, then why didn't they also go to the lit portions of other doors on the ship? It must be a combination of the light *plus* the banana-like odor of the paint that was attracting them.

Whatever the energy from this combination might be, the insects seemed to be resonating to it, not with their eyes or with their olfactory senses, as most entomologists maintain, but with their *vibrating antennas*. Could it be, he wondered, that the ship's fifty-cycle light was heating the door and causing the scent molecules of paint to oscillate as if they were little transmitters of infrared radiation? That was the way the sun was supposed to work: affecting all gases in the atmosphere, at a molecular level, with its infrared radiation, pumping them with energy which they re-radiate like tiny radio stations. Was the light bulb a low-energy sun and the paint smell a gas? If so, how and why did it affect the moths?

Next to radio waves, higher up the spectrum, come the extra-high-frequency emissions used in navigation-assisting radar, detectable with metallic *dish*-shaped antennas. Of higher frequency than radar beams are *microwaves*, used for long-distance telephone communication, which are sent, or seized, by antenna shapes reminiscent of trumpets and other brass instruments, consequently known as *horns*.

Higher still in the spectrum are rays of visible light detectable, not by metal devices, but by the rods and cones on the retina of the eye, which allow humans to see the world around them so long as a light source is available. Just beyond these frequencies are the so-called ultraviolet ones, which the eyes of insects, made of multiple lenses—imaginable as foreshortened and rounded antennas—can easily pick up, allowing them to "see" a nighttime world denied to human beings.

In between the band of frequencies captured by metal antennas and the band sensitive to the visual sensors of humans and insects lie the still not fully-explored frequencies of the world of infrared, an understanding of which requires a special knowledge not only of antenna design, but of the physics of optics. Just as in Dobler's day, a decade or more earlier, no sensors for filtering out these infrared frequencies from the ether, as radio receivers do for radio, had yet been pinpointed when Callahan attacked the problem. To Callahan, the antennas of insects, whose unaccountable behavior he had been observing around the world, seemed somehow to function as if, like their man-made counterparts, they were receiving strange signals to guide them to objects of desire: food, members of the opposite sex, or, tragically, a source of fire. Unlike higher-frequency radiations, these seemed to require nonmetallic, insulating antennas, made of horn or wax.

If, as Callahan was beginning to suspect, insect antennas were indeed receptors of infrared signals, the principle, duplicated with man-made transmitters, might enable man to control their comings and goings, especially those that cause depredation to vegetation, and in so doing avoid the pestilence of chemical pesticides. No entomologist had ever addressed the problem, perhaps because its solution required not only an in-depth knowledge of the *behavior* of insects but a detailed description of their *morphology*, or physical structure, and the *chemistry*, or organic makeup, of their antennas. The task was all the more daunting because it required a familiarity with many other branches of knowledge, including electrical engineering, which most scientists, preferring the safety of sticking to their specialties, were not willing to undertake.

Using an old microscope, Callahan took a closer look at the antennas of certain moths that preyed on farmers' fields, particularly the one known to be most destructive to corn, cotton, tomatoes, and a host of other agricultural crops: the corn earworm moth, which, together with relatives in Europe and Asia, was estimated to have destroyed billions of dollars of food plants all around the world.

Painstakingly, he brought into focus the tiny spines jutting from the antennas of corn earworm moths. It was to take him a decade and a half before he was sufficiently confident to publish a paper meticulously illustrating and describing in full detail the complexities of this microscopic system of communication in the infrared band.

Moth antennas look very much like TV antennas, with long bars or spines for long waves, and short spines at the tip for shorter waves. They are even organized mathematically as log-periodic antennas, with bars closer together as they grow shorter—implying that nature not only geometrizes but mathematizes.

While the paper was in press, a new technological marvel was invented at Cambridge University in England, the scanning electron microscope. As soon as the first sample was delivered to the United States,

Callahan, accompanied by his corn earworm moths, rushed to California, where it had been installed. To his immense satisfaction he noted that, at powers of magnification and resolution many orders of magnitude greater than he had been able to attain with his own microscope, the huge, clear images of antennas and their sensilla—tiny sense organs in the form of spines, plates, rods, cones, or pegs, each composed of one or a few cells with a nerve connection—exactly corroborated the drawings he had made. What had taken him fifteen years to accomplish could now be done in the same number of days.

Appointed to teach at the University of Louisiana in Baton Rouge, Callahan used its excellent laboratory facilities to discover that female moths, aroused from their quiescent daylight period, engage in a peculiar behavioral sequence. Moving their antennas forward, they spread their wings to vibrate so fast it is difficult to see that their wings are actually moving. This wing vibration raises their body temperature about eight degrees higher than the usual 65 degrees Fahrenheit through which male moths fly at night in search of mates.

As the female heats up she begins to radiate waves in the infrared. With a complex instrument known as a bolometer, Callahan could see these signals emitted by the corn earworm moth just as easily as if it were a firefly. And the fact that the signals were coded by the moth's wing beats—like a blinker signal that chops a light beam into segments to send messages from a ship's bridge—supported his theory that the moth was broadcasting a *unique navigational message.*

Using an oscilloscope to read out both the intensity and amplitude of the signal from any point around a moth, Callahan was able to determine the moth's position in space. He could tell whether a male detector moth was located above a female emitter moth, or, as aircraft pilots put it, approaching from two o'clock high at 50 degrees compass azimuth.

Yet, no matter how excellent the characteristics of the moth's modulated infrared signal might be as a navigational *guide,* it was not likely, in Callahan's eyes, to also include the *attractant* signal. A male moth would be induced to home on this unique signal only if some other message told it the signal was actually coming from a female of its own species *and* that she was in the proper state to mate with him.

Like a detective in search of further clues, Callahan began to ponder the shape of moth antennas. More comparison between man-made antennas and those of various insects caused him to come to the startling conclusion that every single antenna shape designed by electronics engineers could be found at much smaller scale in the insect world. Insects had indeed anticipated man as the designers of radio antennas. Callahan's microscopic studies, and the photos they produced, revealed a

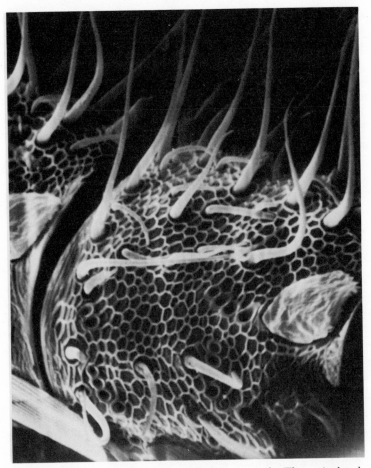

One segment of the antenna of the pink bollworm moth. The reticulated scatter surface and sensilla (spine) antennae are evident at this magnification (4000 ×). They are of different lengths to resonate at different wavelengths. (Credit: Philip S. Callahan)

world of exquisite detail. A tiny green peach aphid, or plant louse, on a desmodium stem, had dome-shaped sensors on its antennas surrounded by curving spines. The spines on the antennas of a red-banded leaf roller moth rose like pickets on a fence. A species of gall midge, to whose extended family belongs the Hessian fly that destroys wheat in the Midwest, has loop-shaped spines. A yellow jacket, one of a family of small wasps, has pyramidal and corrugated helical sensilla reminiscent of highly directional loop antennas.

The corrugation on antennas, he realized, served to amplify the incoming signal by making it tap-tap like a drum.

As stunning as it was to Callahan that what electrical engineers call log-period antennas had already been anticipated in insects, his entomologist colleagues, to a man, kept criticizing this radical notion because, as everybody knew, for insect spines to be able to collect frequencies in the infrared band, to act like radio receivers, the emitted wavelengths would have to be *coherent*, i.e., gallop abreast like horses in a cavalry troop, rather than stampede off in all directions, as they do from light bulbs. Without such coherency, insect antennas could not possibly "tune" to resonate to them, as one tunes a crystal radio to one's favorite station.

Because his colleagues, like the majority of university men everywhere, kept to the placid view that chemicals were the only way to assure

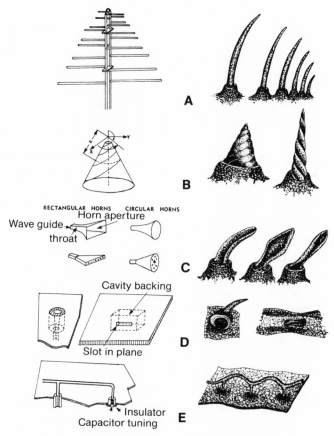

Comparison of insect dielectric and man-made metal antennae. (A) Log periodic spacing; (B) Conical spiral type; (C) Horn and lens types; (D) Cavity types; (E) Loop types.

a defense against insect crop depredation, Callahan resigned his professorship at Louisiana State. Hired by Dr. H. C. Cox, Director of the U.S. Agricultural Research Service's Southern Grain Insect Laboratory in Tifton, Georgia, he found his new co-workers in total agreement with his point of view that chemicals do nothing but destroy ecology.

Drawn back to Ireland, Callahan made another discovery about insects and infrared radiation as he hiked down from Coragh Patrick, Erin's sacred mountain, to spend an evening in a pub, listening to a tenor singing ballads. In the course of an IRA lament called "The Siege of Venice" he noticed a tiny moth making endless circles plumb in front of the singer, directly below a single hanging light bulb and directly above a glass of Guinness stout. The balladeer, annoyed by the moth, tried unsuccessfully to grab it, but the insect kept dodging to resume its circular flight above the pint of ale.

Why, wondered Callahan, did the moth not spiral up to the light bulb, or down to the brew, but remain gripped by some invisible force just in front of the singer, much like the wavering ghost moth at Pullan Brae so many years earlier?

At last it became clear to him: the molecules wafting up from the ale were being energized by the light of the bulb while their amplitude was being raised by being jostled by the singer's voice, causing them to emit laser-like bursts of infrared radiation. These the moth picked up with its antennas, attracted and imprisoned into fixed gyrations.

The next clue came during a hike across Devonshire's wild Dartmoor, where the ghostly hound of the Baskervilles had spread nightly terror in Sherlock Holmes's tortured tale. Pausing to eat lunch near Watern Tor, one of a collection of towering masses of stone strangely sculptured and weathered by the fierce winter winds that blow across the moor, Callahan became aware of a cloud of small insects hovering against the sky over the tor in a dance for which entomologists had no explanation.

An accomplished rock climber, Callahan scaled the steep side of the tor to put his hand on its sun-warmed topmost surface of black rock. On his downward climb, he became aware of a mysterious mist, full of scent, coming from the heather, bog cotton, mosses, and lichens on the surrounding moorland.

It came to him, on his way to Cranmere Pool, that as the scented vapors drifted across the tor their molecules would be bent and rotated by the hot rays from the rock, stimulating them to oscillate at many unknown infrared frequencies. Had he "infrared eyes," he mused, he would have been able to give names to those infrared "colors," as easily tuned to by an insect antenna of proper design as visible colors are distinguished by normal human vision.

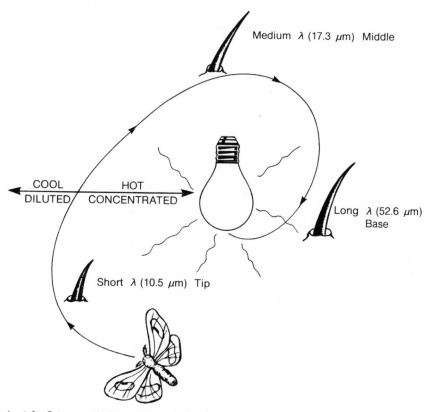

Medium λ (17.3 μm) Middle

COOL
DILUTED

HOT
CONCENTRATED

Long λ (52.6 μm)
Base

Short λ (10.5 μm) Tip

A night-flying moth flies a log periodic spiral to a light bulb because the moth is not going to the light but to the narrow band (coherent) infrared radiation from the light-stimulated molecules around the bulb. The antenna is log periodic. (Credit: Philip S. Callahan)

Could it be, he wondered, that the flies above the tor were attracted to the infrared vapor "colors" from plant scent molecules in turn stimulated by the heated pile of rock to emit frequencies like so many windblown satellites? Could insect antennas be tuned to these subtle frequencies? Trudging on, he was increasingly filled with certainty that the frequencies existed and that insects were tuning in to them. But how did they do it?

An electrical engineer in London's Imperial University, Dr. E. R. Laithwaite, an avid butterfly and moth collector, supported Callahan's assertion that insect antennas might work just like radio antennas. During a walk in the countryside, Laithwaite had noticed male vaporer moths flying with the wind toward females. Since he was sure that suitors did not always track the objects of their affection downwind, he came to the conclusion that they were not attracted by a scent, but by an omni-directional electromagnetic signal that could not in any way be wind-affected.

His opinion enraged an eminent entomologist, H. B. D. Kettlewell—first to discover that female insects lure their males by emitting clouds of hormones known as *pheromones*, from the Greek *pherein*, "to carry," and *hormon*, "to excite"—who angrily countered that electromagnetic communication among insects was claptrap and that only his odoriferous molecules could, sexually speaking, make them tick.

In an attempt to harmonize the opposing views of his colleagues, a physical chemist, Dr. R. H. Wright, held that insects could identify scent molecules through what are called *osmic* frequencies, most likely to be found in the infrared portion of the spectrum. Though how they might do this remained an enigma.

In essence, the conflicting arguments were those of an entomologist saying that insect antennas detect *scent*, an electrical engineer saying they detect *infrared radiation*, and a physical chemist saying they might detect infrared radiation *from* scent.

Callahan was of the opinion that all three were partially correct. Back in his Tifton laboratory, he delved again into the scientific literature to find that he and Dr. Wright were not the first to believe that insects could sense radiation from oscillating molecules. Way back in 1892, C. V. Riley, a noted entomologist, had suggested that insects might communicate through their antennas by *telepathy*: "This power would seem to depend neither upon scent nor upon hearing but rather on certain *subtle vibrations*, as difficult for us to apprehend as is the exact nature of electricity."

Riley's reflection was followed in 1936 by that of the South African journalist and poet, Eugene Marias, who, in his classic *The Soul of the White Ant*, described how a female termite, after finding a suitable spot, would come to rest on her forefeet and lift three-quarters of her hind body into the air, remaining as still and stationary in this position as if she were a termite statue. "What is she doing?" Marias wondered, then replied: "She is busy sending a wireless SOS into the infinite."

Marias also believed the signal might be caused by a scent but, hesitating, qualified the idea: "When speaking of scent, one should also think of waves in the ether. It is false to assume that perfumes consist of gases or microscopic substances. Perfume itself is not entirely a physical substance. One can scent a large room for ten years with a tiny piece of musk and yet there will be no loss in its weight."

No more conjecture on the subject appeared in print until 1949, when an electrical engineer, G. R. M. Grant, published a paper in the *Austra-*

lian Proceedings of the Royal Society of Queensland to theorize that sensory "pits" he had found on insect antennas might be resonators to infrared radiation.

Inspired by Grant's conclusion, Callahan began a search for behavioral patterns among insects that might indicate *both* olfactory *and* an infrared-sensing basis for insect communication. His studies heightened his certainty that this must surely be the case: the insects "smell" odors electronically by tuning into narrow-band infrared radiation emitted both by sex scents and by plants they desire as food. His next requirement was to prove it, indisputably.

On the premise that older knowledge, so often spurned by younger scientists as passé, might still have value, Callahan rebuilt what he called his *Russian Infrared Machine*, an instrument conceived in 1924 by a Russian researcher, Dr. Glagoleva-Arkadieva, for the purpose of detecting infrared frequencies.

When he was finally able to place his moths in front of it, they went into a frenzy of excitement. Believing the infrared emissions emitted from the machine to be the real thing, females tried to lay eggs on it, males attempted to mate with it, extending their genitalia toward its components, adding one more rung to the ladder of proof. With the same machine, by broadcasting different frequencies, Callahan was able to determine that sickening plants signal the news of their impending death to waiting bugs by means of the same infrared radiation, which goes a long way toward explaining why healthy plants remain impervious to pests.

At last it was clear what actually happens with the mating moths. The female emits her pheromone molecules, which are carried on the wind in a plumelike cloud, growing thinner and cooler as they distance from her body. The male moth, flying into this plume of pheromones, uses the sensilla of his antennas to pick up electronically the infrared signals emitted by the scent molecules. If he strays too far from the center of floating scent, the signal fades as the concentration of hormones becomes dilute. He must therefore turn back into the plume, flying a serpentine path, to and fro toward his goal, just like an RAF Catalina following its radio beam.

Two things help the male moth navigate correctly: temperature, and density of pheromone molecules. The further he is from the female the colder and more dispersed are her pheromones, giving off shorter and weaker frequencies, which the male picks up with the shorter sensilla on his antennas. The closer he gets, the warmer and more condensed are the pheromones, giving longer waves, which he picks up with his longer sensilla.

The male also has an additional powerful means of modulating the free-floating scent by vibrating his own antennas at audio frequencies to amplify the signal. All insects vibrate as they move—different species at different frequencies: bees at five hundred cycles, moths at sixty, ants at twenty. This helps the male moth amplify the female signal with his own vibrations. As he reaches his goal, the signals peak, and their matching wavelengths tell him he's right on, and she's turned on.

In the same way moths tune in to the wavelengths emitted by the specific plants their larvae are designed to feed on. The sicker the plant, the more powerful the scent, the easier for the moth to home in on its prey.†

In a chapter entitled "A Blueprint for Insect Control" in his book *Tuning in to Nature*, Callahan makes the final incisive point that his predictions with respect to various characteristics of the free-floating infrared-emitting pheromone open up a whole new way to deal with pests. If an astute group of entomologists, antenna engineers, physical chemists and physicists is assembled, it may be possible—of course with the aid of morphologists—to "produce emissions with enough energy to attract insects or jam their communication systems over great distances."

Such a system would indeed be a boon to farmers. Instead of going to their agricultural coops to buy insecticides, farmers could lease microminiature transmitters either to attract to traps, or, by beating against insect pheromone frequencies, to jam them and prevent mating. The same companies that produce insecticides could produce different chemicals to make solid-state transmitters, even the transmitters themselves. And if, as the Free Masonic founding fathers of this Republic believed so warmly, "there is room at the table of life for one and all," why not room for insects? Of the thousands of species, only a tiny minority are noxious in any way to man or plant. The rest serve multiple and varied vital functions, enlivening the soil, pollinating species, and, in Steiner's clairvoyant view, constituting a vital link with the wavelengths of the spirits of nature, especially the fire spirits he sees so closely tuned to insects.

Surely, says Callahan, any improvement is preferable to the present

---

† More engagingly, the phenomenon poses the question as to whether human sexual attraction may not also be basically reducible to electromagnetic signals. The human body, male or female, lissome or corpulent, is surrounded by an aura of infrared emanations, just as are those of the night-flying moths. The subtle molecular odors that surround human bodies are stimulated to radiation by this infrared emission. Could it be, asks Callahan, that the peculiar chemico-electric signal attractive to mosquitoes and other biting insects, such as gnats, horse flies, deer flies, and the tiny black flies known as *no-see-ums*, is the signal that attracts men to women, and women to men? "Are young people of the present generation closer to the truth of sexual attraction than oldsters when they allude to 'good vibes' as the basis for attraction?"

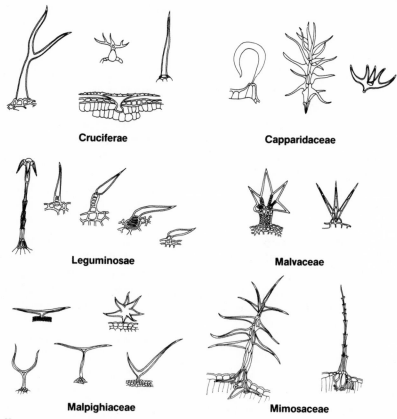

Cruciferae

Capparidaceae

Leguminosae

Malvaceae

Malpighiaceae

Mimosaceae

Different types of triclones on certain plant families. If plants are chemically "talking" to one another, as has been shown, then the triclones must be the antennae for molecular messages. (Credit: Philip S. Callahan)

archaic, indiscriminately lethal, and outdated system of pest eradication. And the ultimate practical advantage of such a new control system, in addition to its ability to attract, bug by bug, only selected species to which it is tuned, is its ability to be turned off—unlike toxic poisons—when not required.

In his geodesic lab in Wichita, with the help of his new infrared spectrometer, Callahan demonstrated for us in scientific terms an overlap between physics and metaphysics, validating the wisdom of the Vedas. Into the beam of infrared light, channeled by mirrors through his machine, Callahan intoned the basic sound of the Hindu mantra "Aum," repeating it several times.

He then pointed to the computerized printout, which clearly showed the effect of his mantra in the form of a group of spikes on the graph. "Above the base line," he explained, "infrared energy is radiated by the molecules jostled by my voice. Below the base line is the energy as it is taken in. When the molecules of whatever is in my breath are energized, they emit laser-like spikes of infrared radiation. And, just as a radio picks up the *Eroica* from your favorite classical music station, this machine picks up the frequency at which the molecules are radiating, and can identify their signal frequencies. If you know the frequency of an atom or compound, you can identify its presence and its amplitude. Here, for instance, is carbon monoxide. You can see it right there on the graph." He adjusted a knob to produce a printout.

"All of this," he said, handing us the evidence, "proves the extraordinary power of the mantra. The more you recite it the more it clears your body of carbon monoxide. See it there? This leads to an altered state of consciousness. As the mantra vibrates the molecules of breath, they give off heightened rays of infrared. These must be picked up by our acupuncture points and fed back to the body, causing an altered state of chemistry, which affects the breath, which again affects your consciousness. The more you chant the more you experience an altered state."

He laughed happily, and offered to let us try for ourselves, promising similar results.

When we asked to what further research the machine was being applied in the center's quest for holistic medicine, Phil surprised us with the blown-up picture of an AIDS virus, which he compared with the picture of an RAF antenna at Heathrow Airport. The similarity was indeed remarkable.

Callahan's object, as he explained it, is to identify the precise frequency given off by the AIDS virus. He then hopes to develop a means of replicating the frequency, and by jamming the virus's signal, to neutralize its effect. The whole of an afflicted person's blood could then be passed through the same sort of machine as is used in kidney ailments, and, with luck, neutralize the disease.

"It's funny," said Phil, as he walked us to our car, "how many important discoveries are made and ignored for decades, or even centuries. The other day I got a jolting surprise when I ran across the works of the almost totally forgotten nineteenth-century Irish scientific genius, John Tyndall, which contained a lost chapter in the history of science. It described how molecules of perfume such as those emitted by pachouli, sandalwood, cloves, lavender, attar of roses, lemon, thyme, rosemary, spikenard, aniseed, and the oils of cloves and laurel, all *absorb* infrared radiations. Tyndall's inspired work spawned a whole new branch of science: infrared spectrophotometry. Yet, though Tyndall also discovered the beneficial effects of penicillin eighty years before Sir Alexander Fleming, and was clearly on the point of inventing the laser a century before that feat was accomplished, his contributions were totally ignored by later chemists and olfaction physiologists." Phil shook his head. "It caused Tyndall's successes to be characterized in the *Encyclopedia Britannica* as "due more to his personality, and his gift for making difficult things clear, than to his original researches."

Equally inspiring to Callahan was the work of Tyndall's great Gaelic contemporary, Robert Lloyd Praeger, whose masterpiece on the natural history and topography of Ireland, *The Way I Went*, described by Callahan as "a field guide to the soul of an entire country," became his admired viaticum.

It was to lead Callahan to a discovery almost as monumental as that of insect communication in the infrared band: proof of the beneficial influence on agriculture of radiations from outer space. These discoveries, made by Callahan subsequent to his years of entomological research, were to validate scientifically the clairvoyant vision of cosmic forces described by Rudolf Steiner, meshing with Julius Hensel's prophetic but discredited conclusion that "animate" rocks, though speechless, may yet rid us of the noxious chemicals with which industry has been greedily and increasingly fouling its nest.

CHAPTER 21

Towers of Power

OF THE SIXTY-FIVE mysterious medieval structures called Round Towers so far located in the lush green countryside of Ireland little is known other than the fact that they are geodetically and astronomically placed and oriented, evidently for some special reason, their windows designed to cast shadows to indicate the day of the solstice or the equinox. That they might be antennas used by medieval Irish monks to capture cosmic waves was Callahan's surprising theory.

During World War II, when he was stationed at Belleek, Callahan often visited one of these Round Towers on nearby Devenish Island, a wild and mystic spot in the middle of Lough Erne. Perhaps the best-preserved of the twenty-five such towers still standing, it is built of finely jointed sandstone, 25 meters high with a circumference of 15.4 meters, divided into five floors with windows specially set.

On one of his trips to Devenish, Callahan asked the fisherman rowing him out to the tower why the local farmers went to such trouble to barge cows back and forth to the island.

"Arragh, man," the fisherman replied, "shar isn't the grass finer out thar than on the mainland!"

Intrigued by the phenomenon, Callahan obtained a map of Eire marked with the location of all its then-known rounded towers. As he studied the map, it struck him that the location of the towers formed what looked like a star map of the northern night sky at the moment of the December solstice, Polaris being clearly marked by an especially magnificent tower on the grounds of the monastery of Clonmacnoise next to the River Shannon on Ireland's central plain.

Could there be a link, Callahan wondered, between the lush green grass around Devenish Tower and the position where it stood under the starry cosmos? Could the round tower be acting as an antenna for some cosmic energy broadcast from the stars? The fact that in 1932 Dr. Karl G. Jansky of the Bell Laboratories had first discovered radio waves from the cosmos, and had measured those from that particular part of the sky as arriving in 14.6-meter wavelengths, seemed more than coincidental. It struck Callahan that the Devenish Tower, precisely positioned and shaped, might be resonating to cosmic radio wavelengths, as well as to some kind of magnetic-field energy. And the fact that it consisted of paramagnetic rock also pointed in that direction.

Irish Round Towers.

Stones, says Callahan, have a secret life that involves two equal and opposing, but very little understood, magnetic forces, the plus and minus of nature, the yin and yang of the Chinese. These are the forces that German and English natural philosophers of the nineteenth century called paramagnetism and diamagnetism—the former attracted by a magnetic field, the latter repelled. As described by the scientists who discovered paramagnetism in the mid-nineteenth century, it is a "weak, fixed susceptibility toward a magnet." By fixed, they mean the magnetic attraction is "inherent" in the substance and cannot be transferred as one does with ordinary magnetism by rubbing a nail or a screwdriver against a horseshoe or bar magnet. Callahan's postulate was clear: the Irish Round Towers, made of limestone, sandstone, or basalt, and therefore definitely paramagnetic, could be massive electronic collectors of cosmic microwave energy as well as giant accumulators of magnetic energy.

To find out, he began to experiment with small-scale models made of paramagnetic sandpaper and later carborundum, using the dimensions published in Professor G. L. Barrow's book *The Round Towers of Ireland*, producing one model to the exact dimensions of the Devenish Island tower. With a high-frequency oscillator called a *klystron* to generate a three-centimeter wavelength of radio energy, Callahan placed his ten-centimeter-wide sandpaper model (fourteen in circumference) within a radio beam and, sure enough, the power meter went up from six to nine decibels of energy, a clear indication to Callahan that the Irish Round Towers were in fact radio-wave guides, acting the way a magnifying lens does to collect and focus light.

The conclusion also resolved for Callahan the mystery of why the doors to the Irish Round Towers were invariably set many feet above the level of the ground, placed there, according to orthodox archeologists, as one of the protections against Vikings. The notion, says Callahan, is quite ludicrous, if only because the first Vikings did not invade Ireland until well after the eighth century, and the height of the doors would have presented no obstacle to their determined attacks. There had to be a better reason, and the anomalous fact that, within the base of differing towers, the monks had compacted the earth to varying levels indicated to Callahan just what that reason might be.

No matter how mathematically accurate electrical engineers design their antennas, says Callahan, their figuring seldom provides a sharp enough resonance. Antennas need to be shortened or lengthened by trial and error to conform with an incoming wavelength. To tune their stone antennas to the night-sky radiation, says Callahan, the monastic tower builders merely needed to fill each of the interiors with dirt up to a level at or near their door heights until they obtained the right incoming frequency.

To strengthen his contention that the Irish Round Towers are indeed paramagnetic antennas, Callahan realized he would have to be able to show magnetic field-force lines on one of his models similar to those shown by iron filings on a sheet of paper under which a magnet is placed. In theory, these would appear as rings at different levels. To make them visible he decided to soak a carborundum model of the Turlough Tower in County Mayo in a solution of Epsom salts for forty-eight hours in the hope that this diamagnetic salt would be light enough to be affected by the extremely weak forces involved. He was elated to see the model become circled with a spiral of white lines from its pointed top all the way to its base. Even more astonishing, at each level of the model, where a floor and window were located on the larger tower, a particularly broad, strong band of white Epsom salt appeared.

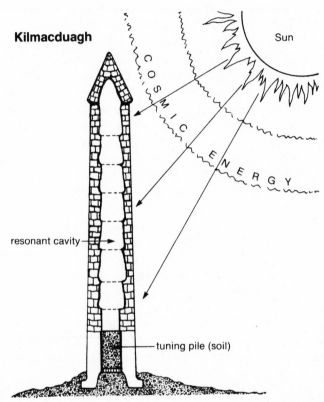

The tower at Kilmacduagh in County Galway, Ireland, thirty-four meters in length, is the tallest of the Irish Round Towers, which served, according to Professor Philip S. Callahan, as antennae to receive cosmic or magnetic energy. Part of the base below the doorway of such towers was filled with dirt. They could thus, according to Callahan, constitute a dirt "tuning pile" for shortening or lengthening the paramagnetic rock antenna. (From *Magnetic Antennae and Ancient Agriculture*)

These lines of force were similar to the standing waves of energy that can be measured on a resonant electromagnetic radio antenna, called electromagnetic *modes* by electrical engineers. In other words, it looked as if the towers were designed so that the strongest "mode" lines were concentrated on the floors of the towers where the monks would be observing the stars through the small tower windows, chanting invocations, as Callahan reconstructs the scene, to increase the inflowing energy.

To what purpose all this effort? Over millions of years, says Callahan, man and all of nature's plants have evolved under low-energy microwave radiation that constantly bathes not only our own bodies, but all of our agricultural crops. For high frequencies the best antennas, says Callahan, are not made of metal conductors, but of dielectric substances— electrically nonconducting—such as Plexiglas, wax, or stone. He believes our agricultural forefathers, the Celts, knew how to "tune in to nature," using the Round Towers as silicon-rich semiconductors rather than using metallic conductors—creating a huge resonant system for collecting, storing, and relaying meter-long wavelengths from the cosmos. The towers, he says, were "tuned" magnetic antennas, massive electronic collectors of cosmic microwave energy and giant accumulators of magnetic energy. As such, they "contributed beneficial cosmic energy to fertilize the fields of those ancient, low-energy and stone-handy engineer monks, centuries before the word *electricity* had come into common parlance."

What the monks were doing, says Callahan, was collecting cosmic paramagnetic energy, and focusing it with their Round Towers onto the earth in which they planted their crops—"doping" the plants with that energy. Infrared paramagnetic forces would radiate in waves from the bases of the towers to increase the attracting paramagnetic properties of the surrounding soil, rather than directly affecting plants, which have a diamagnetically weak, fixed force, repelling a magnetic field, a fact discovered by the English scientific genius Michael Faraday, and confirmed by his Irish contemporary John Tyndall, who tested some thirty different species of trees for diamagnetism.

The towers, huge, well-designed stone waveguide detectors of microwave radiation from the cosmic universe, would resonate, according to Callahan, by day to the magnetic energy transmitted by the sun, and by night would capture the 14.6-meter wavelengths emitted from that starry region of the universe to which they were carefully aligned at the winter solstice. And, because plants exhale highly diamagnetic oxygen during sunlight hours, they are diamagnetic by day, but, as unrecognized by Faraday and Tyndall, become paramagnets at night when they concentrate on breathing out highly paramagnetic carbon dioxide.

Aware that form as well as length is necessary for strong resonance in radio antennas, and that the same is true of stone paramagnetic anten-

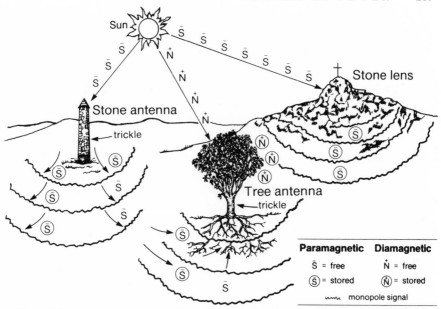

Place of the magnetic poles in nature. A circle around the monopole means it is stored in the soil, stones, or plants. Starting at the sun's magnetoelectric dipoles, the dipoles are torn apart by flare activity and free magnetoelectric monopoles, south and north, head for earth across space. Most reach earth where the Ss are being adsorbed and stored by stone Round Towers, stone mountains such as the sacred mountain (shown with cross), and by paramagnetic soil. The N monopoles are adsorbed by plants such as the tree shown. Once adsorbed they are stored as in a battery (circled monopoles).

nas, Callahan wondered what part the characteristic tapering of the Round Towers might play in their paramagnetic properties. His models showed that tubular structures with conical pyramidions, or dunce-cap covers, were ideal conductors. Experimental plants growing around them invariably bowed toward the central tower, with seedlings sprouting more quickly on the north side, leaving those on the due south side to grow to only about half the size.

The first serious large-scale field test of Callahan's theory was carried out in 1986 in Arkansas, on land north of West Memphis, belonging to Thomas C. Quackenboss, a gentleman farmer. Operating on the premise that, if the Irish Round Tower system worked for Callahan with his small-scale models, it should work just as well with medium-sized models built to the same proportions as the Round Towers of Ireland, Quackenboss and his family raised three six-foot-long and one-foot-in-diameter terra-cotta drainpipes filled with highly paramagnetic granulated basalt to irradiate an area of some two hundred acres. They located the pipes in accordance with Callahan's discovery that his small models radiate the lion's share of their energy in a cloverleaf pattern to the north of where they are situated. To render the terra-cotta towers even more paramagnetic the Quackenbosses topped them with conical pyramidions of basalt mixed into cement. Total investment: $250.

Fields of cotton and soybeans were planted in early summer. By fall the happy result was an increased harvest of more than $8,000 above what had been expected. The following year the Quackenbosses raised twenty-one "towers of power," hoping thereby to increase their revenue. And further to increase whatever advantage Callahan's cosmic radiation might be bringing to the family crops, Thomas's thirty-two-year-old son, John, a natural dowser, made studies of dowsing literature on subterranean water veins, and aboveground "ley lines," with the object of situating his "towers of power" directly over spots where his dowsing rod indicated such lines were crossing or overlapping. The results were surprising from another and novel angle: in an area plagued by drought the Quackenbosses, unlike their neighbors, got plenty of rain, enabling them to harvest 1.8 bales of cotton per acre and over 40 bushels of soybeans, the second best crop in their history, an extraordinary feat in a year of drought such as 1988. The reason for which, says John, may well be attributable to the mysterious powers of the towers, the only extraneous factor.

When two former college classmates heard of John's exploits, they induced their own families to try the experiment. It was thus that in central Virginia, on the land of the well-known Richmond businessman, Jim Wheat, his farm manager, Richard Dix, and his actual farmer, Ray Thomas, had the unprecedented courage to place seventeen towers on a thousand acres of tidewater land in the estuary of the Rappahannock River.

So impressed were they by the first year's harvest with the magical towers, which brought in an added $50,000, they have determined to go a step further and try biodynamics on all of five thousand acres and give up the use of chemicals as fertilizers or as pesticides. A happy move, for when we were there in early summer, just as the corn was being planted in a heavily poisoned field, we found the remains of a beautiful hawk that had feasted on one of three poisoned crows, and on the edge of his burrow, the tracks of a dying groundhog.

That the system of broadcasting cosmic energy with towers, obelisks, or pyramids goes back to the historical developers of such structures, the Egyptians, stands to reason, but the proof of it is very recent, only now come to light through the efforts of Gabriel Howearth, who for many years has backpacked hundreds of miles into the roadless Central American jungle into territory where few white men have ever penetrated, to commune with his fellow Maya agriculturalists.

Six-foot terra-cotta drainpipe filled with basalt and capped with a cone of cemented basalt raised on the New Hampshire farm of a client by John Quackenboss (*right*). The proportions are the same as the Round Towers of Ireland, and early experiments have given significantly increased harvests.

Studying their method of handling the land, he found them to be amazingly sophisticated, with fields of rich produce, sometimes as large as four-hundred-acre lots, richly intercropped—as was the practice before the arrival of Cortéz. "They had permaculture sixteen hundred years ago," says Howearth. "And now they have fruit and nut trees over a thousand years old that are still growing and bearing."

That the Maya were in contact with the Egyptians, as painstakingly documented by Auguste Le Plongeon, and thoroughly corroborated by such eminent epigraphers as Professor Barry Fell, is no longer in question, except perhaps by such establishment skeptics as Martin Gardner and Randy the "Magician," whose subsidized stance, like that of Cardinal Bellarmine with Galileo, is to deride all such wonders as spurious or heretical.

Now the link between Mayaland and Egypt is further supported by Gabriel Howearth's discovery that the Maya used a method of weed and insect control that amounts to a form of what is modernly known as *radionics*. This they married to an amazing science of astronomy-astrology in which cosmic forces are picked up and relayed by tiny pyramids, directed along the local grid of ley lines, just as with Callahan's "towers of power." *

"They know," says Howearth, "which planets affect which insects—such as Venus influencing ladybugs, and Mars influencing aphids—and they have specialists constantly plotting the course of the planets as they move against the stars. To their innumerable and very accurate calendar stones, some carved with glyphs not only of the planets but of the various weeds and insects, the Maya attached movable sections, which enabled their brujos to determine, and then manipulate, as did Steiner, cosmic forces to control unwanted weeds and insects."

Their expertise in these matters, says Howearth, passed down through the centuries, is amazing, and enables them to perform such ecological controls as capitalizing on a grasshopper they know to carry a virus that eliminates a certain weed.

On one unusual calendar stone, said Howearth, the glyphs of the planets run straight down the center, including glyphs for sun, moon, and the asteroid belt. This enables the Maya astronomers accurately to compute and schedule their agricultural activities.

Over the years, as Howearth gained their confidence, the Maya taught him what they knew of the influence of planets, as well as other well-kept secrets not divulged to gringo archeologists. He says they took him out into fields to demonstrate what amounts to a modern technique of radionics. As the planets move into their correct positions, the Maya use small pyramids and special glyphs of weeds or insects to relay forces from the cosmos along terrestrial ley lines, and thus control both weeds and insects.

Howearth says that deep in the jungle, removed from gringo eyes, the Maya practice some of the most advanced agriculture he has ever seen. "Their ability to work the weed problem in the tropics, where the problem is outrageous, is amazing; but they can do it with their glyphs and pyramids. They keep and nurture the weeds they want for their integrated agriculture, but have a subtle method for controlling the ones that otherwise get out of hand. As did Steiner, they will take the ash of a seed and place it in a pyramid. As the influence of the right planets peaks, the effect is radiated to eliminate the weed from where they do not want it. They tell me that in the old days they used to put the smaller pyramids inside the larger ones. They also have inverted pyramid structures, built into the soil, mostly of adobe, in which they can store their seeds indefinitely."

* Radionics, developed in this country by Dr. Albert Abrams at the beginning of this century, and later in England by George and Marjorie De La Warr, is a system for affecting matter by finding and reproducing the exact wavelengths on which it vibrates.

Like Daryl Kollman, Howearth's Mayan friends aspire to rejuvenate agriculture in this hemisphere. And it is Howearth's project to get some of them to come north to help put into practice on his uncontaminated land the ancient wisdom kept alive by natives of America, whether North, Central, or South. These are the natives whom we, of the United States, have continued to enslave or massacre in a tradition worthy of Cortez. Had his extirpator of *malezas*, archpriest Diego de Landa, not destroyed the precious Maya codices, we might have saved our soil the misery of half a millennium of ignorant mishandling.

CHAPTER 22

Cosmiculture

LUCKILY OUR Indian heritage is slow to die. In the highland woods of Georgia, within sight of the Great Smoky Mountains, mystic haunt of the Cherokee, it lives on by the Tallulah River into whose turbulent waters the daughter of a chief once threw herself from a thousand-foot cliff to join her young white lover, sacrificed by her understandably segregationist father. A few miles upstream from this lover's leap, the granddaughter of another Tallulah Cherokee, Sarah Hieronymus, has been tapping cosmic waves. In a laboratory on the shores of Lakemont, not far from the Cherokee reservation, she is carrying on the work of her late husband, T. Galen Hieronymus, running the Advanced Sciences Research and Development Corporation, a nonprofit organization presently devoted to the spread of "Cosmiculture"—the channeling of cosmic energy into the ground for the benefit of plants.

This Steinerian ideal is accomplished with what Galen called *cosmic pipes*, ten-foot-long plastic polyvinyl-chloride tubes, three inches in diameter, which are raised, like an Egyptian *dged* column, to a height of eight feet, their bases inserted thirty inches into the ground. Atop each pipe is a copper electrode designed to absorb the mysterious solar energy Galen called *eloptic*, a combination that obeys "some electrical laws but not *all* of them, and some optical laws, but not *all* of them," and passes it down a wire coiled around a quartz crystal to an underground amplifier, there to be broadcast through the soil for a mile or so in all directions.

"We don't make them any higher," Galen told us in the summer of 1987, a few months before he died, "because the potential increases as you go up; it gets too strong above six or seven feet. All around us is a great sea of energy, cosmic energy, solar energy, lunar energy, planetary energy, and the energy of the earth itself. But, unlike the chemicals sold in commerce, this energy is free, and it isn't toxic; it's highly beneficial. All we have to do is tap it: and that's what we've done. When I saw that chemical fertilizers and patent medicines designed for livestock were making paupers of the farmers of this nation, I got out my early experiments in eloptic energy and adapted them to tap this sea of free energy, and so we devised the cosmic pipe."

For many years an engineer in charge of heavy power distribution in Kansas, Galen liked to quote astronaut Edgar D. Mitchell to the effect that there are no unnatural or supernatural phenomena, only very large gaps in our knowledge of what is natural. For the last fifty years Galen has been pioneering in the exploration of what he calls "subtle energies," energies outside the electromagnetic spectrum, as little understood by orthodox science as are electricity or gravity—the world of energies so lucidly described in the Hindu Vedas, in theosophy, and in anthroposophy.

As early as the 1930s, Galen showed that solar energy could be conducted over wires, and—more difficult—he succeeded in obtaining a U.S. patent for an instrument that did it. Shortly after World War II he developed radionic instruments on the basis of the sophisticated work of Dr. Albert Abrams, a natural genius who did his pioneering in San Francisco. Ever resilient, Galen then discovered his eloptic energy. "We need a new kind of dictionary," he said, "to describe these energies, which are allied to, but are different from those in the electromagnetic spectrum. It's a subtle cosmic energy. It does not attenuate with distance. We can conduct it over wires."

With his radionic instrument Galen mysteriously rid the fields of many a Pennsylvania farmer outside Harrisburg of Japanese beetles and of European corn borers, remotely affecting the fields with a photograph placed in his "black box" many miles away. So successful was this method that a U.S. general helped form a company to exploit the invention. But the Pentagon, quickly realizing that the same system might be beamed on soldiers in the field, did the chemical companies a service by remotely tuning Hieronymus out of the business as effectively as he had tuned out the bugs on the farmers' fields. *

Fingering one of his cosmic pipes in the laboratory of his Lakemont headquarters, appropriately called Oasis, surrounded by innumerable electronic and radionic instruments, Galen spoke of eloptic energy as if it were a friendly djin: "It doesn't like coils. It likes straight lines; but we

---

* See his *The Truth About Radionics and Some of the Criticism Made About It by Its Enemies* (1947) and *The Story of Eloptic Energy* (1988).

can manipulate it with coils and make it radiate. And it moves with the speed of light. We proved this when the astronauts were on the far side of the moon back in the sixties. We knew, fifteen minutes before NASA, that they'd fired their retro rockets. NASA was out of radio communication; but we could plot the astronauts with eloptic energy."†

Well into his nineties, his eye as sharp and his step as lively as that of a Hunzakut, Galen looked forward to demolishing the premise that chemical fertilizers in agriculture can be of any use whatever. "If we don't get 'em in this life," he joked, "we'll have to get 'em in the next."

To help the farmer clean up the poisons already in his soil and balance it for healthy organic growth, Galen developed three other instruments for use in connection with the cosmic pipes. His *eloptic energy analyzer* gives the farmer the means, radionically, of analyzing the soil to find out what's good or bad in it, and how to remedy the situation. The same instrument can also be used to diagnose the health of livestock. Both land and livestock can then be treated with the *treatment instrument*, engineered to restore vitality to the soil, eliminate pests, and cure cows of such diseases as mastitis or leukemia. With the dial set for a particular poison, and the right detoxifying agent in its "well," the instrument is also used to clean poisons out of feed, get the bad algae out of cattle tanks, and even, says Sarah, keep the barn from burning down when it is packed with potentially spontaneously combustible wet hay. The third of Galen's exotic contraptions, a *beam projector*, is designed to transfer a selected energy from one source to another.

If the farmer needs help in analyzing the status of his land, the Hieronymus organization will train him to graph the data so as to understand with their instruments the balance or imbalance of soil, crops, seed, or plants. He can also be shown how to remedy the health of chickens, turkeys, or other animals that have been feeding on poisoned foodstuffs.

The end result of the joint use of these instruments, Galen told us, leads to rejuvenation, enrichment, and revitalization of the soil, earlier fruition, increased yields, higher quality, more nutritious crops, healthier stock, reduction of pesticides and fertilizers, and a consequent increase in profits. Not only is the cosmic pipe designed to take in and redistribute amplified energy, it is also furnished with a "well" around which the descending wire is coiled and into which what Galen called a "reagent" can be placed—in fact, as many as fifty. The reagents modulate the wavelengths of the carrier "eloptic" energy, the same way a radio beam is modulated by voice or music to be broadcast. With their ana-

† See his *Tracking the Astronauts in Apollo 8,* September 4, 1968.

lyzer, the Hieronymuses search for an energy they like, then "transfer" it to a vial of oil or water, place it in the "well" and broadcast it to the land around the cosmic pipe.

Most amazing, the Hieronymuses had recently begun to broadcast the Steiner preps, as well as the barrel compost, by inserting them into the well. "After all," they said, speaking in satisfied unison, "the power of the preps is in the forces they contain, and those forces can be channeled through our pipes and broadcast with eloptic energy, just like any other force."

Sarah Hieronymus and Sara Sorelle (who helped greatly with the research for this book) standing by a cosmic pipe in the garden of the Oasis laboratory in Lakemont, Georgia.

There are now four different-sized cosmic pipes, a small one, which will cover forty acres, costing $500, two intermediates, and a super pipe for $2,500, which will cover 2,500 acres. In the fifteen different states where they have been installed they are also credited with affecting the weather locally, with less violent storms, gentle, more consistent rain, and milder temperatures.

"The cosmic energy manifests in earlier crops," said Galen, "in stronger stems and better quality. A couple of years ago we cleaned up an apple orchard, just about fifteen miles down the road. The land had been so heavily dosed with herbicides and pesticides and commercial fertilizers it had not borne fruit for years. The trees were in bad shape from dry rot. So we put the pipe there in January of 1986. In spring the blossoms were a gorgeous cloud of color, and when the leaves came out they were a lush deep green. The trees are vigorous now, with the most apples I've ever seen."

"The rosy apple aphids bothered us at first," said Sarah, "but we put a reagent in the pipe to resist them. We got hold of a bug and put it in the test tube, corked it and put it in the well of the instrument to find its energy wavelength, then found a substance it didn't like and reduced its vitality to zero. So we treated what it lives on and made it so unpleasant the bugs wouldn't stay. We also had good success with larvae by putting pictures of the trees in the well with something I found they didn't like."

Sarah paused to pick up a small vial of colored water with which to illustrate the process: "We took the leaves of the red geranium and transferred its energy into this vial of water and intensified it about ten times —the equivalent of diluting it and shaking it homeopathically. I put the vial with the cut-up geranium leaves in the well here, and neutralized the water on the plate before turning on the power about ten times. That charged the energy through this plate into the vial of water. To check its effectiveness I put a worm in a jar and began to treat it with the geranium water. The poor critter backed right out and lay down to die. So we did the same thing for army worms and tent caterpillars and gypsy moths, all very successfully."

To demonstrate to visitors the effectiveness of his reagents in disposing of such pests as nematodes, Galen attached a video camera to his microscope to project a living, moving image onto the screen.

"Instead of using pesticides," said Galen, "we can put a reagent in the well of a cosmic pipe that will radiate enough energy to keep the pests away. We put some Shaklee Basic-H in the pipe and got rid of a whole mess of flies. I can show you a barn with dozens of cows standing with their tails hanging down, and not a fly around, maybe three or four on the ceiling. And we've cleaned up the odor. If the cow eats clean fodder, without poisons, its feces soon lose their foulness, and the flies are no longer attracted."

"For a reagent," said Sarah, "we sometimes use the rules of homeopathy that the hair of the dog is good for the bite. We had to work on one place where the thistles were seven feet tall and the cows couldn't get through. So we put a leaf of the thistle in our beam projector and broadcast it to the field. In a year we had the thistles back to size, with a much smaller taproot."

Sarah and Galen Hieronymus standing in their orchard after it had been recon-structed radionically.

Compare with sickened condition of tree before their treatment.

Sarah displayed a photograph of their orchard, loaded with apples. "In our orchard we managed to keep the fruit from being caught by an early frost last spring," she added with pride. "We kept them from freezing by fortifying them with the essence of pine and spruce, which can withstand the cold. And Mark Moller, who has a farm in northwest Arkansas, at Pea Ridge, on top of the mountain where they had that great Civil War battle, was afraid an early frost might kill all his thirty-five acres of blueberries. When he heard a cold front was coming through, he flew down here to Lakemont and we made up a vial of different energies."

Her lips curled into a distant smile: "You may find it hard to believe. It sounds wild, and incompatible, but along with some liquid calcium he put the energy from a picture of the sun taken from the moon, obtained from NASA, and some Agnihotra dust along with its Sanskrit mantra. To get the mantra into the vial Mark played a cassette and ran the sound through a wire into the earphone in the well of the instrument. And at first it looked as if it had worked. By the time the frost arrived, his blueberries had already blossomed and had tiny leaves. All seemed well. Then the leaves failed to mature; so he got worried. When we tested out the leaves in Galen's analyzer we found they'd been subjected to too much heavy stress by the freeze. So we made up a vial with *aqua lithium* —it's produced by a lab in California at San Jose—and when it was broadcast to the plants it relaxed and de-stressed them. They grew beautiful deep-green leaves, and Mark had the best harvest in years."

Every time the Hieronymuses have sent out a pipe they have sent with it twelve reagents, one to increase soil productivity, one to eliminate pests, a soil detoxifier, a chemical detoxifier, and so on. For very cold climates, such as Wyoming, where the temperatures go down to 40 below, they "put their energies" into oil instead of water, so they won't freeze, just solidify without breaking the vial.

"But we never use any poison in the pipes," said Sarah. "Nothing that would lower the vitality of the soil or the plants, or hurt the elementals. The elementals have been displaced from too many places with all the concrete that's being poured, and all the poisons in the fields which cause birth defects in birds and animals and humans. The little fairy folk have had to move away."

The remark opened vistas into Sarah's character, explaining her close association with Galen, who might be described as a "sensitive." Sarah, who travels astrally like Robert Monroe and his cohorts, has accumulated hundreds of tapes of channeling to other dimensions, claims to read past lives, and has the gift of seeing people that have "passed on." She has been told that a special mark on her body denotes her having been a member of the order of the Divinensis among the Albigensians massacred by the church. She says she thought everyone saw "the little people," until her mother scolded her for dreaming. Ever since she was a little girl, she says, watching the raindrops in a puddle she could see the rain elementals, very small, just their outline, silvery and very graceful, misty figures with nothing on their heads like the green and brown ones in the pasture or the woods, different also, she explained, from the ones in the sunlight, or in the leaves of trees, or in the fireplace, changing shape all the time, with bright colors, tiny little things that dance a lot. "You can pick up their thoughts," she said, her Cherokee features distant but warm. "People who can't see the little people think it's all a lot of bunkum. But then they can't see an electron or a proton, yet they believe in them because they're taught that everything that science tells them must be the truth. The smaller creatures in the garden or the grass *are* harder to communicate with. Aeolus, the wind, is a very powerful elemental. With his help you can change the path of a tornado. It's like the healing touch, it's available to anyone, but not everyone knows that it's available."

Sarah said she had asked the elementals for help with the cosmic pipes. "Once you get the pipes set up and spreading energy, the little ones are very happy. They know that you're collaborating. And the energy is what they need to do their job with the colors and the shapes of flowers, with the maturing of the fruit, the gestating of the seeds. There are many grownups now who are beginning to think about these elemental forces and to try to learn to contact them and make a good environment in which we can learn to get along together."

To ascertain whether Sarah and Galen with their cosmic pipes were really onto some old magic, or whether it was all a flight of fancy, we set off to visit an orthodox chemical engineer, replete with the requisite academic background, who has been experimenting with the cosmic pipes for several years.

Nestling in a typical Ohio landscape a few miles west of Lake Erie, just north of the Appalachian foothills, in a flat expanse known as the Firelands—because the British burned the early settlers' homes—a luxurious biodynamic oasis burgeons with healthy produce of all sorts, including sixteen-foot cornstalks growing alongside apple trees that kneel beneath their load of ripening fruit. Where Huron Indians used to play, diesel trucks now roar on their way to Cleveland or Toledo. Unperturbed, Harvey Lisle has set up two Hieronymus cosmic pipes directly over ley lines, one running east-west, the other north-south, with an Indian medicine wheel at the intersection.

A sprightly, outgoing, and lovable gnomelike character with curly gray hair parted down the middle, and a shaggy salt-and-pepper beard, Harvey, in his sixties, has been in biodynamics since the 1970s, when his cunning wife, Louise, placed an anthroposophical text where he could not help but stumble on it. Quit of his job as a chemical engineer for a company producing chemical fertilizers, Harvey now swears by both the BD preps and by the Indian circles of stone he places around each of his many types of fruit tree.

In his cellar, among bushel baskets of early Wealthy apples, and enough glistening jars of fruit to outlast a holocaust, Harvey's lab is set up for the production of sensitive crystallizations and chromatographs—such as were made by Pfeiffer and Kolisko—as a means of checking the forces and qualities in both BD 500 and BD 501. Used either directly in the soil, potentized by homeopathic stirring, or simply transferred to plain water through the Hieronymus instrument, the results are surprising, especially when broadcast via a cosmic pipe.

Ley lines along the earth, he explains, are like nerves in human beings, allowing the energy to move from point to point. "All my stone circles—about a hundred of them on my little eight acres of land—are connected to each other, to the medicine wheel, and to the cosmic pipes. Altogether they make for a powerful center. Energy all over the place! That's why everything grows the way it does, without chemical fertilizer, and only some BD 500 and BD 501 which I put on several years ago. Now I broadcast the preps through the Hieronymus pipes, and it works. But I'm playing around with an even more extraordinary system."

With a V-shaped plastic dowsing rod, held together at the point by a red electrical wire nut, Harvey set off into his jungle of fruits, vegetables, and flowers, to demonstrate how it works, and we had to follow quickly for fear of losing him in the underbrush or having him vanish into one of his own stone circles.

On our way beyond a shallow creek that meanders through the property we came upon a row of blueberries, about six feet high with small cakes of soap attached to them by wire. "Keeps the deer away," said Harvey. "They won't touch the blueberries if there's soap around."

Opening the well of one of his cosmic pipes Harvey revealed its incongruous contents: next to several leaves of poison ivy—by means of which he hoped to rid himself of its luxurious growth, brought on, no doubt, by the excellence of his BD preps—were some baked larvae of the plum curculio, a pest that digs into the heart of plums, but which, thanks to the cosmic pipes, had managed that season to bore into the fruit only about a quarter of an inch before dying. "Next year," said Harvey, "we hope to have the little critters totally licked."

Harvey Lisle with a sixteen-foot stand of corn achieved with biodynamics and Indian medicine wheels of stone laid on ley lines.

From the well of the pipe he also drew a vial about two inches long of what looked like sand. "It's the remains of about a hundred slugs," said Harvey, an impish grin enlivening his eyes. "They were burned to ashes on a charcoal fire. Last year we had slugs by the tens of thousands. This year, after the burning, we had them by only the hundreds. Within a year or two they should no longer be a problem. Funny thing, that's only on my farm. On the other side of the property line, my neighbor still has them by the tens of thousands, just as we used to."

Harvey turned toward a thicket of oak trees in the west. "When the sun goes down," he said, gesturing toward a russet sunset, "the energy goes out of these cosmic pipes, as if someone had turned off a switch. At sunrise it comes right on again."

Inspecting some shriveled leaves on a quince tree, he explained that he was using a Hieronymus beam projector to see if he could improve the quality of his quinces. In one well he had placed the leaves of three quince trees, in the other a piece of his favorite pear tree. As the Hieronymuses claimed to have successfully transferred the flavor of pecans to acorns, Harvey was intent upon trying to transfer the quality of one of his best pears into the quince trees he had planted at the bottom of his garden.

Harvey Lisle placing ground-up slugs and poison ivy in the well of a Hieronymus cosmic pipe together with a reagent to eliminate both slugs and ivy from his land.

"The quince," he complained, "is too hard to eat. But both the quince and the pear are of the same rose family. You can graft either one to the other. Wouldn't it be wonderful to be able to pick a quince—today almost an unknown fruit—right off the tree and eat it like a pear?"

The idea sounded great, but by the look of the quince there seemed to be some resistance still.

To several other trees Harvey had attached "French coils" which he surveyed unhappily, claiming that instead of increasing the trees' energy, as recommended by certain dowsers, the coils interfered with the flow, depriving the trees of energy, so that some of them had developed a blight, only recovering after the coils had been removed.

With a small pendulum Harvey demonstrated how to find the energy "door" to a tree, a door which all trees have. "Block and Davis," he explained, "discovered the neutral spot in a bar magnet. By dowsing we've discovered that a tree has the same thing. A tree is bipolar: the top, with its branches and leaves, is south, or positive; the root part underground is north, or negative, and there's a small neutral block in between, a few inches up from the ground. And there's your door."

A luscious Lodi apple tree swayed gently in the breeze, loaded with succulent fruit. Kneeling by it, Harvey dug to uncover a small plastic film container, half filled with BD 500. "I buried it about three inches, down close to the roots, on the east side of the trunk, because the earth energy runs east-west." Pointing at the bark about a foot above the ground, he explained: "Here I've strapped this tiny crystal made of quartz, pointing up to catch the cosmic inflow. I'm going to take them both off now, and we're going to dowse to see if any energy is coming off this tree. Then I'll put them back, and you can see for yourselves just what I'm talking about. Then I want you to do it too, so you'll know that it's all for real."

The rest of the afternoon was spent in the orchard with a Melrose apple tree experimenting with potentized biodynamic preps and Hieronymus facsimiles of them in vials of plain distilled water. In each case the tree radiated force into the surrounding garden on an approximately equal scale; organic and facsimile appeared to be equally effective.

As predicted, the rod dropped only at a distance of about ten feet from the tree, as it picked up the tree's own energy field. With the 500 buried by the root, the rod went down with force a little over a hundred feet away, and when the 501 was added, about another six feet further. "It looks as if the tree is acting like a cosmic pipe, relaying the forces out around it, just like a Galen Hieronymus pipe.

"The question," said Harvey, "is whether the forces are actually *alive* in every case, the same for the organic substance as for the facsimile in water. We have to know whether the facsimile carries both life and force. And that we can find out only by making some crystallizations of each and every one."

Thirty-six hours later, including one spent stirring the 500 and another stirring the 501, the results, as Harvey predicted, were clear enough to see.

In the crystallization made with the potentized water from the BD 500, not only are signs of energy clearly visible, but also the qualities recognized throughout the years by Pfeiffer, Kolisko, and others, as being attributable to "life." A chromatograph of this same water shows the typical outlines of living organic matter.

On the other hand, the crystallization of the distilled water energized by the Hieronymus instrument does show the force, as can be easily compared with a crystallization of just plain water, unaffected by the instrument. But the crystallization lacks the indications of vitality. It has none of the strong, clear lines radiating from a center to be seen in the organic water. Instead its lines are murky and amorphous.

From all of which it can be deduced that Hieronymus's cosmic pipes may well be broadcasting the energies contained in the potentized BD

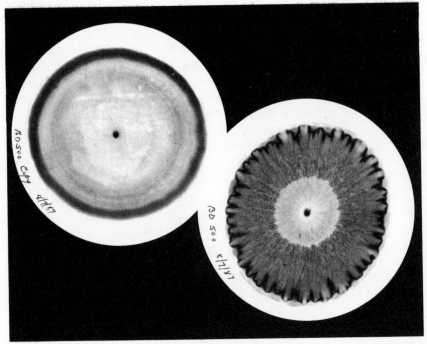

Chromatograph made in Harvey Lisle's cellar from BD 500 stirred for one hour *(top)* and a radionic copy, showing far less vitality. (Credit: Harvey Lisle)

preparations. But somehow in the transfer of these energies via the eloptic analyzer on the beam projector, the "life" in the preps seems to get mangled. Harvey Lisle assumes that it may be because of the electricity used in making the transfer. As the Hieronymuses were well aware, and ready to exploit when they wanted to neutralize a substance, a quick shot of magnetism given to a vial will cause it immediately to lose its power. "The same thing," says Lisle, "may be happening to the 'life' force when subjected to an electric current."

What remains is the fact that the cosmic pipes do broadcast some form of creative energy, picked up with a dowsing rod or with the Hieronymus instruments, as testified to by scores of happy farmers who are growing healthier crops without benefit of fertilizers other than the BD preps, whether they are using the substance itself or the water from it potentized by stirring.

But without making a big issue of it, indeed rather preferring to play the subject down, unless strictly among their own anthroposophs, most biodynamic practitioners admit that a whole mechanic in the transfer of energy into plants, and their successful fertilization, has been left out of the equation: the whole world of nature spirits described, among others, by Paracelsus, Blavatsky, Steiner, Leadbeater, and most enchantingly by Geoffrey Hodson and, at Findhorn in Scotland, by Ogilvie Crombie. All those sensitively involved with the problem attest, without exception, and without the slightest qualification, that in the growth of plants a vital and essential role must be played by the denizens of the elemental kingdom, and that until they too are duly taken into account, as does Sarah Hieronymus, nothing in the world of agriculture, horticulture, or gardening will ever make much sense or progress.

CHAPTER 23

Perelandra

I<small>N THE</small> 1970s the world discovered the community of Findhorn, along with its other world of nature spirits, devas, and the Great God Pan. Some were quick to accept as authentic the messages the sensitive Dorothy Maclean claimed to have received clairaudially from the devas responsible for the remarkable fruits, vegetables, and flowers that flourished so well in that remote and chilly wilderness of Scotland. Others took it as romantic fancy, which might, at best, serve to awaken mankind to a more sensitive approach to the plants it depends on for its life. Actual communication with the nature spirits, a harmless enough performance for the fey entourage of flower people of the sixties, was certainly nothing rational people could pretend to indulge in for themselves.

Now comes a young lady living in the wooded foothills of Virginia's Blue Ridge Mountains who says that anyone can communicate with devas, nature spirits, and the Great God Pan, by the simplest of devices, which she is happy to impart to less spiritually motivated mortals for the asking. It's time, she says, in a more serious vein, that we try, if we want to heal the planet of its ills, to communicate and collaborate with the forces of nature whose function is to do just that.

Her name—as exotic as her pursuits—is Machaelle (pronounced "Michelle") Wright Small and in an open field by her woods she has created an extraordinarily beautiful garden called Perelandra (after the lyrical book by C. S. Lewis), entirely, or so she says, at the direction of her insubstantial teachers. What's more, she now regularly gives workshops to impart to other aficionados of the garden from Oregon to Florida, and from New Mexico to Maine, how to communicate directly with her diaphanous friends.

Her premise, for which there is a strangely growing support around the world, is that only by collaborating with the Third Kingdom, that of the elementals, the world of nature, is there any hope of cleaning up the pestilential miasma that has enveloped this planet, and by healing it, heal all the life that it sustains. The same dream motivated the supporters of such movements as Jose Arguelles's Harmonic Convergence. The same message has been preached by Paracelsus, Blavatsky, Leadbeater, Besant, Hodson, and especially Rudolf Steiner. Only, Machaelle may not have read their books. She gets her information from the *source*, where anyone on earth, she insists, can get it, by simply following her method, though not necessarily the ordeals that led to her awakening.

As a young girl, suffering from dyslexia, with a drunken mother, tyrannical father, and a vengeful stepmother (who threw her into the street in her early teens without money, home, or job), she found herself split into two people: one outwardly somber, trying to survive; the other inwardly joyful, powerful, life-giving.

Meditation led her to astral travel, in which "instead of empty space, I saw before me forms, like a group of people so out of focus that I could barely make them out. I felt I was looking at them through a window, and suddenly I realized that I had come home—my real home. I didn't know where this place was, but I knew, without a doubt, that it was home and that I had left home to come to Earth."

In her astral traveling Machaelle learned that when out-of-body she could be of service to others, giving quick assistance to embodied spirits in distress, such as a priest on a train in Yugoslavia whom she managed to comfort, spiritually, just before the train crashed and he died. "Each time was different, and I learned something new about the amazing, unseen, complex activity that goes on all around us all the time."

Closer to daily life, she helped form the Community for Creative Non-Violence, convinced that ecology was one facet of nonviolence: "the destruction of nature being the destruction of man himself, the quality of man's existence being directly related to the quality of his link with nature."

Moving to live in the woods brought more rapid developments. "I first noticed something different about the woods when I was alone in the house at night. I could feel an energy, and at night it intensified to the point where I would feel uncomfortable about walking in front of a window, or by the double glass doors—especially during the nights of the full moon. My uncomfortableness wasn't fear that I was going to 'get gotten' by something or someone. Rather, it was more a response to being surrounded by intensity—an air of intensity. Nothing hostile."

In 1974 Machaelle discovered books about Findhorn. "All of a sud-

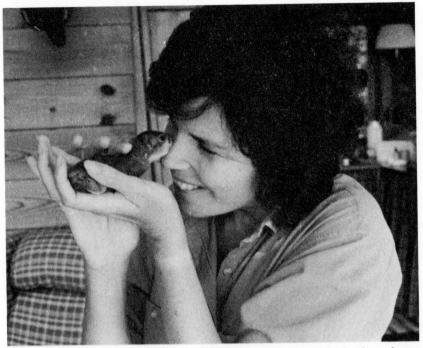

Machaelle Wright at Perelandra with one of the friendly chipmunks that have the run of her garden.

den, I was being told that the vague energies I had felt around me at Perelandra actually had names. Devas. Nature spirits. They weren't created from my imagination. They actually existed! What I had felt in the woods was a life force that now was identified and could consciously be worked with."

By early January 1977 her commitment had been made. "I walked into the woods and announced in a loud, clear voice: I want to do at Perelandra what they did at Findhorn. I want to work with devas and I want to work with nature spirits. . . . Then I left the woods, returned to the house, put myself into meditation and waited."

Machaelle admits that at the time she did not really know what she was doing. But the response was immediate. "Like Dorothy Maclean at Findhorn, a crowd of voices came at me. And only when I asked that they communicate one at a time did they respond, to my amazement. From the meditation standpoint, I found the devic level to be a level of consciousness very high in vibration. As if someone were to hit a bunch of tuning forks and we could distinguish the vibratory difference between them rather than the different sound. . . . It resembled nothing I had experienced in meditation previously."

And so Machaelle's collaboration started with the members of the Nature Kingdom. "I was given instructions. I was told what seeds to buy. What fertilizer to use. How far apart to plant the seeds. When to thin the plants, and how much space to leave between them."

As each deva came into her awareness she says she noticed a slight shift in vibration, until after a while she could recognize them as they entered her awareness. "One day, I felt a very different vibration, and found myself connected to the Overlighting Deva of the Garden."

Machaelle says that because she tends to see the reality around her in terms of energy rather than form she experiences the nature spirits as swirling spheres of light energy. "Out of consideration for me, when the nature spirits choose to be visible, they choose a context with which I am comfortable—energy. I know that they do appear to humans in the form of elves, fairies, gnomes, etc. But I think that's only to people who are comfortable with these concepts. To manifest that way, they make use of our own thought forms."

As we sat in the shade of a young oak by the edge of the field in which her garden thrives, only a few feet from a roped-off sanctuary, off limits to all but the nature spirits, Machaelle laughed openly: "They've taken some bad raps because of our mistaking them for our own thought forms. They are extremely powerful entities, responsible for the existence of all that has form around us, and at the blink of a flea's eye, they can remove that form."

It was then that Machaelle, tall, good-looking, in her early forties, dark hair held back from the midsummer heat, cheerfully showed us how to get in touch with her invisible companions. "It's easily done with a form of kinesiology," she explained. "Hold your left little finger to your left thumb. Into that circle thrust the thumb and index of your right hand, pinched together. Then ask a question, any question, providing it can only have a yes or a no for answer. Push up against the circle formed by your left thumb and little finger. If they part, it's no. If they stay firmly shut, it's yes. You'll be amazed how quickly you catch on, and how quickly you'll be in touch with the world of elementals all around you."

Elaborating on the science of kinesiology Machaelle explained that it is in no way mystical or magical, but that when any "negative" is placed within a person's field—his electrical system, the electrical energy grid contained within his body—it will immediately respond by "short-circuiting," making it difficult for the muscles to maintain their strength and hold their position as more pressure is added. Conversely, when a "positive" is placed within the field, the electrical system holds, and the muscles are able to maintain their level.

"If you ask a question using the yes-or-no format," she added, adroitly demonstrating with her fingers, "the elementals can answer your question by transferring a yes (positive) or a no (negative) into your energy field. You can then read the answer by testing yourself with kinesiology." (Further instructions are in the How To Appendix.)

But our first preoccupation was with finding out about the garden in which she appeared to have outstripped Courtney, Podolinsky, Hieronymus, or even Lisle, producing, as she put it, enough extra food to feed Philadelphia, all from about a quarter-acre plot, without benefit of Carlson, Hamaker, or BD preps, either sprayed or broadcast from a cosmic pipe. Nor does she water her garden, except at the moment of planting. Thereafter, not a drop, even in the summer of 1986 when the Federal Government declared Virginia (and several eastern states) an agricultural disaster area because of the drought, or the equally bad summer of 1988. While all around her well-mulched gardens withered in the sun, Perelandra stayed fresh and bright, leading her neighbors to suspect her of being a witch.

The garden, a hundred feet across, is laid out in a circle, with concentric rows of luscious vegetables and brilliant flowers in the most amazing variety and juxtaposition. In the center, in an area specially reserved for birds to feed and bathe, stands a wire structure in the form of a "Genesa" crystal, an antenna device made of four copper circles, two feet in diameter, designed by plant geneticist Derald Langham "to attract to it the life force energy from all form within its range, cleanse this energy, and spin it back out into the environment." Beneath it lies a gemstone, all on a slate above a Flanagan pyramid tensor. The haven is also surrounded by a circle of stones, like Harvey Lisle's, to create a zone of power.

"The focal point," says Machaelle, "has changed in content, but not in position. It used to have a natural white quartz crystal in the center; but then I managed to get hold of a much more precious jewel."

Three spiral paths lead from the periphery to the center. A fence surrounds the garden to keep out the neighbor's horses and cows, but not the animals of Perelandra, which are free to come and go as they please, including rabbits, moles, deer, and any kind of insect. No foreign repellent or insecticide is used, organic or other.

"I maintain the garden on the principles of energy," said Machaelle, speaking with the intensity and single-mindedness of Bernini's Santa Teresa receiving the arrow of love directly into her bleeding heart. "This gardening, this co-creative energy gardening, is a metaphor for life. As you change your approach to the garden, you will, in turn, change the very fabric of how you approach your life. The Perelandra garden is my

The Perelandra circular garden laid out in the northern Virginia hill country. (Credit: Machaelle Wright)

life, my heart, my breath. It is my friend, my helper, my nurturer, and teacher—especially of myself—my planet, my universe. It is my key to the universe. It gives me access to spiritual truth and universal natural law contained within the universal flow. It is the demonstration of these truths and laws played out before my eyes. It is my proof that what is spiritual truth and universal law courses through all of reality—and this includes a garden."

And some creative miracle must certainly be being played out in the beds of that small garden, for Machaelle's method of fertilizing her soil is more unbelievable even than that of Podolinsky, Carlson, Howearth, Hieronymus, or Lisle. She has made up what she calls her "soil balancing kit," a collection of tiny packages of bone meal, rock phosphate, Nitro-10, greensand, cottonseed meal, dolomite lime, kelp, and essence of comfrey—all on instructions from her devic friends.

"People don't realize that without the proper nutrients the physical aspect of the plant receives no life-sustaining support. It must have nutrients."

These she holds out, one at a time, just a pinch in the palm of her hand, and asks how much of each is needed by the soil. As she gets a response she requests of the appropriate nature spirit that it receive the energy from the nutrient in her hand and shift it in an appropriate amount to the appropriate depth in the appropriate section of the garden.

"You may feel an immediate sensation in your hand or a change in the nutrient you are holding," said Machaelle, quite matter-of-factly, as if describing a recipe for fudge rather than some quintessentially metaphysical feat. "Hold your hand out about ten seconds. Once complete, just drop the nutrient on the ground. Don't try to save it because it is nothing but form without energy, and is no longer useful. At this point, the nature spirit is shifting the energy of the substance you are providing from the form in your hand and using it as the base to expand the energy to the amount needed. The nature spirit will reinfuse that expanded energy into the ground in the proper concentration and at the needed depth. Once the energy is in place, the nature spirit you are working with will shift the energy to the form level. In short, by using the expertise of the nature spirits, you can infuse thirty tons of greensand over a one-square-mile area to a depth of five feet, and it all starts with a teaspoon of greensand in your hand."

Machaelle smiled happily, then gazed again at the beauty of her garden. "When you've done, spend a moment sensing the land you have just balanced. Note any changes. If you sense nothing, don't be disappointed. The effects, because of the nature of this work, may take a few days or weeks to show or be felt. Over a period of time, you will note a change in the garden balance, the devic information received, how everything in the garden weaves together more easily. You'll see the garden rhythms change. Things that an outsider wouldn't even notice, but the very things a person working intimately with the land will pick up immediately."

Machaelle explained that if they wanted to the nature spirits could easily find and move energy about without relying on us. "But that would defeat the purpose of the *game*. They need us as much as we need them in order to have a co-creative endeavor. Although nature is powerful beyond imagination, and humans are powerful beyond imagination, man and nature together hold the promise of many times their individual power. A potential of this union is the creation of the earth's own healing energy grid through its gardening system around the planet. And the healing power which will radiate from the gardens and ultimately from the grid formed by the link-up of the gardens will be equally available and usable to both humans and to nature, because it was created by humans and nature united."

Machaelle's approach to insects turned out to be as amazing as her approach to fertilizing her garden. "If man is to sensitize himself to the communication of the insects, it is important that he view them as messengers of a problem, not the problem itself. When dealing with disease in humans or nature, the insects in question may be conquered or controlled by man with his technology; but the disease itself will not be eliminated until the underlying reason is addressed. In areas where insects appear to be troublesome or out of control, I suggest you draw back and look at the larger environmental picture for the answers you are seeking. If you still have difficulty identifying the real problem, look at precisely how the insects are interfacing with human or natural form: there lie the clues you'll need for the answers to the overall picture."

Machaelle headed toward a smaller circle, about twelve feet in diameter, entirely devoted to tomatoes.

"I have found," she went on, delicately picking a beetle from one of the plants and gently placing it onto the ground, "that in the garden insects function as quick dispatchers of communication. If I see a plant or row suddenly overwhelmed, or seemingly overwhelmed, I'll open to the appropriate deva and ask if the plant balance is off. For example, I may find that a particular rosebush is covered with aphids. When I first inquired, I was told not to panic, just do the monthly fertilizing as planned and that would rebalance the bush. And it's true: once I've done the planned fertilizing, the aphids leave the bush within twenty-four hours."

From the tomato circle she moved to another circle filled with flowers, her tone becoming more serious. "Everything can be going along just fine in the garden and all of a sudden, out of nowhere, there's a horde of something eating three rows of vegetables. I have learned that when this happens more often than not it is because there has been a sudden and dramatic shift in thought, intent, or emotion, either with the gardener, or with the family or community connected with the garden. When it comes to ungrounded, raw, emotional energy released by humans, nature functions in the role of absorber. Even though emotional energy is invisible, it is no less tangible in its effect on the world of form than insects, heavy rain, or drought."

The reason for Perelandra's success with insects became clearer as Machaelle headed for a shaded outdoor table, where some cooling herb tea sweated slightly in the breeze. "Since the beginning of my adventure, I have tithed ten percent of the garden back to nature, just on general principle. To be frank, I don't believe I have ever fully been taken up on that tithing. What I've observed is that where animal and plant interface it is with the softest of touches. At the same time I've felt an air of aggression which hung over the garden gradually dissipate and eventually disappear altogether. Then I realized I had removed my attitude of aggression toward the animal kingdom when I changed my mindset;

and this, in turn, changed the collective attitude with which the animal kingdom interfaced with the garden. They no longer had to fight for life. They could exist within a natural environment without fear of reprisal. What's more, it set into motion the creation of a new balance—one in which the quantity and quality of activity was increased manyfold. By including all members of the chain of life who belong within the balanced garden environment, one encourages and enhances the quality and intensity of life energy within that chain—and within the environment as a whole."

A perfect Perelandra cabbage, radiant with the life forces administered to it by the nature spirits, as photographed and described by Machaelle Wright.

Machaelle's description of how she handles the cabbage worm is wiser perhaps, and certainly kinder, than luring them into sexual traps, however appealing the bait and painless the death.

"When my cabbage, broccoli, cauliflower, and Brussels sprouts," said Machaelle, "became heavily infested with cabbage worms, a common problem in the area, I connected with the Deva of the Cabbage Worm and announced I wished to give one plant at the end of each of the four rows to the cabbage worms, and I requested that the worms remove themselves from all the other plants, except for the four which I had designated."

Machaelle smiled happily: "The next morning all the plants in the four rows were clean of cabbage worms—except for the one plant at the end of each row. What surprised me most was the amount of cabbage worms on the end plants. Each had only the number of worms it could comfortably support; the rest had simply disappeared! Birds, wasps, and various other creatures had been feasting on the abundant cabbage worms and the plants were left to continue their growth without missing a beat. In less than seven days the infested plants had 'healed,' leaving no holes in their leaves; and by late summer even the designated plant had formed a perfect four-pound head."

A tall young man with a thin mustache and the gentle expression of a person who cares for what he does, came wandering up the path from the house a quarter mile away. It was Clarence, ex-Paulist seminarist, who had given up the cloth to live and work with Machaelle. "Then, just as our corn crop began to tassel," she went on, waving to him to join us, "it was attacked—ravaged is the word—by Japanese beetles. They ate the pollen and demolished the silk. But, based on my luck with the cabbage worms, I decided to contact the Deva of the Japanese Beetle. Much to my astonishment, I touched into an energy that I can only describe as that of a battered child. It was an energy of defeat, of being beaten into submission. Yet it still had mixed in with it anger and a manic desire to fight for its life. I was told by the deva that what I was experiencing was not devic but the consciousness of the Japanese beetle itself. So I simply asked that the beetle recognize Perelandra as a sanctuary and invited it to join us so that it could begin to heal. I stated that we would not damage or destroy the beetles. To seal the bargain, I promised to leave unmowed a specific area of tall grass that was a favorite of the beetle."

The sun was almost onto the crest of the Blue Ridge, casting long, cool shadows from the woods that reached almost to the garden while the first of the evening swallows darted after insects in the amber light. "I addressed the issue of the corn," said Machaelle, "still hoping to salvage some. I decided to try to raise the vibration of the individual stalks. Perhaps the ears would fill out despite the Japanese beetle. I spent three days putting my hands on each stalk and loving it. At the end of three days, the nature intelligences had had enough of this nonsense and I was told to leave the corn patch and not return until further notified. Devas and nature spirits do not respond to what they call gooey, sentimental love. Their love is of action and purpose, the kind they desire from us."

Clarence smiled with evident satisfaction. "For three weeks she stayed away from the corn patch, until one morning she was told she could return. Every ear of corn had filled out—not fully, only half. She was told it was all that was needed to feed the birds. A later planting, not damaged by the beetles, would be for our exclusive use. A month later it matured, untouched."

"It was the same with my roses," Machaelle added. "The beetles would take one or two flowers on a bush and leave the other ten for me. After watching this process year after year, I have been able to change my attitude about bugs from focusing on what they attack, or weaken, or damage, or destroy, to focusing on the gift each insect offers to the countless other members of the garden.

"This, in turn, has allowed me to see the insects' right to be part of the plant kingdom. It may sound odd, I'm sure, but I see the insects as not only an integral part of the environment, but also as part of the garden crops. I encourage their health and vibrancy as I would anything else in the garden. And I look to the garden to draw to it and support a balanced and full population of insects, which, in turn, helps to support the overall life of that environment. I was told, for instance, to plant costmary in the herb ring, a herb I personally do absolutely nothing with, though I've heard it is used for medicinal teas. Each year I dutifully fertilize it, and early each spring it becomes completely covered with a billion aphids. About a week later, almost the same number of ladybugs appear in the costmary. Not long after that, there are no aphids, and the ladybugs have scattered about the garden. The costmary in the Perelandra garden is obviously intended as a breeding ground for ladybugs."

Machaelle—like that sensitive and astonishingly literate Scot, R. Ogilvie Crombie, better known as Roc, friend of Findhorn and possessor of a library of arcana in Edinburgh worthy of being the reincarnation of Queen Elizabeth's wizardly astrologer, John Dee—describes making contact with the Great God Pan. As ruler of all the nature spirits, he conveyed to her, as she understood it, that the overriding intent of any garden is to be of service. "Without service," was the message, "there would be no need to create such an environment. Prior to harvest, the gardener is of service as he works to assist the garden into form in balance. During this period, the garden serves humans in return by creating an environment which shifts and heals all it touches and enfolds. This is true healing service. But the garden's full capacity to serve begins with the harvest. At this point, the human experiences through harvest and ingestion the full notion of serving on all levels. And the partnership that was formed from the very instant he set foot upon the planet is celebrated by man and nature on all levels. Those humans who work with nature in the spirit of co-creativity have acknowledged and successfully demonstrated the link they have with the nature world both physically and spiritually, and nature has responded in kind by producing food and fuel for the physical support of the human spirit on the planet."

In unison, says Machaelle, and as a parting shot, the Overlighting Deva of the Perelandra garden, the Deva of the Soil, and Pan all told her that in order for man to be physically, emotionally, mentally, and spiritually supported, as he evolves into the Aquarian Age, he must, in partnership with nature, work toward full ecological balance.

The sun was down as we headed for the house. Machaelle walked ahead slowly, showing the way, thoughtfully kicking a pebble with her sandal. "They're real, the nature spirits," she said quietly. "And they may take many forms; but one thing they never are. They're *never* cute. They seek a co-creative partnership with humans, and they are in the position to accept no less."

Epilogue

IN EARLY MARCH of 1988, as a late-afternoon dusk was settling over Moscow, we emerged from a downtown subway station close to the Arbat, heart of the city. Along intersecting Vesnina Street, we slip-slid for three blocks over powdery snow underlain with a layer of ice so gray it resembled a coating of ashes, to arrive in front of a four-story nineteenth-century Tsarist building in the process of renovation.

Though no identifying sign proclaimed to passers-by what might be taking place behind its massive wooden door, well padded around the edges to seal off blasts of late winter wind, we knew it to be the home of a newly born Soviet "think tank," attached to the U.S.S.R. Academy of Sciences, innocuously and unenlighteningly named *Otdel Teoritiches-kikh Problem:* Department of Theoretical Problems.

A ring of the doorbell admitted us to an all-but-pitch-dark foyer where a dank flight of stairs led up to a spacious, well-lit office and the greetings of a short, wiry, nattily-dressed young man with a wolfish smile. A head of hair the color of dusty straw topped an aquiline face in which the eyes still betrayed the three-hundred-year occupation of the Mongols. Erast Andriankin, a combination of gracious Russian host and wily Soviet administrator, alerted to our arrival via a telephone call from Prague, contemplated us with a mixture of unfeigned curiosity and puzzlement.

The call had come from a Czech philosophy Ph.D. turned experimental scientist, Zdeněk Rejdák, president of the International Association for Psychotronic Research, which, despite myriad Czechoslovak strictures, he had managed to found in 1968.*

---

* *Psychotronics*, a word coined by a French engineer more than a quarter century ago, generically refers to the action of pure mind or *psyche*—regarded by the ancient Greeks as ephemeral as any butterfly—in the same way that its semicognate *electronics* refers to

314

In Prague, Rejdák—who believes it is only modern scientific overspecialization that is blocking new investigations into unrecognized human abilities and unused mental reserves—confidentially informed us that the Otdel in Moscow was harboring the possessor of a mysterious power to increase the growth of plants. "It is time," he told us, "to look deeper into things which might help people come into harmony with the bigger world around us, to cohere with living processes rather than just dead matter. Go to see Andriankin. I think some surprises might await you in his shop. I'll call to say you're coming."

In the Otdel office, half a dozen of Andriankin's collaborators were sitting around a boardroom-style table at the foot of his desk. They included a young and reputedly brilliant math and physics "whiz kid," Andrei Berezin.

Half addressing us, half addressing his collaborators, Andriankin cleared his throat: "Our efforts here may come to an end at any moment. We could be closed down any day. What we are doing is alien to a system built up over the years. Our research is a thorn in the side of senior scientific officials, who look upon it, at best, as nonproductive; at worse, as outright flummery."

With this warning, the assembled company rose and trooped out to another flight of stairs that led to a dark corridor and a half-opened door. Crossing its threshold we entered a high-ceilinged room, no more than ten feet square, suffused with eerily flickering shadows cast by the light of two small tapers. An aromatic trace of incense, burning somewhere in a hidden corner or crevice, wafted into the surroundings to mix with the sonorous tones of an *a cappella* Russian Orthodox church choir broadcast from a large black stereo cassette recorder.

On a low end table next to a three-seat divan lay a heavy black-bound Russian translation of the *Holy Bible* printed in the Soviet Union in 1983. Hanging on all four walls were icons depicting Christ, the Virgin Mother, and saints from the Russian Orthodox pantheon.

At first sight, it felt as if we had been transported by some Wellsian time machine back into the nineteenth century. In the gloom of the strange audiovisual setting, it would not have been surprising to spot the hulk of a black-bearded *svyashchennik* in contemplation of his breviary.

Instead of any clergyman, we found a petite and attractive blond

---

electrically-induced action. Like its precursors, *parapsychology*, and its even older brother, *psychical research*, it welcomes into a kind of scientific foster home a host of phenomenological orphans, among them dowsing, clairvoyance, telepathic communication, and psychokinesis—the apparent ability to mentally move, distort, or otherwise affect material objects in ways inexplicable by known physical means. All these *Wunderkinder* are vehemently denied as bastards by rationalists and scientific pundits, who see in them no more than the symbols of unsubstantiated mysticism.

woman seated at a small desk across from the doorway, her eyes as orientally slanted as those of her obviously Tartar forebears, who must have gone back perhaps as far as the hordes of Batu Khan and the endless waves that swept from the inner reaches of the Gobi Desert all the way to the gates of Krakow.

"I want to introduce you," said Berezin, raising his voice, "to our charming collaborator, Alla Kudryashova, one of our most talented extrasensors and a gifted natural healer."

The title *extrasensor*, another Soviet neologism, defines, more accurately than its Western equivalent, *psychic*, any person endowed with ESP.

Rejdák had forewarned us about an extraordinary woman, somehow connected to Andriankin's effort, who he said had been found, under the strictest official testing, to be able to raise the harvest figures for certain field crops by over 100 percent above normal, affecting them in some way merely by her "presence"—in the extended, figurative, rather than the narrower, material sense of that word.

As our group milled about the room, the muted church music was drowned by a cacophony of several animated conversations taking place at once, making it difficult for a foreigner, who might normally have followed a single Russian exchange between two parties, to make much sense of all the hubbub.

It was then that we noticed huge sprays of wheat and triticale affixed to the walls or standing in urns around the room, their dried heads burgeoning with golden grain. Mixed in were spruce boughs and the sere remains of meadow plants.

"Why the profusion of grain?" we asked.

A collective smile rippled through the room. "It is an example of Alla's incredible power to energize and stimulate seeds before planting," said Andrei Berezin. "It greatly increases their growth."

Bit by bit it was explained to us what Alla could accomplish. A gifted therapeut, or natural healer, her powers were not as publicized as those of Djuna Djugatashvili—reputed to have extended the lives of such failing Soviet leaders as Brezhnev and Chernenko, amassing in the process quite a fortune—but were considered equally effective.

"When did you discover your healing powers?" we asked Alla.

"About ten years ago," she replied. "And it happened quite by chance. My father had an incurable illness associated with low blood pressure and bad circulation. One day, to his horror, he found he couldn't walk. I said to him: 'Daddy, let me massage your legs,' though I really didn't know what massage was. I just moved my hands over his torso and his trouser legs as he lay on his bed fully clothed.

"Only a few seconds had gone by when he exclaimed: 'Alla, my blood feels as if it's rushing along to keep up with your hands.' This was the first time anyone had suggested I was endowed with a special skill."

A week of this treatment and her father was able to take long three- to four-hour walks with his granddaughter.

"It wasn't all that easy," Alla cautioned. "You can't think of it as a machinelike process that turns on and off with a switch. I know a lot about machines. When I was just out of high school, I was plagued with a kind of inner dissatisfaction about my purpose in life, my reason for existing here on the planet. I was tormented by the overall question: What is our human calling?

"Since no answer came, I followed a path of least resistance and entered a metal-working trade school, graduating as a milling-machine operator. I loved working with iron and other metals. My favorite smell was the machine shop; my favorite noise that of machinery; infernal to many people, it was music to my ears." Alla turned her hands this way and that.

"When I was treating my father he would sometimes say: 'Today, my child, your hands are not working.'

"This caused me to reflect upon what my mother had told me when I was still small. 'Whenever you're faced with a problem,' she had said, 'you must seek its solution not outside, but inside, yourself.' My mother was a remarkable individual, a real sunbeam. When she came into a room, she seemed to light it up as if she were a lamp. Another thing she told me that came back to me while I was treating my father was: 'Never talk to anybody or do anything when in a bad mood. If you cook a borscht, for instance, when you're feeling out of sorts, the borscht will be poisonous to those who eat it.'

"So, thinking about why my hands at times seemed unable to help my father's legs, I began to look inside myself for an explanation. I found that whenever I indulged in cheap gossip or gave vent to anger, my hands, as my father put it, 'stopped working.' I realized that nature herself might be wise enough to shut down whatever my hands were emitting, lest the power, modulated by bad or base feelings, be insidious or harmful.

"I began to understand more consciously that I was responsible for my every word spoken or action taken, as well as for the consequences of those actions and thoughts. To walk that new path was very difficult because it required blaming oneself for many things and not foisting the blame off onto others. I saw that motivation was crucial. As I progressed in healing, I understood that the *profit* I was seeking was quite different from the one usually defined by that term; so I took no money for my healing.

"And I remembered still another thing my mother used to say: 'Forget anything you've ever done out of *duty*. All that is undertaken out of a sense of obligation brings only harm. Only things done out of *love* serve the positive and the good.'

"This led me to stop discriminating among people, judging whether they were 'good' or 'bad.' All kinds of remarkable things began to happen when I started to deal with people without passing value judgments on them. When sick people come to my office I treat them without letting any preconceived attitude toward them get in the way."

As a basic explanation for the development of her talent, Alla quoted the gifted Soviet children's writer Samuel Marshak: "Whatever is touched by a human hand is energized, as if illumined, by the living soul of the person touching it."

Alla Kudryashova transferring her energy to jars of water and wads of cotton wool and communing with a bouquet of wheat.

This led Alla to put her powers into ordinary simple substances such as water, oil, or cotton wool, which could be easily tufted and given a desired shape. And so her treatment became known as the VMV method, for the first letters in the Russian words *voda*, *maslo*, and *vata*. Patients seeking succor would send her small sample jars of oil or water or cream and then use them as remedies for their particular maladies.

"Can you explain," asked Berezin, "what goes on in your mind as you treat this or that sample?"

Looking first at him, then at us, Alla measured her words: "One of the things to avoid is any concern with results, or worry about the future. I have come to realize that we must live in the present. How I live, how I am, at any given moment, determines what will happen to me later on. So many people are taught to worry about what they *will become* that they forget they already *are becoming*, that they exist . . . right now! How useless it is for people to think, as they do, 'If I reach a given point, then such and such will happen.' When they project notions into the future about how they should be, they are seldom happy, or fulfilled in life.

"When I am treating people or substances, I admit to myself I don't really *know* anything. I just let myself *be* . . . here . . . this moment. I don't struggle to find the right approach to this or that person, to choose what to say or not to say. I just try to be calm and reach a harmony with the world and its beauty. Most of us never stop to see this beauty. We mask it by making what amounts to a drug of our lives. Life should be *joy*. When I am suffused with joy, hear its voices in my heart, it's *then* that I know I can help people, give them strength that will last a long time."

At that moment, Alla was interrupted by a long telephone call. Her voice, barely audible, was drowned out by two or three other conversations taking place simultaneously; so we couldn't catch what she was saying. But, when she hung up, Berezin explained that she is capable of treating water, oil, or cotton wool over the telephone, no matter what the distance.

"In experiments," said Berezin, "we have found that *distance doesn't matter in the least*. It's like a wireless telephone. She becomes a connecting link and, in that connection, it becomes highly important that she recognize her responsibility with regard to her mood."

Alla, it was explained to us, is sure that it is through "mood" that any person, as a receiving and transmitting station, either attracts or rejects others. The trick, in her words, is to "find oneself," to "learn to attract." It is her belief that this kind of understanding can come only through action, or purpose. Like Alex Podolinsky, she holds that the world wastes far too much time in *criticism*, which does no good because it merely confirms the existing negative. *Action*, or impulse, is the key, not only in the physical, but in the mental sense, because everything external to our being is only a projection of consciousness. To Alla, therefore, whatever is real about the world is in oneself.

It was getting late. Leaving Alla's little lair, we descended to Andriankin's office, where he looked at us with a kind of leer as if to ask: "Well, what do you make of our starlet?"

"Have you anything in the way of *real* attestations for her extraordinary feats?" we asked. "Particularly in agriculture? With animals and plants? Affidavits from directors of the farms on which they took place, documentation signed and sealed by responsible third-party officials, with figures showing the extent of the increases claimed?"

Andriankin's eyes crinkled. "We'll see what we can get together."

Back on the twenty-third floor of our giant Cosmos Hotel, we gazed out at the soaring aluminum parabola of a rocket on its way to space—monument to Konstantin Tsiolkovsky, first Russian to have dared to think about, let alone work on, extraterrestrial travel, in the nineteenth century at a time when his compatriots considered his vision nothing more than fantasy. Was the Otdel, we wondered, exploring a new frontier of mental space? Would hard, substantiated data be forthcoming on Alla's extraordinary powers?

The time was certainly ripe for such endeavors, with Soviet agriculture in its present disastrous state, as authoritatively reported by the Russian agriculturalist Zhores A. Medvedev in his *Soviet Agriculture.*†

In our hotel, the telephone rang, and it was Berezin wanting to know if he could come over with something interesting to show us. Half an hour later he stepped into the room with a stack of papers to validate Alla Kudryashova's remarkable achievements, causing us to wish for what is easily accessible on the corner of any Main and Broad street in America, but so rare in the Soviet Union as to be unavailable to us in Moscow, a copying machine.

On top of the pile was a three-page *akt* dated December 1986 on "The Effect of the VMV Method as Performed by A. A. Kudryashova on the Growth of Broiler Chickens." The work had taken place at the F. E. Dzerzhinsky State Farm (*sovkhoz*), a large broiler-production operation in the village of Mirnoye near the city of Simferopol in the Crimean peninsula.

When we did ask Alla how this work with chickens had come to pass, she replied that it had begun as the result of a fortuitous contact with a former air-force pilot who had learned about her treatment of a small flock of sick chicks owned by a peasant woman. Stricken with an intes-

---

† The book, which covers the subject from before the Revolution to the fall of Khrushchev, with an analysis of what has evolved in the last twenty years, claims the system does not produce enough tractors to plow, or combines to harvest all the land that is sown, and that the machines it does make are so heavy they seriously damage the soil.

Soviet farming he describes as having been in crisis since Stalin collectivized farming methods and deported five million people back in 1929–30, killing seven million in a famine caused by the forced confiscation of all available grain. Today the country cannot feed itself and has become the world's largest importer of grain. It wastes, says Medvedev, seventy million rubles a year subsidizing retail food prices, which are lower than the prices the state pays state and collective farms, which are usually lower than actual production costs. And every effort at reform complicates or simplifies bureaucracy without doing much for farming.

tinal disease that usually proves fatal, all the peasant's birds survived and thrived, thanks to Alla's ministrations.

"Most people don't realize," said Alla, "that chickens specially bred to produce broiler meat have weak systems because a single characteristic is bred into them at the expense of other aspects of their organisms."

When the ex-flier mentioned Alla's feat to the manager of the F. E. Dzerzhinsky State Farm, he invited her to come down from Moscow and run some experiments.

Alla found not just a farm but an associated scientific research institute. "Had I known that in advance," she admitted to us, "I probably wouldn't have bothered to go because I was aware that most professional scientists, unlike farmers, simply can't stomach anything that doesn't fit into the framework of their limited beliefs. And so it was; but only at first. When the institute director called his staff together and I began to explain what I did with respect to treating water and the kind of effects such water could produce in chickens, one of the young scientists began openly and mockingly to laugh in my presence as if all I had said was no more than a crazy joke. And I heard others muttering that their director must have lost his mind to become involved with such a scheme."

Not giving in to any irritation, and sticking to her guns, Alla said she chided the scientists with the statement that she had been taught that true experimental science should regard negative results as potentially as important as positive ones. Were they afraid, she hinted, to repeat experiments that had elsewhere proved successful?

Her challenge reduced them to silence. It was agreed that she be allowed to treat water for over ten thousand birds. But the huge number so appalled her she asked if it could be reduced to one or two hundred. It was then the turn of the chicken specialists to ask if it might not be she who was afraid.

"It was only a lapse on my part, a momentary weakness," said Alla. "I knew that by going there to do an experiment it should be done, no matter the conditions. Mother Nature never gives any of us any more than we can do. And it had always been my belief that one should work on any task as though it were the last thing left to accomplish in life."

The results were amazing. In the months of January and February of 1986, one- to seven-day-old broiler chicks of a "Gabro-6 cross" that drank only water treated by Alla collectively gained three-and-a-half tons more than a similar group of control chicks that drank ordinary water. Furthermore, the experimental chickens ate about 15 percent less feed (see document, page 322).

During her visit in the Crimea, Alla also ran an experiment on the hatchability of eggs stored for periods of up to twenty-one days after

с.Мирное
Симферопольского
района
Крымской области

УТВЕРЖДАЮ:
Директор совхоза
им.Ф.Э.Дзержинского

Г.П.Серый

3 декабря 1986 года

А К Т

производственной проверки применения эффекта
ВМВ научного сотрудника лаборатории ОТП АН
СССР А.А.Кудряшевой на показатели выращивания
бройлеров

Производственная проверка проводилась в январе-феврале
1986 года на бройлерной фабрике совхоза.

В опытном птичнике № 59 цыплята-бройлеры кросса "Гибро-6"
с первого по 7-й день жизни пили воду,энергетизированную
А.А.Кудряшевой.Вода обрабатывалась в деревянных бочках и оцин-
кованных корытах. В контрольной группе в птичнике № 61 цыплята
пили обычную воду. Условия микроклимата в опытной и контрольной
группе были идентичные,цыплята в течение всего периода откорма
получали одинакового состава кормовой рацион вволю.

Результаты опыта

Группы	Кол-во сут. цыплят в нач.опыта гол.	Забито в возрасте 66 дней, голов	Общая живая масса кг.	Средняя живая масса I головы г	Сохране-ние, %	Затраты корма на I кг.при-роста масс к.ех.
опытная	12900	11950	19500	1632	92,6	3,06
Контроль ная	12970	11545	16020	1368	89,0	3,42
+,- к контролю			+3480	+244	+3,6	-0,36

Применение энергетизированной воды позволило в опыте
улучшить все основные показатели откорма. Работы по изучению и
использованию эффекта ВМВ А.А.Кудряшевой желательно продолжить.

Гл.зоотехник-селекционер   В.С.Кромин

Зав.цехом выращивания
бройлеров   А.В.Карташов

Бригадир   Н.И.Севастьянова

Сотрудник лаборатории СТП   А.А.Кудряшова

Official Soviet documents attesting to Alla Kudryashova's remarkable gift for increasing the weight of broilers while reducing their diet, and for increasing the harvest of edible beets (for borscht) by 160 percent.

One document is signed by M. N. Prokhorov, Senior Scientific Collaborator of

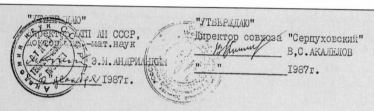

"УТВЕРЖДАЮ"   "УТВЕРЖДАЮ"
Директор ОТП АН СССР,   Директор совхоза "Серпуховский"
доктор физ.-мат.наук   _____ В.С.АКАПЕЛОВ
_____ Э.М.АНДРИАНКИН   "___" "_____" 1987г.
"___" "_____" 1987г.

ЗАКЛЮЧИТЕЛЬНЫЙ АКТ

о проведении производственного испытания
метода биофизического воздействия
на семена кормовой свеклы

II мая 1987г. научным сотрудником ОТП АНСССР Кудряшовой А.А. на основе договоренности о научно-производственном опыте была проведена биофизическая обработка семян кормовой свеклы перед посевом. Семена были высеяны на двух участках: Арнеево (Опытный участок - 20 га, контрольный - I5га) и Нефёдово (Опытный участок - 25га, контрольный - I5га). На контрольных участкахвысевали аналогичные, но необработанные семена.

В течение вегетационного сезона все участки возделывались по одинаковой технологии. До уборки урожая было проведено три обследования опытных и контрольных участков. Обследование проводил старш.научн.сотрудник ВИМ ВАСХНИЛ м.н.Прохоров. При этом выбор квадратов для статистического учета развития проводился рендомизировано. В первом учете оценивались по балльной системе параметры развития проростков, в последующих двух учетах оценивался вес корнеплодов (протоколы прилагаются).

В процессе уборки урожая по участку Арневво существенная часть корнеплодов была убрана без отнесения к контролю или опытк, в связи с чем этот участок из опыта исключен.

По участку Нефёдово с опытной части поля собрано II300ц, т.е. по 432 ц/га; с контрольной части поля собрано 4240ц, или по 283 ц/га. Таким образом, отмечается превышение опыта над контролем в I,6 раза.

Следует отметить, что 1987 год был очень тяжелым экологически: серьёзный недостаток солнечного облучения, большое количество пасмурных холодных дней, поэтому полученный результат представляет несомненный интерес для дальнейшего изучения и внедрения.

Старш.нучн.сотр.
ВИЛ ВАСХНИЛ _____ м.н.Прохоров   Главный агроном   _____ В.М.Кацуро
                                     совхоза

the Lenin All-Union Academy of Agricultural Sciences and by V. M. Katsuro, Agronomist-in-Chief for the Serpukhov State Farm (Moscow Region).

The documents are ratified with the signatures and seals of the Director of the Department for Theoretical Questions, USSR Academy of Sciences.

laying before being introduced into incubators. The longer the storage period, the more the hatchability index dropped. The experimental eggs, over fifteen thousand in number, were put into containers with a layer of cotton wool, specially treated by Alla, covering each layer of eggs. The hatchability index rose for eggs stored seven days by 2 percent, fourteen days by 5.7 percent, and twenty-one days by a huge 21.9 percent. The experiment therefore conclusively proved the effectiveness of Alla's methods in heightening the survival of chick embryos. It also raised speculation about whether—as in the case of her human patients—she was enhancing the immune-defense systems of the chickens.

"The same method was later used to increase the preservation of newly picked ripe tomatoes," Alla said proudly. "That got me to thinking about the vegetal world. I wanted to see if I could be effective in that domain. I was elated by the results obtained on the chickens and described it to my friend who directs a large agricultural institute in Simferopol not far away from the Dzerzhinsky farm. I said to him: 'Vladimir Nikolaievich, why don't we sign an agreement to experiment with plants?' And, finally, an agreement was signed between his institute and our department.

"But then a strange thing happened. During the course of the next year, when we ran into each other several times, he seemed to have totally forgotten that he'd signed an agreement with a branch of the U.S.S.R. Academy of Sciences. The truth I wormed out of him was that he feared what his scientific associates would say about my proposed experiments and tried to make up for this by promising that he would run them by himself, in private, in secret. The academy could have sued him for breach of contract."

Later, on a visit to Moscow, the same director called Alla to make amends and to ask if he could bring over to her office a friend, V. S. Akalelov, who turned out to be the manager of the Serpukhov State Farm near Pushchino in the Moscow District. Far from having any qualms about doing agricultural research with Alla, Akalelov said he was eager to see what she could do on a one-hundred-acre plot of edible beets, specially prized by all Russians as the chief ingredient in borscht, that unique soup, almost a stew, that is rounded out with meat, marrow bones, potatoes, and many other vegetables.

Among conclusive documents attesting to her success was one on "Certain Experiments Indicating Positive Effects Produced on the Seeds of Beet Plants Through Biophysical Action." Signed on December 30, 1987, by the Serpukhov Farm director, it bore the official seals of both the R.S.F.S.R. (Russian Soviet Federated Soviet Republic), branch of the Lenin All-Union Academy of Agricultural Sciences (VASKhNIL), and that of the Department of Theoretical Problems, U.S.S.R. Academy of Sciences (OTP, ANSSR). It reported that Alla's treated beet seeds planted in the experimental hundred-acre plot produced 432 *tsentners* (43,200 kilograms) per acre while nontreated seeds grown on similarly cultivated ground produced only 283 *tsentners* (see document, page 323).

To our query as to exactly what she did, Alla ingenuously replied: "I just sat at the edge of the field where the seeds were to be planted and held my hands against the sacks containing them. I didn't really need to use my hands. I now stress that, in my kind of work, it's best to use a minimal amount of physical contact. I realized this when a farm worker asked me: 'Would your powers disappear if you lost your hands in an accident?' I was taken aback. Would it mean I could no longer be of help to human patients and other living things? I knew the answer straight away: Of course not!

"So I put my hands down and just sat there staring at the sacks of beet seeds and, all of a sudden, I knew that the whole field I was in was a living, breathing organism. In my mind's eye I could see it writhing, tormented by the artificial fertilizers it had been drenched with, the huge tractors and other agricultural machines which compressed its surface, and the coarse language of the farmers, with all the obscene words they use while working the land. It used to be that landsmen had deep respect for the earth, but today, in agriculture treated as industry, that's no longer true.

"I was so upset that my body broke out in a rash and my limbs began to swell. That night, at home, I was running a high fever. For eighteen days I was sick. It was only by going on a rigorous fast, with its cleansing effect, that I purged the disease.

"It took me a little while to come to the understanding that I'd brought the illness on myself. How? By taking on, by internalizing, the woes of that field. I had given in to pity, whereas what I should have resonated with the field was not pity but *sympathy*."

So pleased was Akalelov with the results that on December 30, 1987, he signed an agreement for continued work characterized as "of great scientific and practical interest," during the 1988 growing season. (See document, page 323).

News of Alla's work at the Serpukhov farm traveled all the way through the agricultural grapevine to Central Asia, eliciting an invitation for her to come and work on an experimental field station in Kirghizia near the Chinese border, where five herds of sheep of about one hundred head each had become so inexplicably nervous and jumpy—ewes, lambs, and rams—they had lost a great deal of weight. Treated by Alla, the sheep were quickly calmed and began regaining weight after drinking only water treated by her for less than a week.

"I finally found out," said Alla, "why all the sheep on the farm were doing so poorly, when the director admitted to me that he actually loathed sheep. The animals had become sensitized to his hostile feelings. It's just as with people: real communication never takes place with words, only through feelings."

Berezin looked from Alla to us and then grinned at Alla. "Tell them about the pond," he said, as if to imply we'd heard nothing yet.

Alla smiled. "It was a narrow body of water, only fifty meters wide but a kilometer long. The water in it was rank, muddy, and covered with a thick film of algae and a proliferating growth of weeds. I asked if some of the water could be put in bottles for me to treat, then poured back into the pond. That was done in May and by June the whole pond was clean and has remained so to the present day."

But to a physicist, mathematician, and engineer, like Berezin, the most impressive of Alla's feats involved not animate organisms, living things, but electronic instruments. "She has," he said with pondered emphasis, "been able to repair persistent malfunctions and breakdowns in highly complex instruments, both with the VMV method or solely with her mind. And this raises a fundamental question about consciousness: is it limited to the living, or is it more basic, perhaps a crucial part of the universe since its inception?"

It all started, Alla told us, when her friend Vladimir Vassilyevich Lezhnin, an electronics engineer, called her from Kazan—capital of the Tartar hordes before they were reduced to submission by Ivan the Terrible—where he was working as a senior supervisor for the GNIPI-VI computer center. A delicate instrument used to monitor heart conditions had broken down to the point where a team of repairmen could do nothing to set it to rights. Alla sent Lezhnin samples of cotton wool and a bottle of water, all treated in her special way, and asked him to sprinkle certain key parts of the instrument. Within moments, said Lezhnin, the machine began to function perfectly.

"But in my absence, when my friend tried the same method on another machine," said Alla, "it broke down completely. He was going about his work in a mechanistic, formal, indifferent, and heartless way, with no *soul input*; and when I told him so it helped him diagnose what was the matter with other instruments. He found that by concentrating lovingly on his maintenance work he could, in many cases, get the malfunctioning machines to run again smoothly. He is a very open-minded person, unlike so many scientists I have met in laboratories who believe that only their kind of theory works, not my kind of practice. Some of them have only to come near a sensitive instrument for it to begin acting up."

When the Kazan computer center began having serious trouble with some of its most sophisticated computers, Lezhnin again turned to Alla for help. Involved were two similar computers, models SM-1403 and SM-1600, which work at speeds of 400,000 operations per second. They have adjunct equipment such as memory blocks to increase the reliability and quality of their output, and drum-type alphabet-numeral printing devices capable of producing 500 lines of 132 symbols per line per minute.

Much in demand, the machines operated twenty-four hours a day in a "multi-terminal regime" on shared time. As a result of their intensive use, the number of malfunctions and breakdowns, as carefully recorded in the engineering log, began to rise sharply. Information transcribed on disk packets could not be consistently read out, and there were "floating failures," when the machines would not supply what was needed.

On Alla's recommendation, the machines were wiped free of dust with cloths impregnated with her treated water, and tufts of cotton wool drenched in the same water were applied to key places in their construction. In the words of Lezhnin's official report: "The quality of the magnetic-tape and disk readouts greatly improved as did the morale of the work force responsible for maintaining the machines in proper working order."

In Lezhnin's analyses of what seemed to be miraculous goings-on, he referred to a complex cybernetic system's being significantly affected by the "climate of the operator-machine relationship." It reminded him of the father of a boyhood friend who could repair television sets despite only three years of elementary schooling. With no idea of how the sets functioned, Lezhnin said, the man could unerringly select, from an array of parts, whatever tube, condensor, or resistor was necessary to his task.

The most satisfying explanation for this elusive concept—as well as Alla's effects on animate and inanimate objects "transmitted over distance"—has been advanced by an inventor, neo-natural philosopher and cosmologist, Arthur Middleton Young. Exasperated with the ravages inflicted on the human psyche by increasingly reductionist thinking in science, and long before the birth of the so-called "new-age consciousness movement," Young set up, in 1952, his Foundation for the Study of Consciousness, first of its kind in the U.S., to delve into the mysteries of clairvoyance, precognition, and other forms of "seership" as well as the apparent ability of the human mind to affect matter at a distance. Objuring any reliance on "fields" to explain psychotronic abilities, Young points to the power of intent, harnessed to will. ‡

In the garden of his summer home in Downingtown, Pennsylvania, Arthur Young told us: "You can't explain *intent,* the very basis for Alla's success, in terms of something else. You explain the 'something else' in terms of *intention.* That's what's done in everyday life. If you see someone digging a hole outside your window, you go outside and ask: 'What are you digging that hole *for*? What's your *intention*?' We only have all our physical paraphernalia because we had *intentions.* We have cars and planes and space shuttles because we have the intention to travel at higher speeds."

Young's is a sobering thesis, but one that is now supported by Robert G. Jahn, professor of Aerospace Sciences at Princeton University, and his colleague, Brenda J. Dunne, manager of an "anomalies research laboratory" at the same university. They have pursued ten years of rigorous research evidencing the power of mind over matter, as described in their recent book *Margins of Reality: The Role of Consciousness in the Physical World.*

Among two dozen extraordinary insights by leading physicists dug up by Jahn and Dunne and quoted in their book is one by the English astrophysicist Sir James Jeans. In his *Physics and Philosophy,* Jeans wrote that the theory of relativity showed that electrical and magnetic forces are not real at all but merely mental constructs resulting from our misguided efforts to understand the motions of atomic particles. "It is the same with the Newtonian force of gravitation," added Jeans, "and with energy, momentum and other concepts which were introduced to help us understand the activities of the world. All of them prove to be mental constructs, and do not even pass the test of objectivity."

And Alex Podolinsky in one of his recent lectures to Australian farmers took up the same theme: "We have not even the justification to speak of a particle picture of reality, neither have we the justification to speak of a ray picture. These are all working hypotheses with which we try to explain what is in the background of matter. But as such there is no matter; it has been exploded away. And if there is no matter there certainly is no space in the sense of matter, and so there is no time. These three pillars of Kant have vanished."

For Jahn and Dunne, it was Duke Louis Victor de Broglie (whose predictions that the electron particle had wave properties won him the Nobel Prize in 1929) who may have hit upon the most unsuspected

‡ See Arthur Young's *The Reflexive Universe: The Geometry of Meaning,* and *Which Way Out?*

explanation of the power behind the seemingly inexplicable capacity of consciousness to meet and relate with, or affect, matter at a distance. "If we wish," wrote de Broglie, "to give philosophic expression to the profound connection between thought and action in all fields of endeavor, particularly in science, we shall undoubtedly have to seek its sources in the unfathomable depths of the human soul. Philosophers might call it *love* . . . that force which directs all our actions, which is the source of all our delights and pursuits. Indissolubly linked with thought and action, love is their common mainspring and, hence, their common bond. The engineers of the future have an essential part to play in cementing this bond."

Flying away from Moscow, we compared Alla's extraordinary mental powers with Machaelle Wright's extraordinary communication with the devas and the nature spirits, wondering what possible connection there might be with the sonic, supersonic, and very high frequencies of thought communication. It reminded us of Kudryashova's modest statement in her candlelit office that her doings were really based solely on a feeling of *purity* and *love*—the same ingredients as Steiner's Spiritual Science. Could it be, we wondered, that these very best intentions were the ultimate explanation for Alla's telekinetic powers? If, dominating the projection of her "mind powers," they can produce such a bonanza of wheat and borscht, might they not produce for the world at large the even greater prosperity of peace between the two planet's superpowers, brutish, bellicose giants, deprived, like Wagner's Fafner and Farsolt, of the benefits of Freyja, goddess of love?

APPENDIXES

APPENDIXES

APPENDIX A

Light from the East

To appreciate the agricultural wisdom of Rudolf Steiner and Alex Podolinsky requires an insight into the hermetic tenets of a spiritual movement Steiner first belonged to in Germany at the turn of the century, and of which he became general secretary, the Theosophical Society, founded in New York in *anno Domini* 1875. Derided as crazily fey or far-out, members of this century-old institution have most recently been credited by orthodox scientists with stunning and prophetic breakthroughs in the abstruse science of particle physics. This has led to a reevaluation of the tenuous overlap between the domain of physics and the realm of metaphysics.

During the second half of the nineteenth century a vein of ancient knowledge from India and Tibet, tapped by a handful of adventurous Europeans, startled the Western world into realizing that certain secluded Orientals, whose notions of science had been ignored or derided, might, after all, have understood more about physics than Newton, more about electricity than Faraday, and a lot more about agriculture than Justus von Liebig. Millennial Hindu and Tibetan doctrines threatened to lift a veil of matter from the eyes of Victorian Englishmen enough to shock them with a glimpse of life disrobed, as exquisitely spiritual as the present age of Kali Yuga is materially gross.

Occult philosophy, a system of knowledge cultivated in secret since remotest antiquity and handed down to initiates "selected for their strength of character and purity of purpose," showed a comprehension of the forces in nature quite different from—and in many cases in advance of—those described by contemporary European and American science.

Occultists could deal not only with such unusual physical phenomena as "anti-gravitational devices or machines which could fly beyond the atmosphere," but with the inherent and powerful capacities of the human spirit, itself capable of examining these marvels of nature from within or without the human body, at will.

Some of the newly rediscovered wisdom of the Orient, already partly familiar to the West, had been kept alive by the magi of Chaldea, the priests of Dynastic Egypt, aesthetic Jewish Essenes, Christian Gnostics, and third-century neo-Platonists who combined elements of Oriental mysticism with Judeo-Christian concepts. But the bulk of a new wave of disclosure coming from the East seemed stunningly original.

The reputed masters of this arcana were said to be a group of highly advanced beings known as *adepts*, credited with being able to communicate with each other or with chosen mortals from either in or out of a physical body, by telepathy or through actual materialization. Organized into such secret societies as the Great White Brotherhood, they had functions that were described as "the education of mortals in the secret, ageless wisdom of the past." Acting as spiritual teachers and inspirers of mankind, they were held to constitute an inner government of the world.

Such adepts were credited with superhuman powers over nature, the ability to assume or discard a physical vessel at will, using bodies as mere vestures. Other adepts were said to be able to remain in the flesh for vast spans of time, mostly in secluded retreats in the Himalayas, Tibet, China, Egypt, Lebanon, the meridional Carpathians, the jungles of Yucatán, even in "England's mountains green," appearing to whomsoever they chose.

To become such a teacher, many previous lives were deemed a prerequisite —as many as several hundred; after which the adept, having achieved perfection of body and soul, could become "immortal" and have access to "all knowledge" and to the entire "akashic record"—or cosmic history—without the need for laborious scientific research.

Reputed adepts in Tudor England were Sir Thomas More, beheaded by Henry VIII for his defiance in upholding the Roman Catholic Church (which canonized him in 1935), and Thomas Vaughn, alchemist and mystic, author of *Anthroposophia Theomagica*, whose Rosicrucian tendencies foreshadowed those of Steiner. Another famous "master" was the historical and legendary Compte de Saint Germain, supposed reincarnation of the English baron and philosopher Francis Bacon, essayist (and possible co-author of Shake-Spear texts), later player of a mysterious role in the New World, whose seventeenth-century body, believed to be buried in Williamsburg, Virginia, was unaccountably missing when the vault was opened.

To pass on the tenets of this ancient wisdom to maturing humanity, the system required that lesser humans be chosen by higher adepts to act as vehicles for transmitting the "Secret Doctrine." One such individual, according to her own account, was an aristocratic Russian maiden, Helena Petrovna, fathered in the North Caucasian city of Ekaterinoslav in 1831 by an army colonel, Peter von Hahn, whose ancestors had settled in the Romanov domains during the reign of Catherine the Great, and whose mother, a Russian princess, descended directly from Ryurik, the Norse adventurer who founded the first of Russia's ruling houses.

From childhood, Helena claimed to have lived simultaneously in two worlds, one physical, the other spiritual. In the latter she claimed to have been accompanied by invisible companions, and to have been "contacted" by a Tibetan master, whom she named Koot Hoomi, who told her that he would appear to her in the flesh when she was ready.

In her autobiography, written many years later, Helena recounts that one night in London, when she was sixteen, strolling by moonlight along Hyde Park's Serpentine, she ran into an Indian Rajput prince whom she instantly recognized as her master—also strolling in his earthly form—who told her she would ultimately come for training to his ashram beyond the Himalayas.

Back in southern Russia, aged seventeen, Helena was married by her family to Nikifor Blavatsky, an army general more than thrice her age, at that time vice-governor of the province of Yerevan in Armenia. Kept in ignorance—as was then the custom—of the physical details that marriage might entail, the young bride, rather than submit to her aged husband, ran away in disguise, never to return. Her autobiography colorfully describes a quarter century of wanderings in Europe, Egypt, America, Mexico, India, and Java, performing feats of "apparent magic" from drawing rooms to circuses. But corroborative evidence is scarce, as is evidence of her reputed journey to the forbidden reaches of Tibet. There she claimed to have spent four years in the vicinity of the famous Lamasery of Shigatze on the Brahmaputra River, near where the oldest known text, *The Book of the Secret Wisdom of the World*, was said to be in the safekeeping of the Teshu Lama.

Helena P. Blavatsky at "Maycot," Norwood, London in 1887.

The inference is clear that it was from this book—described as set down on palm leaves treated by some long-forgotten processes to make them impervious to air, water, or fire—a work far older than the ancient sacred Hindu writings and the canonical Vedas, and one credited with having given rise to fourteen volumes of commentaries, that Helena wished one to believe she had acquired her knowledge of arcana. True or not, from some such source she did acquire an encyclopedic grab bag of astounding data. Back in the West, with what she felt was a life mission to perform, but no clear idea of how to begin, "because a *chela*, or student disciple of a master, far from being treated as an automaton, is left to perform suggested tasks in the light of his or her own sagacity, in perfect freedom, unlike soldiers in a military hierarchy," Helena settled in London, where she became involved in spiritual séances. Her intention, she declared, was to found a society with the central purpose of investigating spiritualistic phenomena—at that time a popular fad—but with the higher aim, as she defined it, of leading its members "beyond those limited interests to a real knowledge of the inner nature of man and the universe." For her efforts to reveal to mankind the dreadful materialism into which she saw it sinking, she says, she was at once attacked by the spiritualistic movement in the first of many campaigns to calumniate and misrepresent her efforts, attacks which were to last throughout her earthly life.

In Paris in 1873, she says she received from her masters instructions to go to America. Penniless, and with no great prospects, she arrived in New York City to take a job embroidering neckties until she was clairaudially informed to proceed to Chittenden, Vermont, to view some fantastic spiritualistic happenings reportedly occurring in the home of a family called Eddy. The reading public, avid for stories about current mediums and their feats, was at that time agog with the mysteries of spiritualism, especially since Horace Greeley had attested as genuine in a column in the New York *Tribune* the materialization of spirits by the Fox sisters of Hydesville, New York.

In Vermont, Helena Blavatsky's spiritual match appeared in the flesh in the form of a Civil War veteran and experienced journalist, Colonel Henry Steel Olcott. Captivated by the current vogue for spiritualist séances, he had gotten himself assigned by the New York *Graphic* to cover the events at the Eddy home largely because of his reputation for integrity and for his known accuracy in reporting sensational events. A graduate of Harvard, Olcott had already by the time he was twenty-five made himself such an expert in experimental agriculture he had been invited to become Director of the United States Agricultural Bureau in Washington. Founder of the first scientifically-based American experimental farm in Mount Vernon, New York, Olcott had also published the first textbook in America on Chinese and African sorghum, and from his European travels to study the latest developments in farming he had produced an impressive report published in the *American Cyclopedia.*

A promising career, it had nearly ended when he was twenty-seven in 1859. As agricultural editor of the New York *Tribune* and foreign correspondent for the London *Mark Lane Express,* Olcott was sent to Harpers Ferry, Virginia, to cover the hanging of that fire-and-brimstone abolitionist "old Brown of Osawatomie," better known to history as John. Arrested as an intrusive reporter, young Olcott was almost hanged from the same gibbet, only spared when his captors discovered him to be a fellow Freemason.

Back in New York City following the Civil War, Olcott became a successful lawyer specializing in customs, internal revenue, and insurance cases. As special commissioner for the War Department to investigate that even then perennial plague on the body politic, corruption among army contractors, he was awarded the rank of colonel for his successful services.

From his Vermont experiences with the Eddy brothers, veteran Olcott was to produce a remarkable book, *People from the Other World,* in whose nearly five hundred pages, thoroughly illustrated with apparitions and apportations, he did his best to validate the amazing mediumistic feats he witnessed, including the levitation of bodies, the rising off the ground of heavy objects without human or other physical contact, the appearance of luminous objects, either self-luminous or visible by ordinary light, phantom forms and faces accompanied by various weird sounds, including exotic musical instruments, and the voices of materialized spirits, recognized by the audience as speaking in their own living tones. Most impressive to Olcott was the fact that for Helena Blavatsky's benefit the Eddys had summoned spirits authentically dressed in Russian costumes who could converse with her in authentic Russian dialects, and play Armenian folksongs familiar from her childhood, something the Eddys could not possibly have known.

Only a year younger than Madame Blavatsky, of whom he wrote, "In the whole course of my experience, I never met so interesting, and, if I may say it without offense, eccentric a character," Olcott found in her a spiritual master whose efforts at spiritualizing the world he would support for the rest of his life, abandoning his former mundane pursuits, becoming a teetotaler and vegetarian in pursuit of adeptship.

Most impressed by Blavatsky's own extraordinary talents as a medium, which he described as totally differing from any other he had met—"instead of being controlled by spirits to do their will, it is she who seems to control them to do her bidding"—Olcott teamed up with the rebellious lady to form an occultist society in New York City to "diffuse information concerning those secret laws of nature which were so familiar to the Chaldeans and Egyptians, but are totally unknown by our modern world of science." He proposed that the society study mesmerism, spiritualism, the "Odic Force" of Baron Karl von Reichenbach, and the universal ether, or "astral light" of occultism. For her part, Madame Blavatsky saw the society as a vehicle for imparting to the world the Ancient Wisdom she believed it was her calling to present anew on instructions from her bodied and disembodied Indian and Tibetan avatars.

Thus was engendered the Theosophical Society, whose declared object was "to collect and diffuse knowledge of the laws which govern the universe." Its early members included Thomas Edison, inventor of the electric light bulb, and General Abner Doubleday, supposed originator of the game of baseball. The term *theosophy*, or "divine wisdom," was used to refer to the strain of mystical speculation associated with the Kabala and the writings of such earlier occultists as Agrippa, Paracelsus, and Robert Fludd. Olcott spoke of "freeing the public mind of theological superstition and a tame subservience to the arrogance of science."

In the heart of Gotham, Blavatsky and Olcott lived together in a Greenwich Village suite on Irving Place over the entrance to which presided the stuffed head of a fanged lioness. With its Bohemian atmosphere, the flat became known as "The New York Lamasery," a mecca for Kabalists, spiritualists, platonists, and seekers after the marvelous, a haunt where William Q. Judge, secretary to the society, wrote of "amazing feats of magic, hundreds of which I witnessed in broad daylight or in blazing gas-light from 1875 to 1878." Into this menage Olcott's sister was obliged to move to maintain the appearance of propriety, Olcott, the father of four children, having by this time been divorced by his wife.

For two years Olcott hovered over Blavatsky as she sat at her writing table, chain-smoking, to compose a fifteen-hundred-page book containing startling theories concerning the evolution of humanity and of religion, all of which she claimed were conveyed to her by direct astral communication from her master in Tibet. A veritable encyclopedia of occult wisdom, displaying a lapidary command of English prose, well beyond Blavatsky's normal vocabulary, it was written so fast, according to Colonel Olcott, that, as the pages were cast to the floor, he was unable to keep up with them, and so got the order mixed, a volume fascinating in its content, but somewhat random in its presentation, the monumental *Isis Unveiled*, "A Master-key," as its subtitle infers, "to the Mysteries of Ancient and Modern Science and Theology."

Olcott, convinced that the book—which postulates man as a spiritual being, and chronicles the human race through eons of karma—was inspired astrally or telepathically by highly evolved masters, described how Blavatsky's pen would fly over the paper until she would stop, "look out into space with the vacant eye of the clairvoyant seer, shorten her vision as though to look at something held invisible in the air before her, and begin copying on her paper what she saw."

Ten days after the publication of *Isis Unveiled* in 1877, it had sold out its first edition, and three subsequent printings were gone within the next half year. Later described by California occultist Manly P. Hall, founder of the Philosophical Research Society, an encyclopedic writer on the arcane, as "the most vital literary contribution to the modern world," the book created a world-wide interest in the newly-formed theosophical movement. Contemptuous of both scientific materialism and the weaknesses of a Western religion incapable of uniting Christian peoples in peace, the book was either pilloried or ignored by orthodox scientists and religionists. It also engendered well-documented accusations of gross plagiarism, to which Olcott replied by admitting that Blavatsky's writings did contain "a large number of citations from other authors without giving credit," but claimed the act was not willful or conscious plagiarism because she drew the material from "the Astral Light," a metaphysical record available to all.

Ill-at-ease in New York, and convinced that India and Tibet were the true sources and reservoirs of hidden and immemorial wisdom, Blavatsky and Olcott heeded the suggestion of their spirit master, Koot Hoomi, to move to India, where, in 1878, accompanied by two adherents, they settled, first in Bombay, then in Adyar, near Madras, taking over an old summer estate on the shores of the Indian Ocean.

Immersed in the ancient wisdom of the subcontinent, Blavatsky soon launched into another massive work, based, she said, on information divulged to her by master Koot Hoomi, information which "appeared before her as writing, pictures, and symbols." Begun in 1885, the work was published in 1888 as *The Secret Doctrine*, soon to become the bible of Theosophy, offering to the Western mind an interpretation of the sacred writings of the Hindus and their predecessors, "passed on," said its author, "by generations of seers from higher exalted beings."

In two volumes, totaling fifteen hundred pages, the book claimed to be based on the "Stanzas of Dzyan," a mysterious ancient religious text, unknown in the West. The first volume, *Cosmogenesis*, deals with the creation of the universe; the second, *Anthropogenesis*, deals with the history of the earth and the evolution of humanity through a succession of "root races."

To Blavatsky and Olcott the "secret doctrine" contained the "alpha and omega of universal science," and was the keystone to all knowledge both ancient and modern. "What we desire to prove," wrote Blavatsky, "is that underlying every ancient popular religion was the same ancient wisdom-doctrine, one and identical, professed and practiced by the initiates of every country, who alone have been aware of its existence and importance."

As analyzed by Bruce J. Campbell in his *Ancient Wisdom Revived*, the secret doctrine was based on three fundamental principles: (1) the existence of one absolute Reality, the infinite and eternal cause of all; (2) the periodicity of the universe: its appearance and disappearance in cycles; and (3) the identity of "all Souls with the Universal Over-Soul, and the pilgrimage for every Soul or spark through the cycles of incarnation."

Goals of the Theosophical Society—branches of which were being organized in other countries—were described as "the study of comparative religion, philosophy, and science, as applied to the investigation of the unexplained laws of nature and powers inherent in man with the ultimate aim of forming the nucleus of a Universal Brotherhood of humanity."

In Adyar, the two founders were joined by Charles W. Leadbeater, a Church of England clergyman who had been attracted to theosophy in the early 1880s while serving as curate in a Hampshire parish. Admitted into the society's London Lodge together with Sir William Crookes, the celebrated English physicist, discoverer of the element thallium and editor of *Chemical News*, Leadbeater wrote a letter to Blavatsky's master, Koot Hoomi, asking to become a *chela*, or student, to which, after a long waiting period, he received an answer that contained little specific advice or instruction but ended with "Our cause needs missionaries, devotees, agents, even martyrs, perhaps. But it cannot demand of any man to make himself either. So now choose and grasp your own destiny, and may our Lord the Tathagata's memory aid you to decide for the best."

Giving up his pastor's post, Leadbeater sailed to India to devote himself wholly to the movement. As a leading member of the society he was soon describing how under Koot Hoomi's supervision and the direct help of another master, Djwal Kul, he was able to break through to continuous "astral consciousness, with the body awake or asleep," and thus was able to investigate "the constitution of superphysical matter in the structure of man and the universe, and the nature of occult chemistry."

In this state of consciousness Leadbeater claimed to be able to see the gross physical body of a plant, animal, or man as being held together by an "etheric" counterpart, a sort of energized sheath that serves as a blueprint for the organization of the gaseous, liquid, and solid elements of the physical body, "ether," in theosophical parlance, being matter in a finer state than gaseous, usually invisible to normal sight, but still definitely physical.

During these periods, Leadbeater testified: "I have on many occasions seen the masters appear in materialised form at the Headquarters in Adyar. The materializations were frequently maintained for twenty minutes, and on at least one occasion considerably over half an hour."

That Leadbeater was an authentic disciple of the master Koot Hoomi was attested to by another English theosophist, Annie Besant, who declared she had constantly met Leadbeater "out of the body and seen him with the master." An ardent free thinker and Fabian Socialist, Mrs. Besant had become a theosophist overnight when asked by William T. Steed, editor of the *Pall Mall Gazette*, to

Annie Besant in 1885.

review *The Secret Doctrine*, of which he could make no sense. Devouring its fifteen hundred pages, Annie Besant was "dazzled, blinded by the light in which disjointed facts were seen as parts of a mighty whole, and all my puzzles, riddles, problems seemed to disappear. . . . In a flash of illumination I knew that the weary search was over and the very truth was found."

Less appreciative, the British Society for Psychic Research, after an adverse investigation of Blavatsky's supposedly magical practices, called her "one of the most accomplished and interesting impostors in history." Undismayed, a captivated Besant sought her out to become her acolyte, her lieutenant, and, after her death in London in 1891, her successor as head of the Theosophical Society.

Behind her Blavatsky left a large following of more than 100,000 devoted adherents with branches all over the world, who acclaimed her affectionately as "White Lotus Lady"—white lotus in India being the flower that rises from murky depths to break above the water into sunlight, symbol for the illumination of the spirit freed from the prison of the body.

APPENDIX B

Seeing Is Believing

From Madame Blavatsky, and from her successors in the Theosophical Society, the world was to receive detailed descriptions—purportedly obtained through "second sight," "hearings from adepts," or "readings from scriptures"—of the subtle makeup of human beings, animals, plants, and even "inanimate" objects, all of which they claimed to be able to see surrounded and interpenetrated by "fields of formative forces."

Separately and together Annie Besant and C. W. Leadbeater, both adeptly clairvoyant, clairaudial, and clairsentient, were to spell out in a series of remarkably lucid books and articles the tenets of theosophy for those who could not peer beyond the limited spectrum of "normal" human eyesight. To share their vision—and later that of Rudolf Steiner and of Alex Podolinsky, with their unorthodox recipes for revivifying agriculture—an inkling of their notions of physics and of metaphysics is essential. This means visualizing the human body as composed of seven interpenetrating bodies or "sheaths," each one more refined as it approximates a divinely guiding "spark."

The three finest sheaths, known to theosophists as *Monadic, Buddhic*, and *Causal*, constitute a tripartite segment of the whole, postulated by all the great religions as the essence of an immortal being. The more materially mortal sheaths, following *grosso modo* the Hindu pattern, are described—proceeding from the finer to the grosser—as *mental, astral, etheric*, and *physical*. These disintegrant bodies are seen to fade away at death, one after the other, leaving the immortal trio to reincarnate in another set of "material" sheaths, at least until their bearer has become wise enough not to be viced into such a dying game.

In humans, animals, and plants, occult science maintains that the etheric body serves as the matrix for cell tissue, interpenetrating the physical to produce its shape, and extending about one-sixteenth of an inch beyond the skin. The function of this etheric body is visualized as the force that molds and holds the body in shape, and is best understood when it is remembered that in a relatively brief space of time, in a rhythm of about seven years, most if not all of the physical substance composing the human body will have ceased to belong to that body, transient molecules of matter, marshaled by the etheric power into cells, organs, and entire limbs, only to be scattered in the life process to the other kingdoms of nature, to be replaced by newer substance. Yet the body, constantly transmuted in its components, keeps whole its unity of structure as perceived by

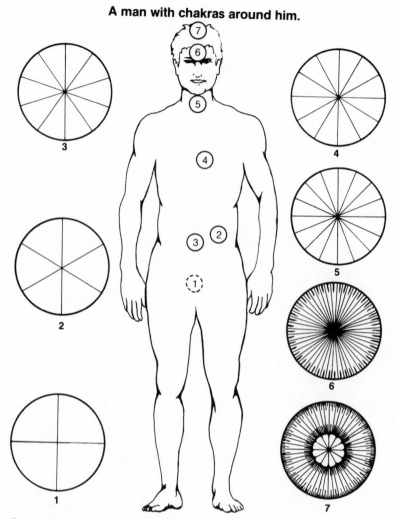

A man with chakras around him.

The Seven Chakras. (Credit: From *The Chakras*, Charles W. Leadbeater, The Theosophical Publishing House, 1927)

human eyesight. To occultists, that which renders possible this vision of the human body, eternally metamorphosing, is the etheric forces—cosmic in origin —which interpenetrate and interweave the physical.

In humans the etheric body is seen by clairvoyants as mainly lavender in color, with other colors running through it, of which orange predominates. Its vital energy is viewed as coming from the sun in the form of prana. A concept basic to both Hindu and theosophical thought—and not all that easy to assimilate —prana is said to be a sustaining, apparently "intelligent," highly active "life force," radiated by the sun. Dual in nature, prana is described as both superphysical and physical, the two forms meeting in every cell. Within the cell, the etheric double is seen as being responsible for the absorption, specialization, and distribution of this vital force.

Blavatsky speaks of the sun as the storehouse of prana, operating in seven different forms in the seven different planes. On the physical plane, she says, prana builds up all the minerals and is the controlling agent in the chemico-physiological changes in protoplasm that lead to differentiation and the building of the various tissues of the bodies of plants, animals, and man, all of which show the presence of prana by the power of responding to stimuli. By means of the etheric sheath, prana is described as running along the nerves of the physical body, enabling them to act not only as carriers of external impacts, but also as the motive force that originates from within. Without prana, say theosophists, the body would be nothing but a collection of independent and disorganized cells.

Clairvoyants view prana as entering the human body through the seven flower-shaped vortices, familiar to the New Age as "chakras," the word meaning wheel or revolving disk in Sanskrit—about four inches in diameter, close to the surface of the etheric body, connected by what appear to be "stalks" to the seven glands of the physical body at the sacrum, spleen, solar plexus, heart, throat, brow, and crown. Nerves and veins are seen as having etheric sheaths, which constitute etheric canals along which the prana can flow.

The functions of the etheric body, in theosophical terms, is not only to hold and control growth, keep the true shape of the body when it is injured or needs repair, but to function also as a connecting link with the next-finer body, the emotional or astral, and through it to the mind and ego of the person, the etheric body's chakras forming the link between physical and superphysical conscious-ness. Impacts on the physical body coming from without are conveyed by vibra-tions of prana as *motion* on the physical plane, as *emotion* or sensation in the next-finer astral body, and as *perception* in the still-finer mental body.

When a thought or perception causes the throat to tighten with self-con-sciousness, or the stomach to sink with dismay, these emotional vibrations in the astral body connect through the etheric chakras to the physical glands and nerve centers, resulting in physical sensation. Likewise, the etheric double can be shocked or damaged by noxious "astral" emotions such as anger or jealousy. The etheric also becomes separated from the physical body by death, accident, anes-thesia, or hypnotism, though a thread is said to connect it firmly to the physical body until the actual moment of demise.

Etheric matter, as described by theosophists, though invisible in ordinary light, is still considered purely physical, and can therefore be affected by heat and cold, or by powerful acids. Being subject to densification, etheric matter can also reflect light so that it can become visible, affect a photographic plate, and even be tangible, according to theosophists.

In persons who have lost a physical limb, but can still feel pain where the limb should be, clairvoyants see the etheric counterpart still persisting, much as Kirlian photography shows a ghostly etheric outline where a segment has been cut from a living leaf.

According to Leadbeater, many human beings are endowed with etheric vi-sion, and almost everyone can develop it. Full and controlled possession of etheric sight is described as enabling man to see through physical matter: a brick wall has the consistency of a light mist, the contents of a closed box can be accurately described, and a sealed letter read. With a little practice, says Leadbeater, it is possible to find a specific passage in a closed book. Remarkable psychics, such as the painter Ingo Swann, and many others, have effectively demonstrated this ability in controlled experiments at Stanford Research Institute in California.

When the faculty of etheric sight is perfectly developed, it is completely controllable, say theosophists, and can be used or not, at will, making it as easy to change from ordinary to etheric vision as to alter the focus of the eyes—the change being in reality a focusing of the consciousness of the inner, or "real" self.

With etheric vision, the earth becomes somewhat transparent to a considerable depth, as through fairly clear water. A creature burrowing could be seen, or a vein of coal or metal could be distinguished, if not too far below the surface. Etheric sight also descries several entirely new colors in the etheric band, quite different from, and more splendid than, those in the spectrum as we know it, presumably in the ultraviolet and infrared frequencies, to which normal eyesight cannot tune.

With etheric sight, the physical bodies of humans and animals become transparent enough to reveal the action of the internal organs, and, in many cases, allow for the diagnosis of disease.

Even more extraordinary is the theosophists' claim that etheric sight can make visible to sensitives many other "entities" with etheric bodies, such as those frequenting the lower orders of nature spirits, including fairies, gnomes, brownies, and so on, in great numbers and varieties.

At death the etheric body is described as withdrawing from the physical body like a violet mist, which gradually condenses into the counterpart figure of the expiring person. Many ordinary bystanders watching someone die have noticed it as a gray mist. Occultists say that the cloud of etheric matter at first does not go far, but usually floats over the denser body of a person for a few days before dissolving.

At the turn of the century, a French medical doctor, Ippolite Baraduc, obtained some extraordinary photographs of a light cloud hovering over the just-dead body of his beloved wife. Many successors, also dabbling in the medium, have produced equally remarkable prints.

All objects, animal, vegetable, and mineral, from a planet to a fly, from a cloud to a grain of sand, say theosophists, are interpenetrated by a counterpart body of etheric matter. These etheric bodies are then vivified by the next-finer sheath, the astral or desire body, which imparts to them sensation.*

---

* Not only the animal and vegetable kingdoms but even minerals are described as having etheric doubles, with life currents playing through them to awaken out of latency the finer astral matter in their atomic structures, producing what theosophists call a "thrill of desire." In minerals, says Leadbeater, desire expresses itself as chemical affinity. And certainly the attraction between positive and negative poles is basic to chemical reactions, forming bonds as powerful as any. And it may be—judging from the latest discovery of "monopoles" (free energy)—that all of physical matter consists of nothing but monopoles, variously attached and conjoined.

More lasting than either the physical or the etheric, the astral body appears, in clairvoyant sight, as ovoid, usually some eighteen inches larger than its physical counterpart. Astral bodies are described as being in continual motion, "with clouds of color melting into one another, rolling over one another, appearing and disappearing as they roll, the surface of the luminous mist resembling the surface of violently boiling water, the varied colors, corresponding to human emotions, feelings, passions, seldom pure."

In an undeveloped man, the colors are viewed as dark and gross, often so dense they hide the outlines of the physical body, being responsive to stimuli connected with the grosser passions and appetites of life.

Leadbeater says that, whereas a developed man has only five rates of vibration in his astral body, the emotionally unstable can have as many as a hundred, the whole surface being broken into a multiplicity of whirlpools and cross-currents battling each other in a mad confusion of unnecessary and weakening emotion.

In the case of a spiritually developed person, clairvoyants see a much larger astral body composed of the finest particles, in brilliant sparkling colors. The aura of the Buddha is said to have extended three miles in radius; and a friend's clairvoyant view of the present Dalai Lama showed a great dome of golden light covering several city blocks. In principle, there is no limit to the extent of a highly spiritualized aura.

During sleep, the astral body of the theosophist—and, in their lexicon, of all humans—serves as an independent vehicle for conscious action; after death, it is said to separate itself from the physical body to move freely in the astral plane.

Whereas every person is deemed to possess an astral body and many to use it regularly in sleep, few are clearly aware of the phenomenon or can consciously control its function. Others have repeatedly reported becoming fully awake and vividly conscious on the astral plane, leading active lives in their astral bodies, alternating between the physical and the astral plane. Literature abounds in appealing descriptions of the Topper-like night life of astral travelers.

Some describe the astral world as diffused, with a luminosity not clearly coming from any direction, all astral matter being in itself luminous. The passing of a physical cloud in front of the sun would make, they say, no difference on the astral plane, nor would the shadow of the earth at night.

Astral bodies are seen to be transparent, so they make no shadows; and astral sight differs from etheric sight by being more extended, four-dimensional, so that an object is seen from all sides at once, every particle in the interior of a solid being as plainly open to view as the outside, free from the distortion of perspective. A rock seen with astral sight would be no mere mass of stone, the whole of its physical matter being visible instead of just a very small part of its outer shell.

Inexperienced visitors to the astral world are warned they may find it difficult to understand what they are seeing, especially as the entire astral body is endowed with perception, so that one may perceive with any part of the body in any direction. The effect is easily achieved with the use of mescaline, but with it comes the consideration that it is both more practical and more sensational to see things one at a time with "normally" limited eyesight. More is not always better.

Leadbeater describes inhabitants of the astral plane as having the power of changing their forms with protean rapidity, and of "casting practically unlimited glamor over those with whom they choose to sport."

Communication in the astral world, as described by its inhabitants, is not by thought transference but by formulated thought—halfway between thought transference and the concrete speech of the physical world. It is therefore necessary, says Leadbeater, silently to formulate the thought into words, for which two parties must have a common language. Other travelers speak of purely non-verbal communication as even more effective.

A prominent characteristic of the astral plane, as described by those who can retain a memory of it, is the ease with which two of the densest of the astral bodies can pass through each other. People on the astral plane describe passing through one another constantly, as well as through "fixed" astral objects.

Leadbeater says that on the astral plane one never "touches" the surface of anything; one does not feel hot or cold; but on coming into contact with the interpenetrating substance one would be conscious of a different rate of vibration, which might be pleasant or unpleasant, stimulating or depressing.

To the astral body the densest rock offers no impediment to movement: "Truly one may leap from the highest cliff, plunge into a raging volcano, sink into the deepest abyss." As part of the astral scenery there are said to be materializations of the past, with a living photographic representation of all that has happened, reflected from a higher plane. The full records of the past, say occultists, are actually on a higher level and are only imperfectly reflected on the astral plane, fragmented and distorted as reflections on a ruffled surface of water. Which may account for the sometimes garbled versions of history obtained by psychics such as Edgar Cayce, a mixture, hard to separate, of what may be fact and fiction.

Suddenly-disembodied humans on the astral plane are said to be likely to find a Dantean scenery peopled by the ghosts of the dead, the astrally-traveling bodies of sleepers, a world of nature spirits, and a confusion of the astrally materialized thought forms of both sane and aberrated humans.

"A visitor to the astral world," says Leadbeater, "will be impressed by the ceaseless tide of elemental essence, ever swirling around him, menacing often, yet always retiring before a determined effort of the will; and he will marvel at the enormous army of entities temporarily called out of this ocean into separate existence by the thoughts and feelings of man, whether good or evil. Hosts of them advance threateningly, but return or dissipate harmlessly when boldly faced."

These "false" elementals are to be distinguished from what occultists call true elementals, those consisting of the four elements of fire, air, water, and earth, which they describe as forming a hierarchy of nature spirits, the lower ones with only etheric bodies, the higher ones with astral and mental bodies, responsible for the construction of forms in the mineral, vegetable, animal and human realms, translating vital energies into plants, and vital energies into the cells of animals.

Animals are said by theosophists to possess all three of the lower bodies—physical, etheric, and astral—but as yet only germinal egos of individuality. Plants, less developed than animals and humans, are said to have only physical and etheric bodies, their astral or emotional bodies being only partly evolved. To flower and reproduce, says Leadbeater, plants must have the astral brought to them by insects, birds, and nature spirits.

Animals with astral bodies apparently stay only a short time on the astral plane, the vast majority, except for certain domestic pets, not having yet fully "individualized."

When an animal dies, according to theosophists, the monadic essence, which has been manifesting through it, flows back into the animal "group soul" whence it came, bearing with it such advancement or experience as has been attained during its earthly life. Leadbeater, in an impassioned antivivisectionist plea, like that of his sensitive contemporary Bernard Shaw, maintains that the astral plane is filled with the shrieking, terrified forms of millions of animals butchered in slaughterhouses or killed for "science" or for "sport." Shaw quipped that when he died he would be "followed not by mourning coaches but by herds of oxen, sheep, swine, flocks of poultry and a small traveling aquarium of live fish, all wearing white scarves in honor of the man who perished rather than eat his fellow creatures."

The astral world of the theosophist—all around us, above, below, within, but intangible, they say—is normally imperceptible to the physical body that imprisons us because our physical particles are too gross to vibrate under the action of astral matter. †

What motivates the astral body to sensation and thus to action in the physical world is what occultists call the "mental" body. This body is seen by clairvoyants as also ovoid, its matter not evenly distributed throughout the egg but partly gathered more densely within the physical frame. When thought waves from the thinker strike this mental body, the vibrations tend to communicate to the astral, etheric, and physical matter in waves, as a bell communicates to the surrounding air.

By means of this quatrain of bodies—physical, etheric, astral, and mental—the spiritual thinker, the real person, can, according to the ancient Indian philosophy (in which theosophy is rooted) operate in the world of the physical, his thought producing emotion in a kind of "quantum of action," which produces force, which in turn moves the body. As Leadbeater and Besant interpreted the matter, were it not for the interaction of these interpenetrating bodies there would be no means for exerting physical action and, conversely, no connection between impacts on the physical body and their perception by the mental body —as in anesthesia.

Next comes the triumvirate of truly "spiritual" bodies, Causal, Buddhic, and Monadic: immortal vestures for the "I" of the being.

The "causal" body, first and most easily available of the trio, is described by theosophists as a storehouse—a kind of "bank account"—for the totality of man's experience in various incarnations. It is considered the cause of all the effects

---

† Ancient sources maintain that just as an involucrum of skin protects the human body from being overpowered by cosmic rays, so a web composed of "a single layer of physical atoms, closely woven, much compressed," acts as a barrier to unwanted forces moving in either direction between the astral and the half-physical etheric body. This web is said to be intended as a protection by nature to prevent premature opening of communication between the astral and the physical plane. Otherwise, theosophists warn, all kinds of experiences from the astral plane may overwhelm the consciousness of an individual operating on the physical plane. If astral existences seek to introduce forces that men or women on the physical plane cannot cope with, this atomic shield serves, they say, as a safeguard against emotional shock, which could drive them mad until they have become initiates.

that manifest on the planes below: its thinking directly motivates the mental body, which in turn affects the astral, the etheric, and the physical. It too is represented as ovoid, surrounding the physical body and extending about eighteen inches beyond it, a mere film, just sufficient to hold itself together and make for a reincarnating entity. Vibrations aroused in this causal body are said to manifest as colors of the most delicate hues, "beautiful beyond all conception." In the spiritually developed, it becomes a glorious iridescent film filled with lovely colors, typifying higher forms of love, devotion, and aspirations toward the divine. It is also depicted as being filled with "living fire" drawn from a still higher plane—the Buddhic—with which it appears to be connected by a quivering thread of brilliant light.

Whereas in the "causal" plane one recognizes divine consciousness in all human beings, in the "Buddhic" the "I" is no longer seen as separate from the "you," as if all were but leaves on a single glorious tree. Finally, the "monadic" sheath is described as "pure spirit," associated, like the "Buddhic," with attainments far above those accessible to normal human beings.

Thoughts generated by the causal body can have a life of their own, called by occultists "thought forms." Thought, they say, gives rise to vibrations in the matter of the mental body, which throws off a vibrating portion of itself shaped by the nature of the vibrations—much as fine particles laid on a disk are thrown into a form when the disk is made to vibrate to a musical note. This thrown-off mental matter is said to gather from the surrounding atmosphere "elemental essence of the mental world," thus becoming a temporary "living" entity of intense activity animated by the idea that generated it. Elemental essence is described as a strange semi-intelligent life energy all around us, vivifying matter of the mental plane, which readily responds to the influence of human thought. That is why occultists sometimes call thought forms "artificial elementals" to distinguish them from sylphs and undines and such like, which to them are actual nature spirits made of the elements of fire, air, water, earth.

Anyone who can "imagine," say the theosophists, has the power to create thought forms that persist and can be very powerful—"good" thought forms perpetuating as active beneficent powers, evil ones as maleficent demons. Thought forms are said to be visible to a hypnotized person, who can see and feel them as actual objects: they can also be projected onto a piece of blank paper or a film.

According to occultists, many thought forms, the result of popular fancy, have a semipermanent existence, having coalesced from the products of the imagination of countless individuals. Man, they say, is continually peopling space with a world of his own, crowded with the offspring of his fancies, desires, impulses, and passions, a current which reacts upon a sensitive or nervous person in proportion to its dynamic intensity.

Contact with the astral world, says Leadbeater, has a salutary effect on human beings in that it brings home to them the enormous responsibility they have for their own thoughts and emotions, which can have such a powerful effect in the world. He says it is significant that many of the higher types of thought forms assume shapes closely resembling vegetable and animal forms. From this he deduces that the forces of nature work along lines somewhat similar to those along which human thought and emotion work, the whole universe being a mighty thought form somehow called into existence by some great mind.

Elementals and nature spirits can, in this fashion, create etheric bodies—facsimiles of Goethe's archetypal plants—to be quickened by pranic outpourings from the cosmos and given feeling by the incorporation of astral matter. The vegetable kingdom, says Leadbeater, is much more developed than the mineral in its capacity to use the lower astral matter. "Plants are quick to respond to loving care and are distinctly affected by man's feeling towards them. They delight in and respond to admiration: they are also capable of individual attachments as well as of anger and dislike."

Leadbeater describes the sensations of plants as being a diffused feeling of well-being or discomfort, recognizable in most plants as the result of growing activity of the astral body, which causes them to probe and grow toward what is pleasant to them, such as sun, air, rain, music, or steer away from that which they consider unpleasant.

A great number of "eyewitness" accounts of experience in the astral world have been published in the West by persons who claim to be able willfully and deliberately to move in the "second body."‡

A thorough treatment of his own astral experiences has recently been presented by Robert A. Monroe, a former businessman and president of two corporations dealing in cable TV and electronics. In *Journeys Out of the Body* (Doubleday, 1971) and *Further Journeys* (Doubleday, 1985), Monroe details a series of trips starting with limited excursions to the ceiling of his bedroom, from which he confusedly looked down to see a stranger sleeping next to his wife, a shock that brought him scrambling back into his sleeping body.

Through a series of exercises, Monroe gradually learned to project himself even further, accurately describing activities of his friends and neighbors, observed while in his independently astral state and thus invisible to them. Soon he was experiencing an increasingly complex series of phenomena in what he calls the "Second State," including his discovery that the sex act is a true intermingling or fusion, "not just at a surface level, and at one or two specific bodily parts, but in full dimension, atom for atom, throughout the entire Second [or etheric] Body." The moment, says Monroe, induces "unbearable ecstasy, then tranquility."

Better to analyze the extraordinary phenomena of out-of-body travel, Monroe organized his own research institute on a hundred-acre farm near Virginia's Blue Ridge Mountains, where he developed a stable of sensitive "explorers" who, under strictly controlled conditions, describe their adventurous explorations.

But, by dint of repetition, Monroe and his explorers encountered whole new worlds of other beings, of a less-material matter than the physical, usually more

---

‡ One of the earliest, Oliver Fox, an Englishman, gave detailed reports in his *Astral Projection*, first published in *Occult Review* in 1920, of his out-of-body experiences, but they were thought too fanciful for his post-World War I contemporaries and consequently attracted little attention beyond London's "psychic underground."

Slightly better known in his time was Sylvan Muldoon, who, together with the psychical researcher Hereward Carrington, wrote in 1936 *The Case for Astral Projection*.

Robert Crookall, a British government geologist and lecturer in botany at Aberdeen University, spent more than a quarter century collecting 160 case histories of natural out-of-body experiences as well as those enforced by anesthesia, which he presented in 1960 as *The Study and Practice of Astral Projection*.

advanced and more ethical beings, with whom Monroe found he could discourse with a nonverbal form of communication, "a quantum jump beyond a talking moving color picture . . . direct, instant experience and/or immediate knowing transmitted from one intelligent energy system and received by another."

All of Monroe's and his explorers' contacts are described as benevolent, and the *Weltanshauung* they developed as a result of their wide-ranging explorations makes for a more cheerful future, devoid of the gruesome hells concocted by clerics, a cosmos in which other entities are loving and compassionate, and in which all souls are eventually saved as they develop and improve spiritually through a series of incarnations in this earthly school of very hard knocks.

Here then is the magical world of immortal spirits described by Blavatsky, Besant, *et al.*, validating the ancient *Secret Doctrine*, a world in which, as Monroe puts it,

> We can create time as we wish or the need arises, reshaping and modifying within the percept itself. We can create matter from other energy patterns, or change the structure thereof to any degree desired, including reversion to original form. We can create, enhance, alter, modulate, or eradicate any percept within the energy fields of our experience. We can transform any such energy fields one into another or others except for that which we are. But we cannot create or comprehend our prime energy until we are complete.

APPENDIX C

Three Quarks for Muster Mark

The most stunning validation of the power of clairvoyants to probe and accurately describe realms where science dares not tread was produced, at the turn of the century, by the two theosophical leaders Annie Besant and Charles Leadbeater. With intensive yoga training under expert guidance, they claimed to have acquired the faculty known in the extensive literature of Indian yoga as *siddhi*, or psychic power. It enables the yogi to develop an inner organ of perception with which to display "knowledge of the small" in a visual form.

For many millennia, Oriental yogis have used this form of perception, described as "magnifying clairvoyance," the equivalent of modern "micro-psi." The technique is described as not consisting in the actual magnification of the small object under observation, but, conversely, in "making oneself infinitesimally small at will."

With this faculty of extrasensory perception the two theosophists claimed to be able to see and describe the inner makeup of all the then-known chemical elements.

The results of this amazing investigation they incorporated, some fifty years ago, into a large illustrated volume entitled *Occult Chemistry*. Replete with detailed drawings of the constituent subatomic particles of the then-known ninety-two chemical elements, the theosophists' opus accurately describes them all, from hydrogen to uranium, along with many of their yet to be discovered isotopes.

Early twentieth-century scientists who came across *Occult Chemistry* felt justified, after a mere cursory inspection, in rejecting it as fantasy. But recent discoveries in particle physics show that Leadbeater and Besant were evidently describing in accurate terms the inner makeup of elements now theorized to consist of such particles as quarks and sub-quarks, considered to be the very basic substance of the physical world, not even postulated before the 1970s.

To validate this apparent flight of fancy has taken the determined efforts of an English physicist, Dr. Stephen M. Phillips, authority in particle physics, equipped with all the requisite degrees from recognized academies—a B.A. and M.A. in theoretical physics from Cambridge University; a M.Sc. from Cape Town University; and a Ph.D. in particle physics from the University of California.

In his recently produced *Extra-sensory Perception of Quarks* Phillips vindicates the amazing claims of the theosophical couple. Carefully documented, clearly illustrated, and strictly scientific, his book analyzes, on the basis of modern theoretical physics, the extraordinary clairvoyant talents of Leadbeater and Besant.

It was 1895 when Leadbeater first directed his clairvoyant vision upon the atoms of the chemical elements, soon joined by Annie Besant in what was to become a long series of investigations, lasting on and off for thirty-eight years. In their task they were aided by a colleague, C. Jinarajadasa, later president of the Theosophical Society, who acted as recorder during the experimental sessions, while their friend and fellow theosophist Sir William Crookes, the renowned chemist and inventor of the forerunner of what today is the television tube, provided specimens of elements.

Starting with the lighter elements, hydrogen, oxygen, and nitrogen, the investigators gradually extended their field to cover all those known at the time, plus several others, as yet undiscovered by science. They also analyzed a number of typical inorganic and organic compounds, depicting them in admirably aesthetic detail.

Annie Besant and Charles W. Leadbeater in 1900 when they were both fifty-three.

Often sitting cross-legged on a rug with a pad in her lap, Annie Besant would sketch, while Leadbeater, lying prone under the ministrations of a masseur, would visualize the interior of his atoms.

When some of the rarer elements were located for him in a museum, Leadbeater even found it unnecessary to carry out his detailed clairvoyant examination of the element *in situ* in the museum. Once he had memorized the position of the specimen, he could find it again later by visiting the museum in a "more subtle than physical body," yet still be able to dictate his observations to Jinara-jadasa.

In his *Inner Life*, Leadbeater describes how "etheric sight" can be used for magnification. "The method is to transfer impressions from the etheric matter of the retina direct to the etheric brain: the attention is focused in one or more etheric particles, and thus is obtained a similarity of size between the organ employed and some minute object being observed."

By 1907 some sixty elements had been analyzed, with variations being noted in neon, argon, krypton, xenon, and platinum, showing, as science developed, that what the theosophists were seeing were isotopes—the same element but with a different atomic weight—the existence of which was not yet postulated. Only in 1913, five years after their published results, was the name *isotope* first given to an atom differing in mass from its basic element. Neon (atomic weight 20) and a variant meta-neon (atomic weight 22) were described in *The Theosophist* six years before Frederick Soddy introduced the concept of isotopes to science.

As members of the scientific community were obliged to admit, there was no scientific reason in 1908 for suspecting a second variety of neon and no earthly purpose in fabricating it.

That same year the two theosophists published their interesting discovery that there were nucleons of two types, protons and neutrons. This was three years before the nuclear model was even proposed, and twenty-four years before the neutron was actually discovered.

By 1909 Leadbeater and Besant had studied twenty more elements, including what they called *illinium*, only recognized by science as promethium in 1945. By 1932 they had described elements 87 and 91 subsequently recognized as astatine and prototactinium.

To appreciate the difficulties in relating these researches to the contemporary state of science, says E. Lester Smith, D.Sc., F.R.S., who wrote the introduction to Phillips's book, it is sufficient to consider hydrogen. "Its atom as 'seen' by E.S.P. contained eighteen of what Leadbeater and Besant called *ultimate physical atoms*, grouped into six spheres of three apiece, spheres that appeared to be arranged at the corners of interlacing triangles." No subatomic particles, says Smith, were known then—or indeed are known now—eighteen of which could make up the greater part of a hydrogen atom, possibly its nucleus.

As Phillips tells the story, one day he was browsing in a bookshop in Los Angeles, where "by chance" he saw and bought a copy of an old theosophical book, Kingsland's *Physics of the Secret Doctrine*, which contained a few of the *Occult Chemistry* diagrams. Back in England Phillips found a copy of the third edition of *Occult Chemistry*, and, as he puts it, was hooked.

The first pointer to a possible reconciliation between what the theosophists described and what modern physics concedes, says Phillips, came when quarks were first postulated, requiring subdivision of the proton into three bound quarks. But the theosophists had further subdivided their atoms into six smaller units connected by a "very thin line of lighted force." As Phillips points out, "Between three and eighteen there still remained a factor of six to be bridged."

Smith remarked that this unsatisfactory state of affairs involved Phillips in some abstruse mathematical calculations, which led the particle physicist finally to suggest that quarks must be subdivisible, each into three subquarks: new particles to which he gave the name of omegons. If his theory were accepted, the number of ultimate particles per proton would be tripled to nine, still allowing for a factor of two. The major breakthrough made by Phillips was to realize that the theosophists' diagram of a hydrogen atom represented not one atom but some kind of compound nucleus containing two protons, a mirror-image phenomenon, which resulted from their having to slow down the atom sufficiently to be able to observe it.

When the quark idea was first proposed in 1964 by Murray Gell-Mann and George Zweig of the California Institute of Technology, only a very few people took it seriously. Believing in quarks, said Professor Harold Fritsch, seemed to require the acceptance of rather too many peculiarities, not only the unconventional electric charges of quarks but quite a few other mysterious features as well.

In view of this attitude, what was later to be accepted by science as one of the great theoretical breakthroughs of the century had to be first ushered in as a joke during an amateur cabaret show at Aspen. As Barry Taubes reports the story in *Discover*, Murray Gell-Mann jumped up from the audience on cue and babbled wildly what seemed like nonsense about how he had just figured out the whole theory of the universe, of quarks, of gravity, and of everything else. As he raved with increasing frenzy, two men in white coats came on to drag him away, leaving the audience in laughter.

Even the manner in which the new particles were named was enough to incite ridicule. The word "quark" in German describes a special kind of soft cheese, and is also synonymous with "nonsense." Gell-Mann claimed it was the number three that led him to introduce the word, there being a passage in James Joyce's *Finnegans Wake* which reads:

> Three quarks for Muster Mark!
> Sure he hasn't got much of a bark
> And sure any he has it's all beside the mark.

Reaction in the theoretical physics community to the quark model was also far from benign. "Getting the CERN* report published in the form that I wanted," wrote Gell-Mann (later to receive the Nobel Prize for physics), "was so difficult I finally gave up trying. When the physics department of a leading university was considering an appointment for me, their senior theorist, one of the most respected spokesmen for all of theoretical physics, blocked the appointment at a faculty meeting by passionately arguing that it was the work of a "charlatan." To which Gell-Mann added, modestly: "The idea that hadrons, citizens of a nuclear democracy, were made of elementary particles with fractional quantum numbers did seem a bit rich. The idea, however, is apparently correct."

Gradually the quark model developed from a bold hypothesis into a viable theory, until by the end of the 1970s six different kinds of quarks had been established, plus corresponding antiquarks. To add to the Alice-in-Wonderland atmosphere of particle physics, each kind of quark was called a *flavor*, with names such as *charmed, strange, up, down, top,* and *bottom*. Furthermore, each flavor came in three different colors, though physicists were quick to point out that the

---

* European Center for Nuclear Research, near Geneva.

colors they had chosen—red, blue, and green—had nothing to do with our ordinary perception of color; the choice was just another one of those "whimsical concepts applied to unfamiliar aspects of the microworld."

Then John H. Mauldin in his *Particles in Nature* pointed out that it looked as if there were eighteen distinguishable quarks (or thirty-six, counting antiquarks), adding that "the supposed simplification of elementary particles seems in danger of extinction."

Thus, says Phillips, not without satisfaction, "the new patterns derived by application of the rules of theoretical physics tally perfectly with the diagrams which illustrate *Occult Chemistry*."

To explain how the theosophists accomplished their remarkable feat, Phillips described the altered state of mind in which they experienced visual images of objects too small for human sight to discern. Their perception, says Phillips, was from a point of view in space that gave the illusion that the observer had shrunk in size commensurate with the objects he was viewing.

Accordingly, the experience of a person in this state is not that of passive spectator peering down a microscope, but "is characterized by a vivid, subjective sense of actually being in the microcosmos, of being suspended in space amid particles of great dynamical activity."

While functioning in this state the investigator, says Phillips, retains full control of his intellect and can converse normally with people around him, able to describe to them what he "sees." But he has to apply certain techniques of yogic meditation as he concentrates on the interior of some specimen substance that is placed in front of him. With eyes open or closed, the investigator experiences images, says Phillips, but in practice his concentration is aided by keeping his eyes closed so as to eliminate distracting images due to his normal sight. The images experienced are three-dimensional, may be colored, and usually exhibit rapid, complex motion.

The first chemical atom selected by the theosophists for examination, that of hydrogen (H), carefully scrutinized was seen to consist of six smaller bodies contained in an egglike form. The atom rotated with great rapidity on its own axis, vibrating at the same time, its internal bodies performing similar gyrations. The whole atom spun and quivered and had to be steadied for exact observation, forming two triangles, each containing three smaller bodies, described by the observers as "ultimate physical atoms" which could not be further reduced without disappearing from the physical plane into what they casually referred to as "the astral matter of another dimension."

The two theosophists claimed they could not only observe such atoms but could disintegrate them into their constituent bodies, and these in turn into smaller groups, until everything was finally broken up into what they described as free *ultimate physical atoms*, "not the atoms of which the chemist speaks, but the ultimate atoms out of which all of the chemists' atoms are made." For these, Phillips coined the term *omegons*.

The modern scientific notion of a chemical atom is that it consists of a nucleus, into which nearly all the mass of the atom is concentrated, composed generally of positively charged protons and neutral neutrons. The nucleus is minute in relation to the whole atom; the rest of the atom is almost empty, containing only the very small and light electrons circulating in their prescribed

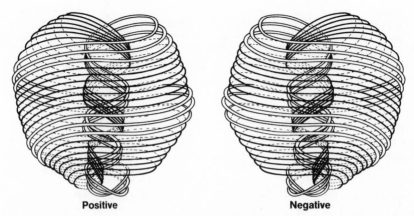

Positive   **Negative**

An ultimate physical atom or UPA as seen and described by Besant and Leadbeater, positive and negative images of each other. The UPA is described as having ten whorls or currents flowing in parallel, spiral-shaped, closed, continuous curves. Three brighter whorls flow currents of different electricities. According to the two Theosophists, force pours into the heart-shaped depression at the top of the UPA and issues from the same point, changed in character by its passage. The UPA is seen to pulsate regularly, incessantly spinning on its own axis like a top. (From *Occult Chemistry*, by Leadbeater and Besant)

orbits. The lightest element, hydrogen, is simpler than this, having a single proton or hydrogen ion as its nucleus, and one orbiting electron. As Smith points out, this picture contrasts starkly with the occult diagram depicting eighteen ultimate physical atoms in the hydrogen atom, with no electrons in evidence.

To Leadbeater and Besant, their ultimate physical atom appeared as "a little miniature sun," the basic constituent of all physical matter, dual in nature, having a positive and negative mirror image, ovoid, with ever-diminishing spirals, or "spirillae," consisting of millions of dots of energy whorling in and out from a fourth-dimensional astral plane, entering the male or positive atom, exiting the female or negative atom. Each atom, as observed, was seen to have three proper motions of its own: spinning on its axis like a top, describing a small circle with its axis like a top, contracting and expanding like a heart. To examine these atoms, says Leadbeater, required making an artificial space for them by willfully "pressing back and walling off the matter of space."

In most cases, "bright lines" or "streams of light" were seen to enter and to leave each ultimate physical atom. These lines were regarded by the investigators as "lines of force." Force, they said, "pours into the heart-shaped depression at the top of the ultimate physical atom and issues from that point, and is changed in character by its passage." This was noticed in both + and − ultimate physical atoms, mirror images of each other, but definitely not, according to Leadbeater and Besant, electrons.

Because their atoms were in such vigorous motion in all conceivable modes, it was necessary for the investigators to slow them down by a special effort of will power (or psychokinesis) before observation and counting of components were possible. Both Leadbeater and Besant claimed they could sufficiently retard the inner and external modes of motion of atoms and molecules to be able to observe them indefinitely. As observers they found they could facilitate the examination by applying further psychokinesis to dismember the atoms, in stepwise fashion, into smaller groupings of ultimate particles, but that at each step a considerably greater power of "magnification" was needed.

"The object examined, whether an atom or a compound," wrote Besant and Leadbeater, "is seen exactly as it exists normally, that is to say, it is not under any stress caused by an electric or magnetic field."

Phillips says the investigators could vary the sizes of the images at will, and that there appeared to be no limit to the level of magnification attainable, although a practical limit was set by the ability of the viewer and by the strain felt when viewing magnified objects. Unlike some other forms of extrasensory perception, the state could be induced or terminated at will.

The much-larger atoms of chemistry Leadbeater and Besant classified into seven basic types according to whether they displayed the shape of spikes, dumbbells, tetrahedrons, cubes, octahedrons, bars, or stars.

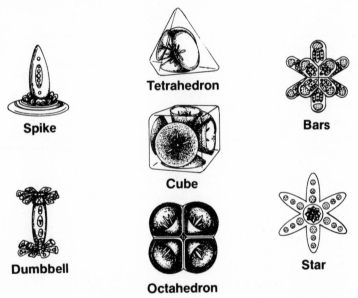

Tetrahedron

Spike

Bars

Cube

Dumbbell

Star

Octahedron

The seven fundamental forms of the elements as "seen" and described by Besant and Leadbeater in their *Occult Chemistry*.

As Smith points out, the five Platonic solids, the only completely regular solid geometrical figures, are all to be found in these archetypal atoms and molecules. Their external shapes follow those of the three simplest Platonic solids, tetrahedron, cube, and octahedron.† And in *Occult Chemistry* Besant and Leadbeater also discerned the infinite Fibonacci Series and the related Golden Section favored by Greek philosophers as representing the ideal proportion.

† For the groups of divalent, trivalent, and tetravalent elements respectively.

When the investigators disintegrated chemical atoms, they reported that their constituent funnels, bars, spikes, and star arms always separated and disappeared, releasing the contents as free bodies. When this happened, says Phillips, the "walls" enclosing groups of particles might change in shape, but individual groups did not break up or suffer any deformation, although their arrangement in clusters of groups might change.

Both Leadbeater and Besant subscribed to the theosophical view that, in the physical plane of our normal reality, physical matter exists in seven distinct states: the solid, liquid, and gaseous ones of which man has sensory awareness, and an "etheric" state, visible only to psychics, consisting of four substates, as different from one another as are solids, liquids, and gases. These they described as: the etheric, which acted as a medium for electricity; the super-etheric, a medium for light; the subatomic and atomic, the latter a medium for transmission from brain to brain.

Never having had any concept during the time of their investigations (1895–1933) of atomic nuclei composed of protons and neutrons, Leadbeater and Besant, says Phillips, believed that what they were studying when they disintegrated chemical atoms was what they called the "etheric" state of these atoms. Accordingly, they named the various states represented by the ever-smaller bodies released during successive stages of disintegration "ethers" 4, 3, 2, and 1, the last being the final state of freed "ultimate physical atoms." ‡

As analyzed by Phillips, on the basis of recent developments in theoretical particle physics, what the theosophists were seeing in the first stage of disintegration—in what they called ether 4—was the chemical atom being separated into its two component protons. Next, in ether 3, they were seeing each proton broken up into a diquark, or strongly bonded pair of quarks. In ether 2, the diquark was being split into two free quarks; finally, all quarks were broken up into free omegons, each embedded in a spherical domain of superconducting vacuum. These omegons, or ultimate physical atoms, according to Phillips, are trapped inside quarks by the same mechanism that confines quarks inside observable protons and neutrons. Thus he describes the ultimate physical atoms or omegons as being identified with those elusive entities previously postulated by physicists but never actually observed, namely magnetic monopoles—magnets with only one pole, north or south, but not both. In 1931 Paul Dirac suggested that magnetic monopoles might exist as pointlike entities with a single magnetic charge, analogous to electric monopoles, notably the electron, so far unobserved by orthodox science.

The investigators found that it was not essential for them to have the elements in a free state. By an exercise of will power they could sever the chemical bonds in compounds to release their constituent atoms. Thus common salt (NaCl)

‡ According to Phillips, when a person with micro-psi powers examines the interior of materials in the solid or liquid state, he reports generally that he observes many chemical atoms in close association, instead of single, isolated chemical atoms, which may be either bound together in regular lattice arrays extending beyond the range of his vision, or be separate, belonging to distinct groups. The composition and pattern of these groups seem to be characteristic of the compound under examination, inasmuch as the same combination of chemical atoms always becomes visible to the observer whenever he examines a sample of the chemical, irrespective of whether its identity is known to him. The chemical atoms belonging to these groups are those of the elements present in the compound.

provided specimens of both sodium (Na) and chlorine (Cl). It would appear, says Smith, that Leadbeater had very sensitive control over the "strength" of the special will power he exerted. This in turn could explain the selectivity that he appeared to exercise when looking for specific elements—a kind of tuning effect resembling the tuning of a radio to a particular station.

Most extraordinary, according to Smith, Leadbeater could use his abilities on occasion to perform apparent chemical changes of one molecule into another by sheer will power (psychokinesis in modern parapsychology) and even to transmute one element into another.*

Each ultimate physical atom was seen to be composed of ten parallel strands of "coiled coils," each of which formed a closed loop that twisted spirally around and down the surface, returning to its starting point via a narrower helix at its core. The coiled formation of each strand or whorl was found by Leadbeater to consist of precisely 1,680 turns or "spirillae," the ten whorls divided into groups of three and seven, the three major whorls being thicker and brighter than the others, carrying "currents of different electricity," each whorl consisting of billions of tiny light sources or bubbles. Further analysis, with their greatest power of magnification, revealed the bubbles to consist of nothing but empty space.

Jinarajadasa described Leadbeater's inexhaustible patience in counting the number of turns in each of the coiled strands, always coming up with 1,680, or 16,800 per atom. Annie Besant, less interested in such counting, concentrated upon the lines of force playing between the ultimate physical atoms in their smaller groupings. To Phillips's surprise, her sketches showed whorls virtually identical with diagrams published in the 1970s by theoretical physicists showing the "lines of force," or *strings* as they are now called, between quarks in smaller groupings.

In recent, unpublished research, Phillips has shown that his family of ten omegons, or subquarks, corresponds with the psychic's ten whorls, that Leadbeater's numbers can be derived from his omegon theory, and—adding further substantiation to the theosophists' insight—no possible alternatives to that theory can give rise to these numbers.

One of the strongest pieces of evidence supporting the primary claim that Besant and Leadbeater were indeed able to describe the composite character of quarks and protons through their use of micro-psi vision, one of the eight siddhis of yoga, says Phillips, was their representation of the elementary particle baryon, with noncomposite quarks as end points, an observation that was to precede the latest breakthrough in physics known as the string theory. A string model was first formulated in which strings were tightly knit bundles of so-called "color flux" (lines of force analogous to the magnetic field around a magnet) with a quark at one end of the string and an antiquark at the other. More significantly, and surprisingly to Phillips, in 1984 it was discovered that a theory of superstrings formulated in ten-dimensional space-time would be free of anomalies, provided that the symmetry of the nongravitational forces acting between superstrings is one of two possible types: one has superstrings with free end points, which inter-

---

* That animals and plants are capable of *transmuting* elements is the thesis, based on years of research, of the eminent French scientist C. Louis Kervran who, before his death in 1985, wrote *Proofs in Biology of Weak Energy Transmutations; Proofs in Geology and Physics of Weak Energy Transmutations;* and six other books on the subject.

act with one another by joining their ends to create a new superstring of the same type; the other, closed superstrings, which interact by touching at some point and join to form a new closed superstring. For some time, says Phillips, the second type was not thought to be a physically realistic model because it interacts only gravitationally. But in 1985 a new kind of closed superstring was discovered, called the "heterotic superstring," which has not only become the most studied model by physicists, but has very remarkable similarities with Leadbeater's description of an ultimate physical atom.

"Psychic observations of the structure of the basic constituents of matter," says Phillips, "are shown to be in striking agreement with superstring theory, which has emerged recently as a potentially consistent quantum-field theory unifying gravity with strong and electro-weak interactions." To which Phillips adds an explanation that wanders into the multidimensional space of modern theoretical physics: "They provide clues to the solution of the 'compactification problem' by revealing the topology of the curled-up, six-dimensional space in which superstrings must be embedded for their consistency with quantum mechanics. Most importantly, the observations indicate how the rapidly-evolving theory of superstrings should be developed."

When the micro-psi observer concentrates on the interior of a crystal or a metal, a specimen of which is placed before him, another revealing phenomenon occurs: he often claims to notice a gray mist or fine haze of light enveloping it, sometimes followed by discharges of rays or streams of "points of light" shooting out from the material in all directions. With greater magnification, the mist becomes particulate, being found to consist of myriads of similar points of light in chaotic motion.

As Phillips sums up the situation: the work of Leadbeater and Besant in the first two decades of the twentieth century, in which they described ultimate physical atoms as "spinning like a top," as having a "regular pulsation and changing shades of color," as "wobbling when external magnetic fields are switched on," as "arranging themselves in parallel lines," as "having external electric fields, and surrounding magnetic fields," and finally as "being connected by a very thin line of lighted force," bears such a striking resemblance to current ideas (some only a few years old) about "spinning, colored magnetic monopoles that are confined by strings or bundles of color flux lines, squeezed into a narrow tube by the ambient, superconducting vacuum," that the only conclusion he could come to, strange as it might seem, is that the work carried out by Annie Besant and C. W. Leadbeater over fifty years ago is in fact consistent with many ideas currently being discussed by physicists, despite the fact that most of the theosophists' work was completed well before the era of modern nuclear physics.

"How," asks Phillips, "can two individuals describe the microcosmos in a way that is found, fifty or more years later, to harmonize with both experimental and theoretical elementary particle physics?"

It was a question which he, as an academic scientist, was unable to answer. But the mere fact that he was compelled to ask the question should be compelling enough to encourage a closer view of the world as seen and described by clairvoyants with the benefit of their "etheric" or "siddhi" vision.

APPENDIX D

Steiner and Anthroposophy

Just as Madame Blavatsky's physical body was being lowered into the grave in London on a warm spring day in May of 1891, in Vienna a young Austrian of thirty was obtaining his doctoral degree by taking issue with the philosopher Immanuel Kant in a thesis entitled *The Fundamental Problem of a Theory of Knowledge*.

The fledgling doctor of philosophy, who was described as "having the appearance of white blood streaming through the veins of his dark skin," was to marry Goethe's concept of levity-gravity with the theosophists' concept of "etheric force" and produce with the result a startling new picture of the earth, its inhabitants, and especially its plant life, flourishing in an interplay of cosmic and terrestrial influences, that make for a living soil.

Clairvoyant from childhood, Rudolf Steiner, who came to be called a "scientist of the invisible," had a hard time convincing his family that beyond the visible world of matter there lay another whole world of the spirit, which he could see without difficulty. From the age of seven Steiner had been able to distinguish between things "seen" and "unseen," both of which were equally real to him, validated by his inner experience of the abstract relationships of geometric forms, which he found quite as real as the table at which he worked—giving substance to the Platonic idea that the deity geometrizes, and to Helena P. Blavatsky's statement that "nature ever builds from form and number."

At eighteen, in the summer of 1879, as he later recounted in his memoirs, Steiner met a man on his daily train ride to Vienna from the small town where he lived, with whom he could at last share his spiritual world, Felix Koguzki, an herb gatherer who collected medicinal plants in the woods near Vienna to sell to apothecaries in the city. With this untaught peasant Steiner was able to share the secrets of his "other world," the solitary one in which he had lived so long. He also learned from the herb gatherer more of the lore of nature, which had persisted for centuries in that region of Austria, unaffected by modern materialistic trends in thinking, including, perhaps, the art of tone singing.

That same year, 1879, Steiner enrolled as a student in the Technical College in Vienna, to study, among other subjects, biology, chemistry, and physics—determined to acquire a thorough training in science, if only to validate his inner vision. In his spare time, devoted to studies of optics, botany, and anatomy, Steiner discovered Goethe the scientist, as distinct from Goethe the poet and

361

philosopher. It was then that he realized that science, as taught in the academies then as now, negating the spirit, could understand in nature only that which was dead, never the living process. Goethe's scientific writings about nature indicated to Steiner a way of research into the organic realm by means of which a bridge could be built between nature and spirit.

By *spirit* Steiner did not mean anything to do with spiritualists, mediums, or "channelers," as they are termed nowadays. Like Madame Blavatsky, he considered the average spiritualist approach to be mediated through a dulled, trance-like condition of the soul, in which all those contacted were merely discarded astral "shells" of the defunct, whereas his own way of communing with the spirit world brought him to a heightened and enhanced state of consciousness in which he could see *beyond* the astral. But, when he tried to share his experience of the spiritual world, few at first seemed interested in listening, preferring to talk of spiritualism. "Then it was I," says Steiner, "who did not wish to listen. To approach the spirit in this way was repellent to me."

Steiner insisted that a new, exact, and scientific form of clairvoyance would have to be included in man's approach to science if the half-truths of materialism were not to drag the world into a materialist and mechanist disaster.

Steiner found that Goethe in his scientific investigations, which were accompanied by intense and imaginative contemplation of what he had meticulously observed, was able to deploy a method of cognition which transcended Kant's limitations of knowledge.*

By this time Steiner had become a theosophist, then under the leadership of Annie Besant, and had been made General Secretary of the German Theosophical Society. But, when Mrs. Besant put forward a plan to present a young Indian boy, J. Krishnamurti, as the reincarnation of Christ, Steiner felt he could not follow, and even when a package deal was offered him in which he was to be presented as the reincarnated John the Baptist, Steiner indignantly refused. Failing to obtain his support, Mrs. Besant expelled Steiner and the entire German section of the Theosophical Society, including all its branches, canceling the diplomas of more than a thousand members. Fourteen German lodges remained loyal to Besant; the rest went with Steiner.

With these, Steiner set up instead an organization of his own known as the Anthroposophical Society, to practice what he called Spiritual Science. By Anthroposophy—a word coined from the Greek for "man" and "wisdom"—Steiner meant that man in *this* life could achieve a state of consciousness in which he could *experience* the spiritual world of which everyone is a member, witting or unwitting. His new society gained rapid influence in Germany and grew to a large membership throughout Europe.

Anthroposophy differs from theosophy, but it would be as idle to enter into the differences as to argue the relative values of Christianity as interpreted by

---

* In 1883 Steiner was selected by the Berlin publisher Joseph Kuershner as the best talent available for preparing an edition of Goethe's scientific writings on nature. And three years later Steiner produced his own first work, *Theory of Knowledge Implicit in Goethe's World-Conception*, an essay which went deeply into Goethe's whole manner of thinking.

Another three years, and Steiner was asked to go to Weimar to the Goethe-Schiller archives, where new manuscripts had become available, to take part in collaboration with a group of eminent scholars in the preparation of a new edition of Goethe's complete works.

Rudolf Steiner.

differing denominations of that faith.

Basic to Steiner's vision is the fact that the inner, spiritual man, can live—as occultists have been aware for many millennia—in complete detachment from the physical organism, perceiving and moving within higher realms. Clairvoyance, as developed by Steiner, was to reveal spiritual facts to those with spiritual vision as clearly as men's ordinary senses reveal to the intellect the facts of the physical world. This special clairvoyance—which Steiner claimed could be developed only by modern man, was to be no "hangover from an ancient past—dreamy, unclear, atavistic, dying in the present-day world"—but a clear-cut faculty capable of providing the means by which lucid answers to scientific problems could become readily available.

The tenets of anthroposophy concerning spiritual life were developed by Steiner into an impressive array of literature and lectures. In twenty-eight published books and some six thousand lectures taken down in shorthand by his followers, Steiner detailed "cosmic history" and the "wisdom of the world." His object was in no way to keep spiritual facts concealed, but to open up the "secret" and the "occult" to the whole of mankind. His intention was to deepen man's understanding by showing how both man and the world have originated from a divine-spiritual cosmos. "There slumbers in every human being," said Steiner, "faculties by means of which he can acquire for himself a knowledge of higher worlds."

It was clear to Steiner, as it was to theosophists, that the spirit of man, for its ethical development, must live repeated lives on earth, and that the deeds of earlier lives bear fruit—through the laws of Karma—in later incarnations. But he also taught that the highest spiritual development of man is that which leads to a modification of his physical body, arguing that one cannot be an adequate philosopher if one cannot efficiently chop wood. And he maintained that so long as one feeds on inferior food from unhealthy soil the spirit will lack the stamina to free itself from the prison of the body.

Unlike the easterner's pursuit of Nirvana, Steiner's Spiritual Science was directed at *this* world: it aimed at developing the spirit for the upbuilding of both man *and* the world, leading to an increased, not a decreased, valuation of life in the physical now.

Then why, with such a genius in our midst, do only devoted anthroposophists appear to have access to what the "master" really taught or meant? A short passage from Colin Wilson's biography of Steiner gives a clue:

> Of all the important thinkers of the twentieth century, Rudolf Steiner is perhaps the most difficult to come to grips with. For the unprepared reader, his work presents a series of daunting obstacles. To begin with, there is the style, which is formidably abstract, and as unappetizing as dry toast. The real problem lies in the content, which is often so outlandish and bizarre that the reader suspects either a hoax or a barefaced confidence trick. . . . The resulting sense of frustration is likely to cause even the most open-minded reader to give up in disgust.

Colin Wilson, who in the end concluded that Steiner was "one of the greatest men of the twentieth century," and that it would be impossible to exaggerate the importance of what he had to say, twice refused to write his biography, and was only prevailed upon to do so when a replacement author found by his publisher committed suicide rather than finish the job.

Wilson suggests that like Gurdjieff Steiner may have written and spoken in a deliberately complicated style to force the reader to make enormous mental efforts. But the trouble lies more in his use of words, the recondite meaning of which differs from the conventional, a meaning which it is possible to parse only after much scattered exploration of endless Steiner texts—running perhaps to a million pages. Had he, like L. Ron Hubbard, simply invented new words to describe "spiritual" concepts unfamiliar to his listeners, anthroposophy might have spread as rapidly and as successfully as scientology. Even so, as Wilson says, the rise of the Steiner movement between 1900 and 1910 was "one of the most remarkable cultural phenomena of our time."

As a place from which to spread his teachings, Steiner erected—with abundant donations from cosmopolitan proselytes—a stunningly strange structure with twin domes and avant-garde wood carvings at Dornach, just outside Basel in Switzerland, which he named, after his mentor, the Goetheanum.

Steiner's architectural design of the partially overlaid domes, one of which was bigger than St. Peter's, was considered by experts to be the work of a mathematical genius. Large enough to seat a thousand persons, the main building was also used for the enactment of deeply spiritual mystery plays. But unaccountably, and to the dismay of his devotees, the building was burnt to the ground, they say by a disenchanted religious fanatic, to be replaced by an undaunted Steiner with an equally striking structure in concrete, the present seat of anthroposophy. There pupils prepare themselves for initiation by a life of contemplation and ritualistic enactments of Steiner's mystery dramas, in peaceful surroundings, much as in the mystery schools of the pre-Christian era.

All over Europe anthroposophical groups sprang up, to be lectured to by Steiner in places as diverse as Rome, Oslo, Oxford, and Prague. The lectures covered an enormously wide range of subjects, from cosmogenesis and the essence of the Christian myths to the therapy of curative eurythmy and care for the mentally handicapped. But basic to an understanding of the logic of his famous lectures on agriculture is his description of what he called "etheric formative forces."

Throughout this starry universe, in which we dabble as driplets of stardust, science claims the chemical elements to be everywhere the same, built up from the simplest and the lightest—hydrogen—in turn made of quarks, subquarks, and the Ultimate Physical Atoms of the theosophist, stringed, spinning bundles of what Steiner calls spiritual light, congealed.

In Steiner's cosmogony, just as in that of the East, and of theosophy, from which it derives but differs, every living being, plant, animal, or man, has an etheric body, drawn from a universal sea of etheric substance, whose forces carry out the process of life and growth, affected or directed by an astral body, plus, in the case of man, by an "ego," *mas o menos* the causal body of the theosophist.

As Paul Coroze, a French anthropologist, pointed out in the 1950s, the etheric is in a constant state of movement, transformation, metamorphosis, not disorderly, but guided by a regular rhythm, a pulse beat, a cadenced breathing. And every rhythmically repeated phenomenon in nature—movement of the stars, succession of the seasons, alternating of day and night, pumping of a heart —all point to the etheric forces at work in matter.

The astral medium, more rarefied and more sensitive than the etheric, functions to modify the etheric, operating in animals and humans through the inner astral body, which in plants does not exist. What we call the organic life all around us is really the consequence of this interplay of etheric and astral forces. A being solely in the power of etheric forces, and completely untouched by astral forces, would grow on, tirelessly and endlessly, forever repeating identical shapes. It would be a monstrous and overflowing source of life, unfading and deathless. It is the astral forces that set a term to development, check growth, and finally lead to death.

Astrality, says Coroze, brings about the law of rejuvenation, of the perpetual renovation in the universe. If the forms were not perpetually being destroyed, they could not alter, could undergo no metamorphosis. Life would die out in a sort of universal sclerosis. The forms that perish are not rebuilt in identical shapes. Each time a slight change occurs. In this way forms are modified, and can be adapted to their purpose and become more perfect. No progress, no evolution would be possible but for this continual renewing, which at first sight seems, as Coroze puts it, a sort of uselessly Penelopean task.

Etheric, or building forces, and astral, or destructive forces, are ever at play together, one working from without, the other from within the organism.

In the plant world, the astral forces work from outside, on the etheric body, which alone is woven into the plant's physical body. Astrality brings both life and death to plants. Some flowers, says Coroze, are able to move to a slight extent, and they open by day and close by night. They are endowed with sex, many plants having flowers that are either male or female, though not always on the same plant. All have male and female organs with coloring and scent not found in the rest of the plant. In both flower and fruit there are aromatic substances, glucoses, starches, etc., which are not found elsewhere in the plant.

Sir Jagadis Chandra Bose, the great Hindu botanist, found that at the moment of fertilization the temperature of a flower rises considerably. This led botanists to consider the flowering of a plant as a sort of disease, a deadly disease, as many annuals fade and die directly after seeding, and almost all stop growing at that time. Goethe considered reproduction by means of flowers and seed to be abnormal and superimposed. He called it "the original sin of the plant, because it appears to be unduly borrowed from animal sex-life, leading to the arrest of growth and often death."

The astral forces that thrust themselves into the life of a plant at the time when it flowers work there for only a short time, whereas they develop fully in both animal and man. When the etheric body and the astral body are incarnate in the same being, as in animals and man, the being has, in addition to "life," the sensation of sympathy or antipathy. That is where the animal differs from the plant. But, by adopting certain features of the animal world, says Coroze, the vegetable kingdom tends to raise itself for a moment toward the higher realm. "In return, and as a sort of offering, the plant gives these higher kingdoms the results of what the astral forces have done for it"—in the leaf and fruit of its own body. From these, the higher kingdoms draw strength for their own life, for their own etheric bodies, from the very substances that the astral forces have given to the plants. In this way a sort of exchange of gifts takes place between the two kingdoms, and between astral and etheric forces. Coroze explains the role of the plant as being one of self-sacrifice for the benefit of the higher kingdoms of animal and man.

In the purview of Spiritual Science, the physical matter that surrounds us and makes up our bodies is brought into existence out of nothing physical, held together and organized by formative forces so tenuous as to be undetectable with instruments, yet as evident as Newton's falling apple, and as mysterious as the force that raised it to the branch—called by Goethe "levity."

Fortunately for Steiner's reputation as a scientist, he was not alone in seeing and recognizing these formative forces, which he describes as organizing etheric matter—that zesty field of play within which every one of us is momentarily entrapped.

As Coroze put it: "Over and above the world of matter there lives and moves a causative realm, a realm in which the forces of creativity have not yet died the death of matter and become rigidified. Formative impulses from the cosmos press down through the elemental spheres into ever deeper layers until they finally congeal into matter."

To explain the laws of his Spiritual Science, as spirit works its wonders in the physical, Steiner elaborated on the theosophists' astral and etheric levels by postulating four etheric formative forces. These he describes as working in conjunction with the four classical elements of Greek philosophy, fire, air, water, and earth, to jointly produce the physical universe. But by these elements Steiner did not mean, of course, the substances themselves. By earth he means the quality of being solid—anything from cheese to granite; by water he means liquid, any fluid from blood to wine to treacle. And each element clearly contains something of the next: molten lava self-evidently partakes of both the solid and the liquid state, hardening as it cools, yet tending, as it is volatilized by heat, toward the next, more rarefied element of air. By this Steiner means, not just oxygen and nitrogen, but all that is gaseous. The last or most subtle of these elements is fire, with one foot in the physical and one foot in the next dimension, a notion that has persisted since the time of Crookes, who postulated fire as a transmuting substance to the fourth dimension, which some physicists have equated to plasma, or a "fourth state of matter."†

Gases heated beyond a certain point, says Steiner, disappear into the astral plane—precisely as the disintegrated ultimate physical particles described by Leadbeater and Besant vanish from the most refined ethereal state to the least refined or grossest of the astral plane.

In the reverse order, Steiner's "spiritual forces" are forever penetrating the chemical elements to create the substances that make up the physical world. The actual substances, says Steiner, are but the condensed effects of the stars, of the constellations, arrested into elements.

Fundamentally, all substance is nothing but condensed light: light is rarefied warmth, air is condensed warmth, water is condensed air, and earth is a condensation of them all. What's doing the work of spiritualizing matter, says Steiner, is the astral, designed by a thinking ego.

Mind, operating through the astral, produces Steiner's etheric formative forces which in turn bring forth the forms appearing in the physical world. As Sir Oliver Lodge, who may or may not have been familiar with Steiner, expressed it: "The universe seems to be more an immense thought than a giant machine."

To help understand what Steiner means, there are, in the stacks of the Library of Congress, 494 books by or about the author, and though almost every page is illumined by an original or arresting thought, on the whole the going is rough. Many a German anthroposophist in attempting to elucidate the master's arcane and often untranslatable prose has merely added to the complexity of the conceits at hand, darkening already muddied language with further verbiage. Guenther Wachsmuth, who for years was Steiner's secretary, and wrote a difficult biography of his mentor, also wrote a whole book called *The Etheric Formative Forces in Cosmos, Earth and Man*, in which he attempts to explain their nature, but a Swiss anthroposophist, Dr. Dernst Marti, in a more recent publication, claims that Wachsmuth got it wrong. Other anthroposophists, such as Karl Koenig, have produced brilliant insights into the world of Spiritual Science, but for the tenderfoot, groping his way through unfamiliar territory, the path is as mysterious and treacherous as the byways of the Egyptian or Tibetan Book of the Dead.

Unfortunately for most of us, Steiner chose to give to his etheric formative forces the awesomely confusing names of *warmth ether, light ether, chemical* (or *tone) ether,* and *life ether,* never really satisfactorily explaining what he meant by "ether." The best exegesis has it as the vital forces of life wedded to the formative

† Plasma: a collection of charged particles (as in the atmosphere of stars or in a metal) containing about equal numbers of positive ions and electrons exhibiting some properties of a gas, but differing from a gas in being a good conductor of electricity and being affected by a magnetic field.

forces of creation—"the breath of God (or gods)." Not much help. But such is the contribution of these formative forces to an elementary understanding of the forces at play in nature, unexplained by orthodox science—which stills wrestles unsuccessfully with the conundrum: the *origin of life*—that they bear looking into. Working together in all sorts of combinations, their fourfold joint activity constitutes for Steiner what we know as life. To add zest to this variety show, the four ethers—warmth, light, chemical, and life—complemented by the four elements—fire, air, water, and earth—behave in a way quite opposite to the elements with which they are associated. Out of this bipolarity comes the world we know.

The ethers operate as if from the periphery of the cosmos, the elements from the center of the earth. Warmth pours in from the stars, fire diffuses up and out. Light pours down in rays from the cosmos, whereas its partner, air, expands, like any gas, as if from the center of the earth toward the periphery of the cosmos. Light creates space, revealing itself in raying, illuminating, and sucking—as per its power to draw up plants or water. Water coagulates to form itself into spheres, whether droplets or a Pacific Ocean, while its partner, chemical ether, divides, dissolves, forms into octaves, as it does in its tonal quality to create the varying spaces which together make for both chemistry and music. Life ether, allied with, but in opposition to the fixed forms of a solid earth, enlivens and individualizes, develops independent organisms. It creates inner mobility, vitality, as distinguished from mere movement, or liquidity.

The elements, in themselves, are considered lifeless until quickened by the etheric forces, which work together in various degrees of conjunction. In human beings, the elements constitute the physical body; the etheric body gives it life, just as it does with the body of the earth or with the body of the universe. When the physical body disintegrates, its elements are returned to earth, water, air, and fire; when the etheric body dissolves, its constituents rejoin the etheric mass.

The etheric merely shapes and builds the physical body, following the astral blueprint created out of thought. As an example: a flower is created from unformed etheric substance, molded from the astral plane by thought into a colorful and aromatic pulse of life. By analogy: raw paint on a palette has no spiritual effect on the beholder until *enlivened* by a Carpaccio or a Titian into an arresting canvas.

In agriculture, Steiner's fundamental premise is that you cannot vitalize the earth by merely adding chemicals or minerals. Organic matter must first be spiritualized and vitalized by cosmic forces before it can, in turn, organize and vitalize the solid, torpid earth. Life comes from the astral through the etheric into the physical. Hence the development of biodynamic preparations 500 to 508. Potentized with cosmic forces they bring this life back to the soil.

Steiner's whole purpose for agriculture was to give plants the substances they need in a living rather than a mineralized form. Calcium in a living plant, he pointed out, is different from calcium in the solid mineral. To dig calcium into the soil, he warned, does not help the plant because the plant cannot take it in directly. *But calcium which has already been included in the living organism of a plant can be transferred without difficulty to another plant.* Hence the compost inoculated with the "astralizing" and "etherializing" preps.

Why should we use farmyard manure, and especially cow dung? Because, says Steiner, it is rich in *astrality*. The food that enters the cow is only partly assimilated by its organism to develop its own dynamic forces: it is not primarily used to enrich the cow's organism with material substance; most of it is excreted as dung. But having been inside the cow it has been permeated with astral and ethereal forces, in the astral with the nitrogen-carrying forces, in the ethereal with the oxygen-carrying forces.

By giving this mass to the earth, in one form or another, says Steiner, we are actually giving the earth both something ethereal and something astral, which had been in the inside of the belly of the animal. This material has a life-giving influence on the soil. It is best just at the point of dissolution, when it is about to lose both its ethereal and its astral forces. It is then, says Steiner, that the little parasites, the minutest of living creatures, come along to find the soil good and nutritive. "These parasitic creatures are supposed to have something to do with the goodness of the manure. In reality they are only indicators of the fact that the manure itself is in a certain condition. As indicators they are important, but we are under an illusion if we suppose that the manure can be fundamentally improved by inoculation with bacteria or the like. It may be so to outer appearance, but it is not so in reality."

What *are* actually effective, according to Steiner, are the cosmic forces, the astral and ethereal forces brought into play by the cow dung. For many plants, as viewed by Spiritual Science, there is absolutely no hard and fast line between the life within the plant and the life of the surrounding soil in which it is living. To manure the earth is to make it alive, so that the plant may not be brought into a dead earth and find it difficult, out of its own vitality, to achieve all that is necessary up to the fruiting process. All plant life has a slightly parasitic quality, says Steiner; "it grows like a parasite out of the living earth."

Compost on the soil, says Steiner, permeates the earthly element with the astral, without going by the roundabout way of the ethereal, and by strongly astralizing the earth it permeates it with nitrogen to kindle in it "an inner quickness and mobility which is transferred to the plants and to the animals and humans which eat them."

This influence of the astral on the nitrogen can apparently be marred by the presence of a too-thriving ethereal element. But this can be offset if lime is added to the compost heap to absorb the ethereal, draw in the oxygen, and make the astral more "splendidly effective."

Steiner explains that his preparations give the manure the power to absorb "life" from the cosmos and transmit it to the soil in which a plant is growing. The object of the preps is to re-endow the manure with the power "so to quicken the earth that the more distant cosmic substances which come to earth in the finest homeopathic quantities—such as silicic acid, lead, etc.—are caught up and taken in. Silicic acid, arsenic, and mercury are substances the plants suck upward from the soil after they have been rayed into the soil from the cosmos."

With chemical analysis, Pfeiffer found an increase of nitrate in the 500 amounting to about twenty-eight times the original content. Spectrographic tests (made by the Illinois Institute of Technology) showed sizable increases in boron, barium, calcium, chromium, copper, magnesium, manganese, molybdenum, sodium, phosphorus, lead, silicon, titanium, and vanadium. Biological research has shown that traces of these elements, even in a concentration as low as one part per million, are essential to the proper growth and nutritive value of both plants and bacteria. Molybdenum and vanadium are needed for the action of the nitrogen-fixing bacteria such as *azotobacter*. Without these trace elements, the bacteria cannot do their job of absorbing and fixing atmospheric nitrogen and of transforming it into nitrate nitrogen. That they do so in the presence of these trace elements can be seen by the remarkable increases of nitrogen found in spectrographic tests. Pfeiffer further claims that the 500 also stimulates both the production of humus and the germination of seeds by capturing "cosmic properties."

As expressed by a leading biodynamic herb farmer, Heinz Grotzke, in Wyoming, Rhode Island, who adds 500 to a field every time he turns the soil, "the 500 is one way we can allow the cosmos to enter and become part of the field."

Here we appear to be dealing with alchemy: with the transmutation of elements under the influence of cosmic forces, and the actual materialization of elements from the etheric and the astral planes.

The prime function of 500 is to promote root activity, says Steiner, especially of the fine root hairs, to stimulate soil micro-life, and to increase beneficial bacterial growth. Bacteriological tests, made by Pfeiffer during the 1930s, showed that fecal matter in the 500 disappeared while it was buried in the cow horn, whereas microflora, humus-forming bacteria, accumulated, like the castings of earthworms. Five hundred million aerobic bacteria per gram were calculated in the finished product. The preparation also serves to regulate the lime and nitrogen content of the soil, and to help release trace elements.

Having passed through the four-chambered, 120-foot bovine digestive system, the 500 contains a host of digestive secretions, intestinal flora, and what Steinerians call their "beneficial energies." These are explained by pointing to the connection between the digestive forces and the cow's skin. In horned cattle, says Steiner, precisely those points where the forces from within want to radiate outward are covered with a horn cap, and horn reflects the astral forces back into the cow's digestive process.

Buried in the ground, the cow horn also concentrates energies picked up from the soil during winter. In his agriculture lectures Steiner says:

> By burying the horn with its filling of manure, we preserve in the horn the forces it was accustomed to exert within the cow itself, namely the property of raying back whatever is life-giving and obtained from the astral plane. Through the fact that it is outwardly surrounded by the earth, all the radiations that tend to etherialize and astralize are poured into the inner hollow of the horn. The manure inside the horn is inwardly quickened with these forces, which thus gather up and attract from the surrounding earth all that is ethereal and life-giving. And so, through the winter, in the season when the Earth is most alive, the entire content of the horn becomes inwardly alive. All that is living is stored up in this manure.

The reason for stirring the 500 and the 501 in water for an hour before spraying it onto the land is to release an "immense ethereal and astral force" preserved in the cow horn, force that is useless to the plant until released by such stirring. The fettered ethereal and astral forces must be loosened so they can unfold and flow freely. And, because they are fettered sun forces, says the Dutch anthroposophist C.B.J. Lievegoed, "they are loosed by being placed into a rhythmic Sun movement, a spiralling in and out. We stir inwards into the sucking nothingness of the vortex, and stir outwards from the center of the vortex into the periphery."

An explanation is then provided by Lievegoed, almost too arcane to follow, though worth pursuing if one wants to catch a glimmer of the mysterious forces Steiner claimed to be able to see and work with.

> It is important, while stirring [says Lievegoed], to become aware of the essential difference between the two directions: the one accompanies the "light" movement of the Sun, the other is in opposition to the course of the Sun, being the "dark" movement. Through the stirring, the preserving forces are loosed and led into flowing activity. We recognize in the Preparation 500 the dark Winter sun forces seeking their way outwards through the force of the planets nearer than the Sun, past the Earth towards the periphery. This is the Sun force which stimulates germination and spring growth. It is the Winter or Night Sun.

More prosaically, by being potentized homeopathically through stirring, the preparation has its oxygen content raised by about 70 percent. Whether or not this is to be considered alchemical fancy, the results are what count, and they are impressive. For more than half a century farmers have been using the 500, all over the planet, with extraordinary and replicatable success.

Sprayed on foliage, preferably in the morning of a bright sunny day, 501 or silica helps the formation of living plant substance in the green leaves under the influence of sunlight. The purpose of 501, says Pfeiffer, is not to add nutrient elements such as potassium or phosphorus, but to stimulate and regulate what he calls "functions."

His analyses showed that the bacteria count in quartz powder, nil to begin with, ends up, after stirring, with some forty million aerobic bacteria. His spectrographic analyses likewise showed several changes in the plain quartz dust. Ninety percent silicon dioxide to begin with, it miraculously developed small amounts of nitrate nitrogen, magnesium, potassium, phosphates, copper, silver, alum, boron, barium, calcium, chromium, and zirconium. Iron increased five times, to 1 percent, while magnesium increased one hundred times.

Does this amount to creation out of the blue? If so it is a pretty amazing form of agricultural endeavor.

APPENDIX E

Planetary Powers

The earth by itself, says Steiner, is only feebly able to transmit to plants the reproductive process that growth requires; it has no power to do so without assistance from the cosmos. "It requires the cosmic forces shining in upon the earth via the Moon, and in the case of certain plants, via Mercury and Venus. With the Moon's rays the whole reflected Cosmos comes onto the earth, so that the force of growth may be enhanced into the force of reproduction."

All three bodies closest to the earth, says Steiner, affect everything connected with the inner force of reproduction and of growth in plants, producing one generation after the other on the flourishing surface of this planet.

These reproductive forces of moon, Mercury, and Venus, as Steiner explains their effect on plants, work indirectly through the limestone content of the earth and are rayed back upward from below the surface, causing plants to produce what grows annually, culminating in the formation of new seed from which more plants can grow. Direct radiation from the cosmos, on the other hand, is seen as working in the air and water above the earth's surface, so that all that goes on in the plant, throughout summer and winter, is essentially a kind of digestion, which must be drawn down into root and soil for mutual interaction below the surface.

The distant planets—Mars, Saturn, and Jupiter—which revolve outside the circuit of the sun, are seen clairvoyantly to work on the silicious element in the soil, which constitutes 28 percent of the earth's crust, forces that are then rayed up from under the surface. While these cosmic forces working on the silica are drawn up, all that is being digested above ground, says Steiner, must be drawn down into the soil by its limestone content. "Rock, sand, and stone receive the light into the earth and make it effective there. But there must be a constant interaction between what is drawn in from the Cosmos by silicon, and what takes place in the digestive system above ground."

In biodynamic agriculture, what the plant roots chelate from the soil depends on the extent to which "cosmic life and cosmic chemistry have been seized and held by means of stones and rock, which may well be at a considerable depth." The way the soil grows inwardly alive and develops its own chemical processes, depends, says Steiner, above all on the composition of the sandy portion of the soil. "Those plants in which the root nature is important need a silicious ground to thrive in, even if the silicon be only in the depths. All that is connected by way of silicon with the root nature must be able to be led upward through the plant, must flow upward, so that there is a constant interaction between what is drawn in from the cosmos and what takes place in the plants."

It is for this purpose, says Steiner, that clay exists in the soil. "Everything in the nature of clay is a means of transport for the influences of cosmic entities within the soil, to carry them upward from below. Clay is the carrier of the cosmic upward stream."

What comes from the distant planets, says Steiner, works on the plant above the earth to make the plant thick and bulky. And the fruiting results from the forces coming from these planets: "If you have apricots or plums with a fine taste, this taste, just like the color of the flowers, is the cosmic quality which has been carried upward, right into the fruit. In the apple you are eating Jupiter, in the plum you are actually eating Saturn."

A simple but vivid experiment to demonstrate Steiner's notion of the source of color in flowers is described by Al Schatz in his book *Teaching Science with Soil*. There are many species of hydrangeas, but one, *Hydrangea macrophylla*, native of Japan, where it grows to a height of twelve feet, has flowers that can be made to change color, thereby giving information about the soil in which they are growing.

If a stalk with a pink flower is placed in a solution of aluminum sulfate, the flower gradually turns blue. Even more amazing, if the stalk of a pink hydrangea is placed one half in water with aluminum sulfate and the other in plain water, the bloom becomes half blue and half pink. Blue flowers can be turned pink if their stems are immersed in a solution of citric acid. Schatz explains that this is done by chelation: the citric acid binds and holds the aluminum, pulling it out of the blue chelated complex. Schatz explains that the most effective way to change the color of hydrangea flowers is to change the type of soil in which they are planted. Landscapers can change the colors of hydrangeas by varying the pH of the soil.

In primeval times, according to Steiner, man knew how to transform one plant into another. If he wanted cosmic forces not to shoot up into the blossoming and fruiting process, but to be held back in the root, he would place such a plant in sandy soil. "In silicious soil the cosmic is held back, actually caught. But one must be able to see at a glance whatever in the plant is cosmic and what is earthly."

Green plant leaves would not be green unless the cosmic forces of the sun were living in them. It is even more so, says Steiner, when you come to the colored flowers:

> Therein are living not only the cosmic forces of the Sun, but also the supplementary forces from the distant planets. When we contemplate the rose, in its red color we see the forces of Mars. When we look at the yellow sunflower (so-called because of its shape, not its yellowness) we see the forces of Jupiter. Supplementing the cosmic force of the sun, the forces of Jupiter bring forth the white or yellow in flowers. In chicory, with its bluish color we see the influence of Saturn supplementing that of the Sun. So Mars accounts for the red in flowers, Jupiter the yellow, Saturn the blue; the green is essentially the Sun.

There is a great difference, says Steiner, between the warmth that is above the earth's surface in the domain of sun, Venus, Mercury, and moon, and the warmth that makes itself felt within the earth, which is under the influence of Mars, Jupiter, and Saturn.

> In its effect on the plant, we may describe one as leaf and flower warmth, the other as root warmth. The two warmths are essentially different, and in this sense, we may call that warmth above the Earth dead, and that beneath the surface, living. The warmth beneath the Earth is most alive in winter.

The air too, says Steiner, is permeated by a subtle vitality.

> Both air and warmth take on a living quality when they are received into the earth. The opposite is true of water and of the solid earthly element, which become more dead within it, but also more open to receive the distant cosmic forces. The air beneath the surface contains more carbon dioxide, and the air above, more oxygen.

Steiner analyzes the role of nitrogen in agriculture and its influence in all of farm production, saying that men recognize only the last excrescence of its activities, the most superficial aspects in which it finds expression. "Nitrogen as it works in Nature has four sisters, whose workings must be understood to understand its function and significance." The quartet consists of sulfur, carbon, hydrogen, and oxygen; and to Steiner they "represent what works and weaves in the living, or the apparently dead, which is only transiently dead." Together they unite to make protein, from Proteus, the Greek sea god who could assume many different forms.

Here Steiner's anthroposophical spiritual science reverts to a basic concept: that not only the earth and the cosmos but all the physical elements of which it is composed are, to some extent, living and sentient. A far-out conceit, perhaps, but no wilder than Steiner's notion of antimatter, propounded by him over half a century ago, now popular among the orthodox.

"Nitrogen," says Steiner, using the poetic imagery of spiritual science,

> is everywhere moving about like a corpse in the air. But the moment it comes into the Earth it is alive again, just like oxygen. Nay, more, it becomes sentient and sensitive inside the earth. Strange as it may sound to the materialist madcaps of today [he is speaking in the 1920s], nitrogen not only becomes alive but sensitive inside the Earth; and this is of the greatest import for agriculture. Nitrogen becomes the bearer of that mysterious sensitiveness which is poured out over the whole life of the Earth. It is nitrogen which senses whether there is the proper quantity of water in a given district. Nitrogen is conscious of that which comes from the stars and works itself out in the life of plants, in the life of Earth. Nitrogen is the sensitive mediator, even as in our human nerves and sensing system it is nitrogen that mediates for our sensation. Nitrogen is the very bearer of sensation.

Our chemists, says Steiner, speak only of the corpses of the substances—not of the real substances, "which we must rather learn to know as sentient and living entities, with the single exception of hydrogen. Precisely because hydrogen is apparently the thinnest element—with the least atomic weight—it is the least spiritual of all."

Yet hydrogen serves an essential function in the world of spiritual science in that all that is living in physical form upon the earth, says Steiner, must eventually be led back again into the universe, to be purified and cleansed in the universal All. Carbon, hydrogen, and nitrogen, which occur in leaf and flower, calyx and root, are everywhere bound to other substances in one form or another, and can become independent again, says Steiner, only when hydrogen carries them outward into the far spaces of the universe, separates them all, and merges them into a universal chaos. Alternatively, it drives these fundamental substances of protein into the tiny seed formation and there makes them independent, so they become receptive to the inpouring forces of the cosmos. "In the tiny seed-formation there is chaos, and away in the far circumference there is once more chaos. Chaos in the seed must interact with chaos in the farthest circles of the Universe. Then the new being arises." Which is as rational an explanation as any of how an oak tree is manifested from an acorn.

How different, says Steiner, is carbon in its living activity as it passes through the human body or builds up the body of a plant from the carbon we find in nature, as coal or graphite.

> In effect, carbon is the bearer of all the creatively formative processes in Nature. Whatever is formed and shaped, be it in plant—persisting for a comparatively short time—or in the eternally changing configuration of the animal body, carbon is everywhere the plastician. Yet time and again man has the faculty of destroying the form as soon as it arises, by excreting the carbon, bound to oxygen, as carbon dioxide. Carbon in the human body would form us too stiffly and firmly, stiffen us like a palm. But our outbreathing constantly dismantles what carbon builds. In plants it is fastened in a firmer configuration.

Man, says Steiner, needing to create something more solid as a basis and support—a sort of scaffolding—for his existence, yet allow what is living in the carbon to remain in perpetual movement, creates an underlying framework in his limestone bony skeleton. Meanwhile carbon, building the manifold forms in nature, as the hidden plastician, makes use of sulfur in the process, allowing Steiner to produce another of his inspired but imponderable dicta: "It is on the paths of this carbon, moistened by sulfur, that the spiritual being which we call the Ego of man moves through the blood."

Sulfur, says Steiner, is the element that acts as a mediator between the physical and the formative power of the spiritual that is spread throughout the Universe. "Sulfur is the carrier of the spiritual. Hence its ancient name, akin to phosphorus, of 'light bearer.' Sulfur and phosphorus have to do with the working of light into matter. The chemist of today knows little of these substances, little of their inner significance in the working of the cosmos as a whole."

In oxygen, according to Steiner, lives the lowest level of the supersensible world. But, circulating inside of us, the oxygen is not the same as where it surrounds us externally. "Physicists," says Steiner, "know only dead oxygen. That is the fate of any science that considers only the physical: it can understand only the corpse." Somehow Steiner's oxygen must find its way along the paths mapped out by the carbon framework. The mediator, he says, is nitrogen; nitrogen guides the life into the form embodied in the carbon. "Everywhere in the animal king-

dom, and even in the plant world, and in the earth, the bridge between carbon and oxygen is built by nitrogen."

Heady stuff, but it opens a whole new world of physics, dragged in from the realm of metaphysics, one in which the "chemical" elements act out their roles in an ever-changing pageantry, swapping costumes as warranted by script or as choreographed by some invisible director.

In the various Steinerian preps—the half-dozen remedies recommended by Steiner to "quicken" compost with cosmic forces—the elements are enabled to perform as stars as they heal the wounds of Mother Earth.

The function of BD 500 is clear enough: it invigorates the soil and gives sustenance to roots. A lucid explanation of the function of the quartz crystal in 501, ground to dust and buried for a year, is provided by Edmund Harold in his *Focus on Crystals*. According to Harold, the Indians of North America considered the powers of the quartz crystal to be sacred, and to solve the problem of sheltered hillsides with a scarcity of sunlight, which prevented the successful growth of crops, they used the power of crystals. This they would do by selecting a large crystal, which they ground into small particles and placed in large hollow cow horns to be buried in the earth for a year. When retrieved, the fine crystalline powder was scattered over the hillside at the points where the sun's rays did manage to penetrate. This served to magnify and proliferate the sunlight, providing light for all crops subsequently planted. Steiner takes the method one step further, spreading not the crystal powder but its homeopathically potentized essence.

Preparation 502—yarrow blossoms stuffed into a stag bladder and buried during the winter—is described by Steiner as being so powerful, with an effect so quickening and so refreshing, that it will make good again *any* exploited earth. For eons the power of yarrow has been known to man. When Achilles found that with the juice of yarrow he could heal the wounds of his soldiers as they lay in pain on the plains of Troy the flower became known as *stratiotes* (of the soldiers), and by Achilles's Hellenophile Roman imitators as *Achillea millefolium*, because of its many leaves. Its homeopathic sulfur content, combined in a model way with potash, is said by anthroposophists to enable the yarrow to ray out its influence to a great distance through large masses, helping to bring sulfur in the correct proportion to the remaining substances of growing plants. In Steiner's Germanically poetic terms: "Yarrow contains that with which the spirit always moistens its fingers to carry the different constituents to the plant's organs— carbon, nitrogen, etc." The moistening substance he refers to is the subtly quickening and essentially "light-bearing" sulfur.

To prepare the yarrow according to Steiner's recipe one should harvest the part of the plant that is used medicinally, the upper umbrella-shaped inflorescence, press it strongly together, and sew it into the bladder of a stag. Why a stag? Because, says Steiner, the stag is an animal intimately related not so much to the earth as to its environment, to the cosmos and its astral forces. Why a bladder? Because in the processes that take place between the kidneys and the bladder "we have the necessary forces, connected to the forces of the cosmos."

A weed, or wild flower, yarrow grows in pasture and along hedgerows, but

can go for years without flowering, just spreading like grass. It has, as the Latin implies, thousands of deep green leaves finely cut into a multitude of very small parts. When it does flower, its scent is not expansive but contractive; it has a bitter taste, which makes it as useful as hops in the brewing of beer. White or occasionally delicate pink blossoms, growing from a grayish-green stalk, appear from early summer to the middle of September. Tough and full of life, yarrow is so quick to regrow when picked that the Austrians say "it grows under the tooth of the animal that eats it."

Pfeiffer's analysis of the finished yarrow preparation, 502, showed that it acted as a biocatalyst with a stimulating effect on the plants' use of sulfur and potassium. This, in turn, affected the buildup of protein and carbohydrates. The bacteria count (as determined by Pfeiffer by the plate counting method on bee-agar-peptone, after forty-eight hours at 291 degrees Centigrade) was impressive. It showed the raw yarrow blossoms contained 30,000 aerobic bacteria per gram, whereas the finished preparation contained a phenomenal increase of up to 910 *million* aerobic bacteria. And, whereas the microflora of the raw yarrow consisted mostly of dust bacteria, the final preparation had an entirely different flora, mainly actinomycetes and bacteria belonging to the bacillus type, essential for the fixing of nitrogen. The increase in nitrate nitrogen after processing was as much as thirty-five times!

But to Steiner all of this was a secondary, derivative result: the main object of the preparations is to bring into them, and thus to the compost heap, "the creative, enlivening forces of the cosmos," forces he considered much more important than mere substances.

In the Middle Ages, when yarrow was used to purify blood, the herb was renamed *Supercilium veneris*, or eyebrow of Venus, presumably because, as Culpeper points out, yarrow is "under the governorship" of Venus, as, indeed, are kidney and bladder, according to Paracelsus and his masters the Egyptians.

A Viennese physician, Dr. Karl Koenig, barely in his twenties when Steiner gave his Koberwitz lectures, was so swept away by Steiner's suggestions for revitalizing the world's agriculture he called the event comparable to what had happened in the Persian Epoch, when grain was developed from grass, a feat attributed to the incarnate avatar Zoroaster.

A passionate pursuer of both embryology and histology, Koenig was gripped by Goethe's notion of the metamorphosis of form, a phenomenon he attributed, following Steiner, to what they both called "the formative forces in nature," forces which they came to identify with thought. "Outside, in nature," said Koenig, "these formative forces work in such a manner that they let come into existence all organic forms. Inside, in the human soul, they are the formers of our thoughts and ideas."

In his *Earth and Man*, Koenig explains why yarrow needs to be enclosed in the sheath of a deer's bladder, and why the preparation must be exposed to the sun throughout the summer before being buried throughout the winter, to be ready for use only after a full year's cycle underground. All organs, says Koenig, are a kind of scripture—kidneys, livers, hearts, lungs being separate scripts written by the formative forces of nature. "Kidneys are the organs which pave the way for the entrance of astral forces into the organism. The kidneys pull the astral into our body, and the bladder opens up and takes hold of these astral forces."

But why a stag's bladder? The stag or male deer, seen anthroposophically, is an animal in touch with cosmic forces, sensitive through its antlers to the whole cosmos around it. In Kolisko's words: "A stag is not phlegmatic like a cow. It is very nervous; it has great vivacity, awareness. It uses its antlers as a sense organ, far-reaching, with which it communicates with its environment. The slightest noise alarms the stag or deer, whereas a cow wouldn't even turn its head."

In humans the trait is known as hypersensitivity, and the horns on Michelangelo's Moses—or better, his antlers, as they are fluted—may be symbols of the antennas through which he received his commands from higher authority.

As Koenig puts it, with reference to the stag: "You must imagine that the tops of the antlers are continuously piercing through the Maya work, bringing the animal into direct contact with the astral forces all around it."

Antlers are bone, raw living bone, the most-developed organ in the animal kingdom, whose marrow is the source of blood and life. Grown in the spring, antlers are shed in winter, a process that is repeated yearly as they become richer in form, with an additional prong for each successive year. Horns spiral inward, seeking a central point; antlers spread outward, branching into slender ramifications, growing like a plant toward the sun, rhythmically with the season, grained like a tree, exposed to the surroundings when their sensitivity is most in demand.

Protected by skin and fur during winter and spring, antlers are rubbed bare during the mating season, to become highly sensitized to radiations from the cosmos. By means of this "skeletal antenna," says Koenig, the stag can receive incoming forces from the cosmos, forces that are then transmitted to the kidney, which, as Lievegoed reiterates, is governed by Venus—as is the yarrow.

Detailing the functions of the various biodynamic preparations, Lievegoed says the stag bladder is "more able to unfold its Venus activity, to create the space wherein the formative forces of distant planets can manifest their activity, forces which penetrate the element of earth."

During the summer, says Lievegoed, the bladder of the dead stag, stuffed with yarrow, hung up in the open, sucks into itself the forces from the planets beyond the sun, which are active, homeopathically, in the silica process of the warm atmosphere.

It is all part of the same game described by Steiner, in which "spirit" descending into "matter" organizes it and gives it form.

Hugh Courtney, who has become an expert at producing the biodynamic preps, claims that all of them together represent the forces of the solar system brought to earth, and he identifies each prep with one of the system's orbs. The purpose of preparing them and injecting 502 to 507 into compost, says Courtney, is to bring the planetary forces as a whole into what he sees as a unified living substance of compost with which to fertilize the earth. The earth itself is identified with the 500, whereas 501, the light-bearing quartz, is identified with the sun; 502 is seen as the yarrow of Venus; 503 the camomile of Mercury; 504 the nettle of Mars; 505 the oak bark of the moon; 506 the yellow-flowered dandelion of Jupiter; and, as the last of the compost preps, valerian, the purple of Saturn. Courtney identified 508, or equisetum, with the cometary forces, dem-

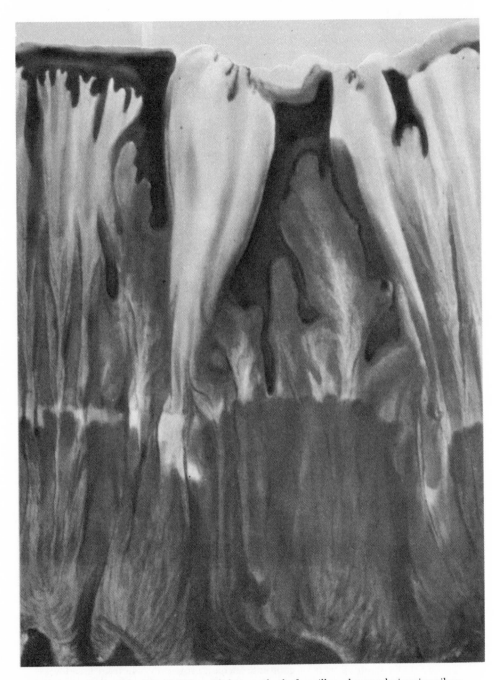

Picture obtained by Lily Kolisko with her method of capillary dynamolysis using silver nitrate and stag's urine. Observers regularly recognized the urine as coming from a stag because of the clearly depicted antler.

onstrating visually—and in the manner of Paracelsus—that a frond of the plant held horizontally does look very much like a tailed comet rushing through space. Courtney also suspects that additional preps may yet be found and developed, which will have the attributes of the unseen planets Uranus, Neptune, and Pluto, as well as of the postulated but as yet undiscovered Ringold in far outer space. There may also be, he says, an intra-Mercurial planet, closer to the sun, though modern astronomy considers it unlikely.

On the more basic side of the question, Kolisko, to discover any difference there might be between unprepared yarrow and that exposed to the forces of the sun in summer and the earth in winter, used her capillary dynamolysis. When yarrow flowers were placed in different animal bladders, her method soon showed that, compared with the bladder of a cow, horse, or pig, the stag produced superior "radiating" forces. The more animals Kolisko studied, the more convinced she became that the stag was unique.

She says she would try to get a stag's bladder as soon as possible after the animal had been killed because she wanted it in perfect condition, with its contents, so as to study the formative forces of the urine by means of her capillary analysis.

Using silver nitrate, she obtained marvelous effects: the stag revealed itself clearly through the formative forces in its urine. Visitors asked to identify from the design on the filter paper the animal whose urine it might be uniformly recognized it as a deer or stag.

Kolisko says the capillary dynamolysis revealed an active connection between the antlers and the deer's excretions, a phenomenon she explained by saying: "The formative force streams through the kidney system, penetrates the urine, is impregnated into the bladder, and is thus conveyed to the yarrow during the process of its fermentation."

Kolisko got a stag bladder with its content of urine still fresh, emptied the bladder, and immediately filled it with yarrow. When she dug it up the following spring a faint smell of stag urine still lingered. To see how long the yarrow preparation retained its potency, Kolisko kept some in its original bladder for all of eight years, noting that it continued to retain a slight smell of urine. Compared with a fresh batch, the best results were from the eight-year-old yarrow. For her compost she would only take what she needed, a little at a time, no more than a pinch between two fingers.

Whereas with yarrow one is mainly dealing with potassium influences, says Steiner, if one wants to get hold of the calcium influences, one needs another plant, "which also contains sulfur in homeopathic substances and draws them into an organic process." By this he means camomile, or *Matricaria chamomilla*, rich in sulfur.

The name *Matricaria*, derived from *mater*, indicated to Culpeper that it should be used to ease the pains of childbirth. It was also a well-known remedy for "all pains and torments of the belly." That, no doubt, is why Steiner says that camomile should be buried, not in a bladder, but in the intestine of an ox.

The pretty white flowers of camomile, which grow everywhere in meadows, blooming from May to June, exhale the characteristic aroma of the familiar tea. At the beginning of October, says Steiner, obtain fresh bovine intestines from the butcher and stuff them with camomile blossoms to make a sausage. These must be buried in good soil covered by snow in winter, but where the sun can shine on the ice crystals to melt them away in order for the cosmic forces to get to the sausage from both above and below the soil. The preparation, says Steiner, helps stabilize nitrogen in the compost heap and later raise a healthy crop. One gram of the resulting 503, dug up in spring, can affect a whole heap.

It also helps plants find the right relationship between silica and potassium, enabling the soil to take in the right amount of silica from the atmosphere and from its cosmic surroundings. Plants affected by the preparation, says Steiner, become especially sensitive to the presence of silica in their surroundings, and have an uncanny capacity to use it in the right way.

For preparation 504, Steiner recommends collecting as many common-or-garden stinging nettles as possible when they flower in June or July, the young shoots not too woody, harvesting the whole plant but for the roots. They are to be allowed to fade slightly before being buried in the soil, just as they are, without benefit of sheath other than perhaps a little peat moss or wire mesh to isolate the plants from immediate contact with the soil. The nettles must stay buried all of one winter and all of one summer, by which time they will be ready for the compost heap, where Steiner says their function is to keep its nitrogen content from evaporating.

Another biodynamic way of using stinging nettles—which Steiner called the jack-of-all-trades of herbs—is by making a liquid manure to enhance the vegetative growth of plants, especially during dry weather. A paste can also be made of nettles as a protective cover on the bark of trees and shrubs. This is done by mixing it with equal parts of fine clay and fresh cow manure. Before the coating is applied, the bark should be scraped and brushed to remove dead, loose parts. Trunks and branches that have received the treatment become smooth and clean after a few days, and the tree grows healthy.

Preparation 505, the spookiest of the lot—oak bark buried in the skull of a domestic animal—is most like medieval witchcraft. A brownish silver-gray, smooth and shiny on the outside and reddish brown on the inside, the bark is extremely rich in calcium, about 78 percent, and in older trees there is even more. When fresh it smells of tannic acid, is slightly bitter to the taste, and is a powerful astringent. Its phosphoric-acid content is about 29 percent. In human disease it was, and still is, used against bleeding, vomiting, and diarrhea.

In the spring, when the skull is dug up, the oak bark will be found to have turned to a crumbly black substance that smells like fresh soil, full of life, teeming with microorganisms, and with a highly activated calcium content.

To establish whether it was really necessary to use an animal skull, Kolisko buried oak bark from the same tree in the same spot, but in an earthenware pot, covered tightly with a fitting lid. Dug up in the spring, the oak bark from the skull came out smelling like living soil; the earthenware pot contained the same reddish-brown oak bark as had been put into it in October.

Kolisko says she tried the skulls of many domestic animals—cow, ox, calf, horse, pig, sheep—all of which worked well, but found it essential to use the skull in a fresh and undamaged condition, never split in halves. The brain, she says, should be pulled out with a wooden stick through the natural occipital opening.

Tested with capillary dynamolysis, the unprepared oak bark rose for Kolisko in the filter paper to produce an insignificant wavy border of light-brown color. The prepared oak produced intense reddish-brown colors. The most beautiful result was obtained with gold chloride: a dark purple, pointed, flamelike form radiated with light purple shades that turned to orange and brown.

One of the most efficacious of the preparations, as useful to humans when eaten *naturel* as when placed into a heap of compost, is the Frenchman's *dent-de-lion*, or 506, the dandelion that opened Hamaker's eyes. With deeply serrated leaves that yield a bitter milky juice if broken, the "weed" grows in meadows and pastures with roots that go deep into the earth. When the seeds ripen, its large yellow-gold flower turns into a small ball with long reddish seedpods beneath it.

Culpeper says dandelion is under the government of Jupiter, that it has an opening and cleansing quality good for liver, gall, spleen, and especially such diseases as yellow jaundice.

Steiner describes the plant as being gifted with the capacity for regulating the relation between silicic acid and potassium in the plant organism. Its ash contains a large amount of silica and calcium. His recipe is to collect the flowers before they go to seed, let them fade slightly, or dry them, keeping them in a cool place until the beginning of October, preferably covered with peat moss. They must then be pressed tightly together to fill the mesentery of an ox—a mesentery being the large sheath of skin that surrounds the internal organs of an animal. Steiner warns that the mesentery must be in perfect condition, without damaged parts or holes. Surplus fat may, however, be cut off, but care must be taken not to break the tender skin. Wrapped around the flowers, the sheath is tied with string, or delicately sewn. When ready, the flowers can be clearly seen peeping through the skin.

The result is buried about eighteen inches below the surface in the fall so that it spends a winter exposed to the strong earth forces, which Steiner says stream through the soil at that time of year.

In the spring the resulting soft friable material is ready to be used in the compost heap. Again, only a minute amount is needed—about a teaspoon for ten tons of compost. It is the forces, not the substance, that is supposed to be at work.

Valerian, which Steiner wants to help the plant find its right relation to phosphorus, is the last of the compost preparations, or 507, and it grows in many different locations, especially where the ground is marshy, near the edge of ponds, on the borders of woods, or between bushes where sunbeams do not penetrate directly. It is perennial, with a brown root as thick as a finger, which spreads laterally. The leaves are highly divided, dentated, or winged. The stalk rises a yard or more and branches at the top, with many small white and purplish flowers.

Culpeper saw the plant as under the influence of Mercury, useful against fevers and distempers. It is alexipharmic (serving to ward off poison), sudorific (sweat-producing), and cephalic (good for disorders of the head); it is also good for hysteria and epilepsy. The dried root, which has a characteristic perfume, has such remarkable therapeutic qualities that in England it is called "all heal." In the Middle Ages branches were hung outside old country houses as a protection against evil spirits, bad witches, and demons.

Steiner wants the flowers placed in lukewarm water and then squeezed out to produce a concentrated tincture that can be kept a long time. Highly diluted, to the seventh or eighth potency, it is sprinkled on the compost heap so that its phosphorus content can be properly used by the soil.

Kolisko says that it is quite obvious from her experiments that the valerian extract has a powerful vitalizing effect on compost, helping in its fermentation. Stored in a glass jar, a dark brown and strong-smelling tincture, it is so powerful it can easily shatter its container.

Last of the Steinerian preparations, 508, horsetail, *Equisetum arvense*, is made into a tea to be used as a field spray or prophylactic to help prevent rust and other fungal diseases. The plant, which looks exactly like its name, is widespread and common on rough ground, but also on cultivated places. It prefers dry locations to moist ones.

Biodynamic farmers say they can hardly use too much of it. Horsetail has pale brown, unbranched fertile stems, which disappear as the plant grows taller and are replaced by green barren ones. These are used for the tea, and can be differentiated from the stems of "marsh" or "shady" horsetail by the places in which they grow.

One collects the barren shoots and dries them as quickly as possible by spreading them in a thin layer in a shady place. The tea is prepared by slowly boiling the shoots in a covered vessel of rainwater, about four ounces of the dried herb per gallon. Sprayed frequently, especially on garden crops, and in cold frames, hot beds, and greenhouses, it is used against mildew, rust, monilia, scab, and soil-born pathogenic fungi.

Sensitive crystallization tests showed a general improvement in the quality of plants as a result of treatment with 508. Relative protein content and ratio of vitamin C were both improved. Kolisko noted that applications made during the morning hours were more effective than those made during the afternoon.

Dr. James A. Duke, of the USDA, has listed the extraordinary quantity of elements and compounds to be found analytically in the various Steiner preps. For a quick *aperçu*, his wife Jane has drawn them as they might appear on an apothecary's shelf. (See page 384).

With the six ingredients 502 to 507, Steiner told his audience at Koberwitz in June of 1924, as he ended his now famous lectures on agriculture, one can produce an excellent fertilizer, whether from liquid manure, or ordinary farmyard manure, or from compost. Then he added with a smile: "I know perfectly well that all of this may seem utterly mad. I only ask you to remember how many things have seemed utterly mad which have nonetheless been introduced a few years later."

He was referring, at that time, to the newly developed electric train. But the list, as it develops, is as long as history, and every bit as unexpected.

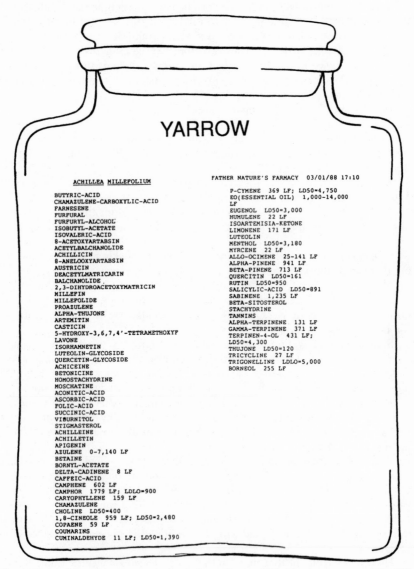

Elements contained in yarrow, the basic ingredients of Bio-Dynamic preparation 502.
(Credit: James A. Duke)

APPENDIX F

How To and Where To

How to Make a Biodynamic Compost Pile

Compost, the primary product of a biodynamic farm, is remarkably easy to produce, especially as the required inoculants, the BD preparations 500 to 507, can be readily obtained from Bio-Dynamic Preparations, Box 133, Woolwine, Virginia 24185.

The following recommendations are an amalgam gratefully taken from the writings of various biodynamic practitioners.

Select a piece of ground that is well drained, and partly shaded, not directly in the sun. The ground should be dug down about eight inches to make a trough of loose earth about four to five feet wide by six long, though it can be any length. The long sides should be oriented north-south so that both finished sides receive an equal amount of sunlight to ensure even fermentation.

The pile should be built up to about four to five feet high, narrowing toward the top. It is constructed in layers, like a cake, each layer being sprinkled with a fine spray of water or valerian-steeped water (see page 402).

The first layer should be made of twigs, about the diameter of a dime, such as might be trimmed from an apple tree. This is to ensure good drainage.

Then comes a layer of hay, weeds, and fresh garden debris, about eight inches thick.

Materials of any length or thickness, such as straw, hay stalks, tall weeds, corn cobs, or corn stalks, should be chopped or shredded if fast fermentation is required.

Next comes a layer of manure—cow, horse, chicken, or whatever. For lack of manure, a good old compost can be used.

Layers of soil should not be more than an inch or so thick. Layers of leaves and grass should be less than two inches thick so as not to compact.

Another layer of fresh green material can follow: cut grass, weeds, and kitchen wastes such as tea leaves, coffee grounds, vegetable trimmings, meat and fish scraps, floor sweepings, vacuum-cleaner contents, old woolen goods, and even old sacks. Dried blood, hoof or horn meal, and pea and bean waste can also be added.

Then add another layer of manure, and so on.

If lime is added to the pile, dolomitic lime is recommended. George Corrin,

385

an English biodynamic expert on composting, recommends only a small sprinkling being usually required, "like dusting icing sugar on a cake."

When the pile is four to five feet high, finish it off with sides sloping at an angle of 70 degrees until it is two feet wide at the top.

Inserting the BD preparations

Take a bar or broom handle and make six holes in the sloping sides (three on each side) twenty inches deep about six or eight inches down from the top and spaced equally apart. Into five of these holes insert a pinch, or better still a pellet, of each of the preparations 502 to 506. Each must go into its own hole, with no mixing. Cover the five holes. (With one set of preparations the most that can be treated is a fifteen-ton pile.)

Stir twenty drops of valerian juice (507) into one gallon of lukewarm rainwater, or good quality water, alternating one way and then the other, for twenty minutes, as with BD 500 and 501.

Divide the gallon in half. Pour one half into the sixth hole. Spray the remainder in a fine mist from a clean sprayer over the entire pile.

Finally, for protection against the sun and against too much rain, cover the pile with an inch or two of straw or peat moss and a burlap bag.

The finished pile should remain moist at all times, holding water like a damp sponge, but not accumulating water so that it runs out or collects at the bottom. Fifty to sixty percent moisture by weight is required for proper fermentation. Each particle of fiber should be almost shining wet.

It is vital that the pile be so constituted that air can circulate in and through it. Its aerobic fermentation means that air, or rather oxygen, can reach all parts of the compost pile at all times. As the activity of microorganisms yields carbon dioxide, this must find its way out to the atmosphere. Only aerobic fermentation yields a product with all the beneficial effects on soils and plants. When fermentation is aerobic, few or no flies will breed on or even approach the pile.

The pile is now ready to start fermenting and decomposing, not rotting.

To test whether the material in the heap is behaving as it should requires some attention—the use of the senses of sight, smell, and touch, and the power of reasoning.

In the new pile an enormous growth and increase in the activity of microorganisms will begin. Their metabolism yields heat, which will develop in the first three days up to a temperature of 150 degrees, or more.

Composts that contain much soil will seldom get very warm; composts with much manure, young plant tissue, and garbage will become very hot. A higher moisture content keeps the temperature lower. It is best to keep the pile between 120 and 140 degrees. If the pile falls to half its size during the first few days, too much air has entered and it has burned like a bonfire.

To reduce heat, make crowbar holes all over the pile to let in air, and dry it out. A gray mold indicates too much heat. To cool the pile, turn the hose into the holes for a while. After a few days, close the holes.

A pile will cool off in fifteen to twenty-one days. If it is turned, so that the top is on the bottom, and the inside is on the outside, it will heat up for another fourteen days. But a pile in proper condition from the start should not need turning.

Earthworms are a sign of good progress. A good pile will be wriggling with worms, which will disappear when the decomposition is complete, leaving their capsules and a tremendous amount of very valuable castings.

Three stages take place before the pile becomes humus:

The original smell disappears and the material takes on a woodsy odor. This may happen in only a few days.
The color becomes uniformly dark brown.
The original texture disappears and the pile looks like rich soil.

In the first phase, the crude source material is broken down by microorganisms, bacteria, and fungi into its original proteins, amino acids, proteids, cellulose, starches, sugars, and lignin. In the second phase, the microorganisms transform the broken-down source material to build their own bodies. In the third phase, as they die, they produce humus of two kinds, lasting or perishable. The stable kind builds up soil; the perishable burns it up.

Best results are obtained from the finished product in about two to three months. Thereafter, the humus will keep indefinitely with protection from sun and drying winds, and with sufficient moisture. Ripened compost can be stored for several months.

If used before stage three, do not plow under, but mix with topsoil so the air continues to have access.

A second pile on the same foundation will generally do better than the first, and have even more earthworms; they will stay in the ground and invade the next pile.

Application of Compost

Compost is applied in the spring and fall to provide a high number of living organisms to the soil with each shovelful of compost.

In the garden one works the compost into the surface layer, two to four inches down, or puts it into rows for seeding, or the holes for planting.

For row crops on the farm, it is worked into the topsoil or broadcast on grasses and hayfields. It should not be dug in deeply, but spread by shovel or muck spreader and worked into the soil with a rake, cultivator, or disk harrow.

Average applications of five to twenty tons per acre are usual every four years in rotation. Ripened composts for basic soil improvement are worked into the upper two- to four-inch topsoil layer in amounts of fifteen to thirty tons per acre. Partly-processed garbage compost for land reclamation and for combating erosion can be applied in a layer half to two inches thick, the equivalent of 40 to 160 tons per acre.

According to Pfeiffer, an old rule suggests that good compost can be applied at any time on any crop and in any amount. But the better the quality of compost, the less is needed.

How to Use and How to Make Maria Thun Barrel Compost

Barrel compost is a homeopathic supplement to regular biodynamic organic compost. Its function is to get the beneficial effect of the preps into soil more often than is possible through ordinary manuring and composting in the normal course of crop rotation. BC is not difficult to make, providing one has access to the required basic materials; otherwise it can be readily obtained from Bio-Dynamic Preparations, Box 133, Woolwine, Virginia, 24185. The current price is under a dollar for a unit weighing a couple of ounces.

One unit, properly seasoned, stirred into three gallons of water, sprayed onto land or garden, will cover an acre. Its effect is comparable to that of regular BD compost, though not so long-lasting. Biodynamic farmers consider it essential to use either compost or barrel compost as a precursor to the spraying of BD 500 and 501.

Barrel compost consists of cow manure, ground-up eggshells, and basalt dust, carefully mixed and inoculated with the Steiner preparations 502 to 507 (see pages 397–402). It is seasoned for approximately twelve weeks in a barrel buried in earth. The purpose of injecting the biodynamic preparations into the compost is to assist in the decomposition process of the manure, and to produce a sweet-smelling substance to improve soil structure. Barrel compost is especially useful for changing from orthodox to biodynamic methods, as it enables one to use the biodynamic preparations more frequently.

For those willing to make the effort, a single barrelful is sufficient to spray onto some two thousand acres. Seasoned, and properly stored in a cool cellar, it will last a long time.

Potentizing the barrel compost in a bucket of water

Thun recommends putting sixty grams (about a handful) of the finished compost into ten liters (two to three gallons) of water, preferably rainwater that has been exposed to the air, to be potentized by stirring in the same manner as the 500 or 501, but for only twenty minutes.

The resulting potentized liquid will cover about half an acre, or a quarter hectare. Thun claims it to be most efficient when used three times in relatively quick succession, about a week apart. It should be sprayed onto the soil in large drops like the 500. This is most easily done with a bucket and a whitewash brush; it can also be done with a regular sprayer.

Before spraying on pasture or lawn, one may lightly scratch the surface with a harrow to aerate the soil and scatter any cow droppings. Spray first during a waning moon, and if possible on a leaf day. Eight to ten days later, spray the 500. Fifteen days later, with an ascending moon, and toward evening, spray 501. Neither the 500 (which helps the growth of roots) nor the 501 (which helps leaves absorb sunlight) goes well on new soil unless it has first been sprayed with barrel compost, which, says Thun, draws the "formative forces from the heavens into the earth in a balanced, harmonious way, making them more accessible and quicker in their action."

Market gardeners use the barrel preparation as a means to apply the equivalent of compost to land that does not receive the actual compost in rotation.

On lettuce beds it should be sprayed at the first hoeing.

Farmers, says Thun, have regularly experienced one-third higher yield as a result of using the barrel compost, and have reported obtaining four crops off the same plot of land within one growing season.

For plants under stress, due to drought, insect attack, or other reasons, a "first

receive two-thirds of the barrel. If the climate is very wet, the barrel should not be buried so deep.

The earth at the bottom of the hole should be as fertile as possible. If you have a poor soil, add some very-well-decomposed compost or fertile soil.

Soil removed from the hole should be banked up around the sides of the barrel to form a slope so that rain will not seep into the hole beneath the barrel. Banking, says Maria Thun, enables the soil forces all around the barrel to work on what is inside.

It is best to dig the hole and install the barrel on a fine day, in the afternoon, preferably during the time of the waning moon. One should absolutely avoid difficult days, says Maria Thun, days of moon nodes (two per month), nodes of Mercury, Venus, and all the other planets.

Filling the barrel

Half fill the barrel with about twenty-five gallons of the manure, eggshell, and basalt mixture. Make five six-inch holes in the surface and place in each one-half ounce—about the size of a cherry—of each of the BD preparations 502 to 506. Add to the barrel the other half of the manure mixture and repeat the dose of BD preps in five more holes. Finally, sprinkle the contents with an infusion of 507, or valerian, five drops to a liter of water, previously stirred for twenty minutes. Any of this liquid left over can be sprinkled on the earth surrounding the barrel.

The barrel should be covered, so that rain does not enter. A piece of slate is recommended, held up a couple of inches by a picket at one side so that air can enter freely.

After a month and a half, thoroughly mix the contents with a shovel for about ten minutes to homogenize it and eliminate unaired lumps.

In about two to three more weeks the barrel compost should be ready.

Barrel compost, which requires much less time to season than regular compost, can be made at any season, but spring and autumn are preferable.

Storage

The finished compost should be placed in a barrel in a basement or cave. It will keep a long time. After two years it looks like a light humus and still gives good results.

How to Use the Biodynamic Preparations 500 and 501

BD 500 is cow manure that has spent the winter buried in a cow horn. The resulting friable humuslike substance is obtainable in the United States from Bio-Dynamic Preparations in Virginia.

Stirring BD 500

For one acre of land stir 1½ ounces (35 grams) of BD 500 into 4 gallons of water in a plastic or ceramic bucket.

The preparation must be stirred by hand or with a stick for one full hour, about twenty seconds in one direction, to form a deep conical vortex down to the bottom of the container, then reversed to form a chaotic mass, which becomes another vortex circling in the opposite direction. The stirring should be done with cheerful intent, not as a chore; and several hands may stir in turn.

Various ingenious methods have been devised for stirring large quantities in more than one container simultaneously. Any system that works will do.

On farms that have mechanical stirring equipment the batches can be larger. The amount of liquid is limited only by the time required to spray it *all* onto the ground. The freshly stirred preparation should be applied within no more than three hours. Maximum effect can be expected within the initial one to two hours after stirring.

Water for the 500 should be lukewarm. Rainwater is best, collected from gutters and kept in a barrel or tank. Clear river water can be used, but it may contain slightly more inorganic and organic substances than rainwater. Never use chlorinated water.

It is advisable to let the rain or river water stand in an open barrel for two to three days to let the light penetrate. The same applies to tap water, which should be allowed to stand for a longer time and should be stirred several times for a minute or two. It should be kept in wooden barrels or glazed earthenware crocks. Oak barrels are preferred. Farms using more water may use large tanks, but they should have a good finish and be free from rust. They should be cleaned with utmost care, using a clean brush, with hot water and a 2 percent lye or soda solution, but no detergents, then rinsed several times with hot and then cold water.

Spraying equipment

Depends on size of farm or garden. Better not use sprayers that have contained other liquids, especially toxic ones.

On small garden patches, liquid can be sprayed by hand with a whisk broom or whitewash brush, flipped with a snap of the wrist in a wide circle, to distribute fine drops evenly.

Small hand sprayers that form a finer mist are commercially available in many sizes.

The next size is a knapsack sprayer, which suffices for a garden or small farm.

There are many larger tractor-drawn or mounted tanks and sprayers that can cover a wide strip. These are arranged so the nozzles face downward, ending no higher than two feet above the surface of the soil; otherwise too much of the fine mist gets blown away.

Sprays applied by pressure sprayers must be strained carefully, or they will clog the nozzles.

For gardens

BD 500 can be applied early in the spring, and again before planting. Soils in cold frames, hot beds, and greenhouses are sprayed before sowing or planting. If possible seed drills and holes should be sprayed before planting.

The rate of application can be higher than on farm land: one portion to cover between two and four thousand square yards.

The stirred 500 is applied directly on the soil, usually ahead of planting; but it is desirable that the application be followed by an early shallow cultivation, such as with a harrow or a garden rake.

500 should be sprayed on in the second half of the afternoon, after a rain, when the soil is slightly moist, never dry, or sealed with a crust. It should not be sprayed around noon, in the heat of the day. It is best if the sky is at least partially overcast (the material then follows the natural daily rhythm of the humidity in the lower atmosphere).

Spraying is not done during rain, and should be postponed if there is possibility of rainfall soon afterward.

During drought conditions a spraying of 500 late in the evening, or (early in the morning) when dew is present, serves to relieve the plants of stress.

BD 500 on the farm

The spray is applied before the last harrowing or cultivating preceding the planting of winter grains or other fall crops, including catch crops. On pastures and hayfields it can be used sometime before a freeze is expected. But more frequently one prefers to treat permanent grasses and hayfields early in spring in order to stimulate growth at the start of the season. On cropland the spray can be applied after the soil has permanently thawed. Mostly one chooses a time during spring cultivation, ahead of planting.

Biodynamic growers prefer to work the soil at intervals in spring in order to control the weeds that germinate from seeds. It is a good measure to include pasture and winter crops in this treatment. It helps strengthen root development of those fields that have suffered from frost or wind erosion.

Storing BD 500

Whether it is made on the premises or obtained from another source, once removed from the horn BD 500 should be stored in a cool and not too dry cellar in an earthenware crock that has a lid of the same material or is covered with a piece of slate or stone. The crock must not be sealed, allowing for the circulation of air.

The container should be placed in a wooden box and be surrounded by a layer of sphagnum or peat moss. The cover of the box, usually made of boards, should have a cushion filled with peat moss on the inner side.

In a dry climate, or during a long dry spell, it may be necessary to keep the peat moss slightly moist, just enough so it does not absorb moisture from the preparation.

In a humid climate, or when the room is rather humid, this is not required.

Those who order the preparations already made should take them out of the traveling container and store them in the same way.

The substances are not dead: if kept under too humid conditions they will go moldy. But they should not be allowed to dry out. Air must have access, but slowly, in order to preserve the microbial life in the preparations.

Root dip with BD 500

This has a soupy consistency. It is composed of equal parts (by volume) of cow manure and subsoil loam, stirred with horsetail tea (one part dried horsetail in twenty parts of water), and stirred 500. Roots and transplants, including vegetables such as cabbages, tomatoes (unless they are already in soil blocks), also the roots of trees and shrubs, will benefit when dipped into this material before they are replanted.

How to Use BD Preparation 501

501 (obtainable from BioDynamic Preparations) is ground-up quartz crystal that has been buried several months in a cow horn. It is to be homeopathically stirred into water for a full hour, the same as with BD 500.

The result is sprayed onto the foliage of growing plants, and acts as a supplement to preparation 500, helping to bring sunlight to the leaves.

The amount to be stirred into water to be sprayed on one acre (or 2,000 to 4,000 square yards in the garden) is a bare 1 to 1.5 grams (1/20 ounce) in 4 to 5 gallons of water.

The resulting potentized spray is applied late in spring or early summer. One sprays in the very early morning when a warm and at least partly sunny day may be expected. The proper growth stage is at the time when the organ of the plants one wants to harvest has begun to form. If applied when the sun is too high, it has been known to cause burns on foliage.

It is important to spray 501 in a fine mist.

To achieve maximum results, spraying of preparation 501 must be preceded by treatment of the soil with preparation 500, which should be preceded by a spraying of barrel compost (see page 388). The effect of all three sprays is stronger than each alone.

501 is said to have the power to stimulate fruit and seed formation; it is also said to improve the flavor, keeping quality, and nutritional value of crops as well as making them more resistant to disease and insects.

Small grains receive the 501 treatment after tilling, when the stalk starts to elongate.

Subsequent applications can be made as they help prevent lodging.

Corn is treated when the stalk starts to elongate and when one can easily drive through the field.

Alfalfa, other hayfields, and permanent grasses receive the treatment not too long after strong growth has started.

Provided there is enough moisture in the ground and one is not in a dry spell, further applications can be made after cuttings have been taken off the field, or a grazing period is finished.

Garden crops that have to be transplanted should not get the spray until they are finally set out.

Lettuces and spinach should receive only one application in the morning; but to prevent them from bolting, another application can be made in the afternoon, tending to drive the forces back toward the roots, as happens normally in that part of the day.

Flowers, tomatoes, strawberries, and fruit should be treated when the flower buds are visible and ready to open.

Potatoes like 501 when the flower formation starts.

Cabbages of all kinds, and cauliflower, broccoli, and other leafy vegetables that grow much bulk, should respond to repeated applications by growing a finer tissue, developing good taste and keeping quality.

Kitchen herbs, soft fruit, tomatoes, melons, and so on, should have their taste improved by repeated applications.

Apple trees get applications during the late stage of flower buds, then again when the fruit is being developed. Spraying 501 on apple trees can be combined with a stinging-nettle spray.

For greenhouses, where such plants as cucumbers, tomatoes, lettuces produce large amounts of plant matter within a short period of time, repeated applications of 501 are recommended.

Carrots should be sprayed on a root day, spinach on a leaf day, and so on according to the desired result.

Steiner recommended the use of biodynamic preps as homeopathic medicine for the living earth to regenerate the soil. Biodynamicists point out that a remarkable change in the condition of the treated soil occurs, that it becomes more crumbly and fibrous, and retains moisture more easily. The appearance of crops is also improved as they become more resistant to drought and infection. Soil and crops show remarkable improvement, even after a short time. The full effect, however, is said to appear in the course of three or four years. It consists in the continuous increase of the fertility of the soil, and an improvement of the quality and flavor of the produce. Both the plants and the soil in which the forces of life are stimulated are found, according to biodynamic farmers, to provide themselves actively with the required substance by attracting them from the surrounding circuit of forces, just as a healthy organ in the body supplies itself actively with what it needs from the circulating blood and the other juices of the body.

How to Make Biodynamic Preparations 500 and 501

Cow horns can be obtained from any cooperating slaughterhouse, and do not have to come from biodynamically fed cattle. But they should be from cows, not bulls or steers. Cow horns are generally thicker and heavier.

To clean, place the horns in fifty-five-gallon drums full of water; cover with plastic to prevent the spread of odors. After a couple of weeks, the thin layer of flesh surrounding the bony core of the horn will rot away; the bones can then be removed, leaving the horns empty. A less smelly method is to leave the horns to dry out. After a certain loss of moisture between bone and horn, the bone will fall out if the horn is struck sharply with another horn. Thereafter the horns may be stored indefinitely.

Manure

In a Northern-Hemisphere temperate zone, cow manure is collected between the fall equinox and the winter solstice. It is desirable that the animals still be grazing, or have part-time pasture or good hay plus some green feed.

In the Western Hemisphere, the horns should be well and tightly stuffed with manure that has been sieved to remove twigs or other foreign objects.

The stuffed horns are buried about two feet below the surface in good rich earth, in the fall, prior to the winter solstice (December in the Northern Hemisphere, June in the Southern), and are left there until the end of winter.

Biodynamic farmer Hugh Lovell, of Blainsville, Georgia, recommends placing the horns tip down, pointing toward the center of the earth, claiming the bovine horn to be an antenna that picks up the telluric forces of winter. Hugh Courtney stacks them in a circle, point down; but in Australia they are laid flat in rows, separated by a thin layer of earth. All methods seem to be effective.

Sara Sorelle stuffing a paste of ground-up quartz crystal into a cow's horn to be buried for a year to produce BD 501.

Quartz crystals

These are obtainable from any good supply source, especially in Arkansas, where there are several mines. A fine powder is produced by grinding the crystals in an iron mortar and pestle, or by any other available method.

Mixed with water, the powder is stuffed into cow horns and buried at the beginning of summer, before St. John's Day. Dug up in early fall, the finished 501 should be stored in a glass jar and placed in a sunny window.

Biodynamic Preparations 502 to 507

These preparations are used to inoculate a regular compost heap or Maria Thun's barrel compost, as described on pages 386 and 388.

The preparations are obtainable from Bio-Dynamic Preparations, P.O. Box 133, Woolwine, Virginia 24185.

Instructions for making your own BD preparations are derived from Rudolf Steiner's 1924 lectures, entitled *Agriculture*, with additional hints from various biodynamic farmers.

502. *The blossoms of yarrow buried for a year in a stag bladder*

Harvest the florets of the common wild white yarrow (*Achillea millefolium*) in spring or early summer, taking the part that is used medicinally, the upper, umbrella-shaped inflorescence. They should then be allowed to dry.

Stag bladders stuffed with yarrow flowers, hanging in the summer sun before being buried for the winter to produce BD 502.

The dried flowers, moistened with a tea made from the same flowers, are stuffed into the bladder of a male deer, elk, or moose.

A bicycle pump can be used to inflate the bladder, or a small balloon can be inserted and blown up with a straw through the ureter duct.

When ready to be stuffed, the dried bladder is moistened with lukewarm rainwater. Moist yarrow flowers, pressed strongly together, are inserted into the bladder stretched to capacity. The hole is sewn with thread, and the bladder hung six feet above the ground all summer in a place as exposed as possible to sunlight.

In autumn the bladder is buried not too deep—ten to twelve inches—in good earth, there to spend the winter.

Dug up after having spent a full year in the bladder from the initial stuffing, the yarrow is placed in a cool cellar in an earthenware pot surrounded by peat moss in a wooden box along with the other preparations.

503. Camomile buried in a cow's intestine

Harvest the wildflowers when they blossom early in the year. As the flowers are small, Hugh Lovell recommends harvesting with a cranberry picker or large-toothed comb to pick several at once. Drying is best done in the shade, on paper or cheesecloth. The longer the blossoms need for drying the poorer their quality.

There is an old belief that the most powerful camomile blossoms for healing are those picked before St. John's day. Afterward the witches are accused of wetting and spoiling them.

In the fall, soak the dried camomile in a camomile tea and stuff them like sausage meat into the intestines of a freshly killed cow. These should spend the winter underground and be dug up in spring.

Biodynamic farmers find a mysterious affinity between camomile and calcium in nature. When the finished camomile preparation is added to the compost it is said to guide the calcium forces in the breakdown of raw material.

CAMOMILE TEA

The definition of this "tea" is one teaspoon of camomile to one cup of water.

The tea is beneficial as a seed bath, especially for seeds that succumb easily to damping-off or seed-borne fungi and bacterial plant diseases.

The seeds are left in the tea for ten minutes and then either directly planted or redried and kept for later use. The same tea can, in diluted form, be used in the greenhouse for watering seedlings, with equally effective results.

Pliny suggested using camomile poultice or bath as a cure for headache and illness of the liver, kidney, and bladder. Today it is still considered a remedy for flatulence, stomach pains, catarrh of the intestines.

504. Stinging nettles buried for a year or more

To make the preparation, the mature plants of *Urtica dioica*, stem and leaves, are picked before they are in bloom, then buried for a full year in a good humus soil.

The stinging nettle is a perennial, which grows as tall as six feet, with a square stem and little hairs; the leaves are opposite and alternate. Little blossoms form like grape clusters in the leaf axils and are pollinated by the wind. The tiny seeds —a thousand of them weigh only .15 grams—ripen in August and can be used the same fall to establish a planting of nettles.

Steiner recommends collecting as many as possible common or garden stinging nettles before they flower in June or July, the young shoots not too woody, harvesting the whole plant but for the roots. They are to be allowed to fade slightly before being buried, just as they are, without benefit of sheath other than perhaps a little peat moss to isolate the plants from immediate contact with the soil. Hugh Courtney favors covering the nettles with a wire mesh, to keep the earthworms from carrying the nettles away and to help find the humus when digging for it the following year.

The nettles must stay buried all of one winter and through the following summer, by which time they will be ready for the compost heap, where Steiner says one of their main functions is to keep its nitrogen content from evaporation.

Underground, the nettles undergo fermentation and turn into dark humus, which is then used to stimulate soil health, providing plants with individual nutritional components such as sulfur, potassium, calcium, and iron. Nettles are able to assimilate iron from the soil and build it into their tissues. Pfeiffer says the most interesting change during the preparation is that the stinging-nettle humus is enriched about one hundred times in molybdenum and vanadium, the trace elements necessary for the activity of nitrogen-fixing bacteria.

The shape of the nettle leaf resembles the heart symbol, indicating to biodynamicists, along with its serrated edge, a relation to the rhythmic forces of the universe. They say that just as the rhythmic system in man, which is centered in the heart, continuously counteracts degenerative forces and strengthens health, so does the nettle radiate healing forces into its surroundings, offering itself as a remedy for many illnesses, including those of plants. In old herbals the nettle is recommended as helpful for ailments of kidney and bladder, skin diseases, and bleedings of all kinds. All herb mixtures sold as blood tonics contain high amounts of nettles.

Another biodynamic way of using stinging nettles—which Steiner calls the jack-of-all-trades of herbs—is by making a liquid manure to enhance the vegetative growth of plants, especially during dry weather.

A paste can also be made of nettles as a protective cover on the bark of trees and shrubs. This is done by mixing it with equal parts of fine clay and fresh cow manure. Before the coating is applied, the bark should be scraped and brushed to remove dead, loose parts. Trunks and branches which have received the treatment become smooth and clean after a few days, and the tree grows healthy.

Hugh Courtney extracting the finished 505 from a cow's skull that was filled with oak bark and has turned to humus after spending the winter buried in a nearby creek. One teaspoonful will be sufficient for ten tons of compost.

505. Oak bark buried in the skull of a domestic animal

Take the bark of an oak, not too old, says Steiner, and break it up into very small pieces. He indicated that in Europe the English oak (*Quercus robur*) was most desirable, but as Lovell points out, this oak does not grow in North America, and there is general agreement among biodynamicists that in the U.S. the most desirable is the white oak (*Quercus alba*).

Take the skull of any domestic animal—though Kolisko recommends that of a sheep and discourages the use of a horse or dog skull—and fill it with the oak bark. With a piece of bone of the same animal, such as part of the jawbone, close the opening of the skull.

Bury the lot during the winter, not too deep, in soil where water has access, and cover with peat moss. Much water must stream over the skull. For lack of a stream, the skull should be placed in a barrel with decaying plant substance exposed to rainwater. The object is to create an ambiance of slimy, rotting mulch.

In the spring, when the skull is dug up, the oak bark will have turned to a crumbling black substance that smells like fresh soil, full of life, teeming with microorganisms, and with a highly activated calcium content.

Lovell points out that the ash of white-oak bark is upward of 70 percent calcium, and the skull of an animal is much the same.

Store the finished product as the other preps.

506. Dandelion flower buried in a cow's mesentery

Steiner's recipe is to collect the dandelion flowers (*Taraxacum officinale*) before they go to seed, let them fade slightly, or dry them, keeping them in a cool place until the beginning of October, preferably covered with peat moss.

Tradition has it to pick dandelion flowers fairly early in the morning, just prior to when they open and are visited by bees, to ensure they are at their highest essence, picking the sunniest spots first to beat the bees.

In the fall they must be pressed tightly together to fill the mesentery of a cow or ox, the mesentery being the large sheath of skin that surrounds the internal organs of the animal. Steiner warns that the mesentery, obtainable from any slaughterhouse, must be in perfect condition, without damaged parts or holes. Surplus fat may be cut off, but care must be taken not to break the tender skin.

Hugh Courtney with a cow's mesentery filled with dandelion flowers that he has just unearthed and that will produce a quantity of BD 506 sufficient for several thousand acres.

Wrapped around the flowers, the sheath is tied with string or delicately sewn. When ready, the flowers can be seen peeping through the skin.

The result is buried about twelve inches below the surface so that it spends the winter exposed to the strong earth forces, which Steiner says stream through the soil at that time of year.

In the spring the resulting soft, friable material is ready to be used in the compost heap. Again, only a minute amount is needed—about a teaspoon for ten tons of compost. It is the forces, not the substances, which are said to be at work. Preparation 506 is credited with establishing the proper relationship between potassium and silica, allowing crops to draw on the substances and forces in the broader surroundings, even within the entire district.

Related to Jupiter, 506 is said to promote functions operative in the liver and regulatory glands and organs of animals. These functions are related to magnetism, fullness, strength, roundness, and attraction.

507. The juice of valerian

The preparation is made from the fresh florets of the valerian plant (*Valeriana officinalis*), commonly known as garden heliotrope. The young blossoms are snipped off on a day in spring, then trimmed from their stems with scissors.

Kolisko wants the flowers placed in lukewarm water, then squeezed out to produce a concentrated tincture that can be kept a long time. Some practitioners use a hydraulic press to extract the juice. Highly diluted, to the seventh or eighth homeopathic potency (twenty to twenty-five drops stirred for twenty minutes into two or three gallons of water) it is sprinkled on the compost heap so that, according to Steiner, its phosphorus content can be properly used by the soil. Related to the planet Saturn, and to flowering, Lovell says it is the burning of phosphorus that is important, that 507 is the warmth giver, and it is sometimes used as a spray to protect against late frosts.

508. Horsetail

Equisetum arvense, or common horsetail, is widespread on rough, bare ground, but also in cultivated places. It prefers dry locations to moist ones. The plant has pale brown, unbranched fertile stems which have disappeared by the time the taller, green, barren ones appear.

PREPARATION

Collect the barren shoots and dry them as quickly as possible by spreading them in a thin layer in a shady place until the herb becomes somewhat brittle.

HOW TO USE HORSETAIL

A tea is prepared by slowly boiling in a covered vessel of rainwater about four ounces of the dried herb per gallon. One can use less water, and dilute the tea, in which case one stirs the solution for about ten minutes.

The tea is used as a prophylactic, mild antifungal agent. One can hardly use too much of it. It is used against mildew, rust, monilia, scab, soil-borne pathogenic fungi. It is a mild agent. One sprays this tea frequently, especially on garden crops. Cold frames, hot beds, and greenhouses are treated before and after having been filled with soil.

The tea can also be added to the water in the watering can. Root dips and tree sprays are made with the tea. During the season when green plants are available one can also prepare an extract by covering freshly picked plants with water and allow them to ferment for about ten days; the liquid is then diluted and used in the same way as the tea.

One uses the barren stems. These can be distinguished from the stems of marsh or shady horsetail by the places in which the plants grow. Marsh horsetail (*E. palustre*) forms fertile stems. Shady horsetail (*E. pratense*) has distinct fertile and barren stems, but the latter end abruptly with the top whorl of the branches. The stems of the common horsetail continue beyond the last whorl of branches; also the last jointed section on the branches is longer than the stem sheath. The plant grows from spores. It has no flower. The barren stems appear in early summer after the fertile ones. The plant is recognizable by its finely marked longitudinal ribs on the stem, the marvelous regularity of the nodes and whorls, its clear, almost crystalline form.

Stinging-nettle liquid manure

This liquid, empirically developed by Pfeiffer, was considered by him to be helpful in enhancing the vegetative growth of plants, especially during dry weather. He recommended it for private gardeners in particular, but also for fruit growers and commercial gardeners who can apply it in various ways.

HOW TO USE

Cut stinging nettles (*Urtica dioica*) at any stage of development, except during the time when the seeds are ripening. Cover about two to three pounds of the green material with water. Place the vessel in the garden where the strong odor that soon develops will not be too bothersome.

Use the liquid diluted or undiluted as a foliar spray, beginning about a week after fermentation has started. The strained liquid can be added to the 501 spray. It can also be mixed with horsetail tea and a small amount of liquid seaweed.

Avoid overuse as the high nitrogenous component can stimulate fungus problems.

Paste for tree bark

Mix equal parts of fine clay (loam-clay type) and fresh cow manure. Dissolve this with 1 percent equisetum tea and one portion of stirred 500 until such a consistency is achieved that one can paint the material on the bark with a brush.

Before being coated, the bark should be scraped or brushed to remove dead, loose parts. The bark of trunks and branches that have received this treatment become smooth and clean after a few years. The trees grow healthy.

ANOTHER FORMULA

1 part dried blood, 2 parts Kieselguhr (diatomaceous earth), 3 parts clay, 4 parts cow dung, mixed with equisetum tea and stirred 500.

These sprays have been in use for sixty years. A wealth of empirical information has been assembled, available in the *Bio-Dynamic Magazine*.

How to Apply "Sonic Bloom"

(A spray and sound combination invented and distributed by Dan Carlson Scientific Enterprises, 708-119th Lane, N.E., Blaine, Minnesota, 55434. Tel: (612) 757-8274.

For backyard gardens

Pour into a one-gallon jug (plastic or glass), one tablespoon (one-half ounce) of Sonic Bloom liquid concentrate. Add tap water with force. (Do not use distilled water.) Insert the tube of a misting spray unit and secure the stopper. Shake well. Turn the misting nozzle to fine.

Insert the cassette into a stereo tape player. If possible, turn treble to high and bass to medium. Set the volume as high as possible, without distortion (or disturbing your neighbors).

Let the sound play ten minutes before spraying your plants with Sonic Bloom. While you hear classical music, the plants will "hear" a hum that invites them to take in more air, water, and nutrients.

Carlson's unique organic foliar feed includes giberillic acid (a plant-growth hormone), a seaweed extract, and some fifty-five trace minerals and amino acids. Aided by the sound, leaves draw in up to 700 percent more nutrients.

While the sound is playing, spray your plants thoroughly, so that both sides of the leaves are saturated, until the liquid drips from them.

For best results, let the sound continue twenty minutes after spraying. Treat house plants once a week, and flowering or fruiting plants twice a week.

Plants absorb the spray best when treated early in the morning (between 5:30 and 9:30).

Do not spray plants when the temperature drops below 52 degrees Fahrenheit. If it is cold early in the morning, spray Sonic Bloom on plants after 4:30 P.M.

Dew, mist, or heavy fog is ideal weather for applying Sonic Bloom.

All young seedlings of vegetables and flowers should be treated once a week for the first three weeks, then twice a week thereafter until harvest.

For head lettuce, wait until the head begins to roll up before spraying.

Fruit and nut trees

To obtain maximum fruit set and size, spray fruit trees:

once before buds form,
once when buds have formed but not opened,
once when flowers are fully opened,
once when approximately one-third of the flowers have fallen off,
once when fruit or nuts have formed (about fifteen days after flowers have
   fallen off).

For commercial application

Use twenty ounces of Sonic Bloom in forty gallons of water (not distilled) at forty pounds pressure, spraying at least five times in a growing season, or about every three weeks.

For farmers

HOW TO TREAT CORN, SOY BEANS OR ALFALFA

To cover one acre of ground, for every six ounces of Sonic Bloom mixture add twelve gallons of water with force for proper agitation. Mount the sound unit, with the speaker facing the rear, on the highest point on the back of your sprayer, using duct tape to attach the unit. The system is designed so the sound waves will assist the sprayed plants to absorb the greatest amount of nutrients. The sound unit must be on while you are spraying. Turn the volume all the way up.

Spray as early in the morning as possible. Each sound unit has a photo-cell activator, so it must be daylight before it can be used effectively. If you wish to start before dawn, there is an override switch to activate the unit. Do not spray when the temperature drops below 52 degrees.

Sonic Bloom sound boxes of different sizes can be rented from Dan Carlson Scientific Enterprises.

How to Garden in Cooperation with the Nature Spirits

Or how to get in touch with the devas that overlight a garden, and the nature spirits that make it what it is.

For details see *The Perelandra Garden Work Book, A Complete Guide to Gardening with Nature Intelligences* by Machaelle Wright Small, Box 136, Jeffersonton, Virginia, 22724.

While waiting for the book you may wish to test your abilities.

Say aloud: "I would like to be formally linked with the devic realm."

Wait a few seconds, says Machaelle Wright Small: "You may feel sensations like a wave of energy gently wash over you. You may feel absolutely nothing and this will mean absolutely nothing because you are going to verify your connection with this level using kinesiology."

If you are right-handed:

Place your left hand palm up. Connect the tip of your left thumb with the tip of your left little finger (not your index finger).

By connecting your thumb and little finger, you have just closed an electrical circuit in your hand, and you will use this circuit for testing.

If you are left-handed:

Place your right hand palm up. Connect the tip of your right thumb with the tip of your right little finger.

To test the circuit (the means by which you will apply pressure to yourself) place the thumb and index finger of your other hand inside the circle you have created by connecting thumb and little finger. The thumb–index finger should be right under the thumb–little finger, touching them. It will look as if the thumb–little finger are resting on the thumb–index finger. This is the testing position.

Ask yourself a yes/no question to which you already know the answer to be yes. ("Is my name . . .?") Once you've asked the question, press your thumb–little finger together, keeping the tip-to-tip position.

Using the same amount of pressure, try to pull apart the thumb–little finger with your thumb–index finger. Press the lower thumb against the upper thumb, the lower index finger against the upper index finger.

If your answer is positive (if your name is what you think it is!) you will not be able to pull apart the top fingers. The electrical circle will hold, your muscles will maintain their strength, and your circuit fingers will not separate. You will feel the strength in that circuit. *Important:* Be sure the amount of pressure holding together the thumb–little finger circuit is equal to the amount pressing against that circuit with your thumb–index finger. Also, don't use a pumping action in your thumb–index fingers to try to pry your thumb–little fingers apart. Use an equal, steady, and continuous pressure.

Play with this a bit. Ask a few more yes/no questions that have positive answers. If you are having trouble sensing the strength of the circuit, apply a little more pressure. Or consider that you may be applying too much pressure, and pull back some. You don't have to strain your fingers for this.

Once you have a clear sense of the positive response of the circuit, ask yourself a question that has a negative answer. Again press your circuit fingers together and, *using equal pressure,* press against the circuit fingers with the thumb–index finger.

This time the electrical circuit will break, and the thumb–little finger will weaken and separate.

Because the electrical circuit is broken, the muscles in the thumb and little finger don't have the power to hold the fingers together. In a positive state, the electrical circuit holds, and the muscles have the power to keep the circuit fingers together.

Play with negative questions a bit, then return to positive questions. Get a good feeling for the strength between your circuit fingers when the electricity is in a positive state, and the weakness when the electricity is in a negative state. You can even ask yourself (your own system) for a positive response and then a negative response. ("Give me a positive response." Test. "Give me a negative response." Test.) You will feel the positive strength and the negative weakness. Now it is just a matter of trusting what you have learned—and practice.

Don't forget the overall concept behind kinesiology, says Machaelle. What enhances your body, mind, and soul makes you strong. Together, your body, mind, and soul create a holistic environment, which, when balanced, is strong and solid. If something enters into that environment which negates or challenges the balance, the entire environment is weakened. The state of that strength or weakness is registered in the electrical system, and through muscle testing it can be discerned.

Once in touch with the devas, says Machaelle, they guide her in how to design and lay out a garden, advising her what to plant, and where, as well as how to deal with insects.

She then shifts her attention to the nature-spirit level, to receive insight and assistance in processing: that is, in how to bring the garden into ideal form, a process of bringing spirit into form.

When not interfered with by humans, says Machaelle, the nature spirits tend to the needs of all physical reality, assuring perfection within form.

Aware that these spirits can take on varied specific forms, visible to humans, Machaelle does not experience them as a phalanx of little elves and gnomes wielding pitchforks and shovels, but as individuated energy presences, without specific form.

This co-creation of a garden in conjunction with the devas and the nature spirits enables her to fertilize her garden with a mere handful of essential elements, the essence or energy of which is magnified and disseminated by the nature spirits, and to keep insects at bay without the use of pesticides by tithing to them a small portion of her produce.

For those wishing to apply these techniques, Machaelle Wright gives seminars at Perelandra during the summer.

RESOURCE LIST: CLIMATE-CRISIS LITERATURE AVAILABLE FROM OTHER SOURCES

Solar Age or Ice Age? Bulletin, Nos. 4–9, 1983–87, by Don Weaver. (A full set of bulletins is an essential source for new information.) Request current prices from Donald Weaver, Hamaker-Weaver Publishers, Box 1961, Burlingame, CA 94010.

Climate Crime, by Barry Lynes, 1985. (Chronicles U.S. government cover-up of climate-crisis information.) $11.95 ppd. from Barry Lynes, Box 5564, Washington, D.C. 20016.

Bread from Stones, by Dr. Julius Hensel. (Rediscovery of soil remineralization in the modern era.) $6 ppd. from Health Research, Box 70, Mokelumne Hill, CA 95245.

Soil Remineralization, A Network Newsletter, published quarterly. (A way to network all those who want to remineralize the soils and regenerate earth's green mantle. A reflection, forum, round table of ideas, experiences, research, etc.) $12 per yr. ppd. from Soil Remineralization, Joanna Campe, 152 South Street, Northampton, MA 01060.

Earth Regeneration Society. (An active and effectively organized network dealing with the climate crisis in its full scope. It has been pivotal in organizing the creation and presentation of key papers, introduction of appropriate legislation, etc.) Annual membership plus newsletter is $25 per yr. ppd. Contact: Earth Regeneration Society, Inc., 470 Vassar Ave., Berkeley, CA 94708.

Institute for a Future. (Created and distributes the film documentary *Stopping the Coming Ice Age.* Available in VHS or Beta for $20 [individuals] or $35 [groups], along with the book *The End: The Imminent Ice Age and How We Can Stop It,* available for $10 ppd.) Contact: Institute for a Future, 2000 Center Street, Berkeley, CA 94704.

Solstice. (A magazine of perspectives on health and the environment. Now carries the regular feature "Climate Crisis Update"—$14.95 for 6 issues U.S., $19.95 Canada/Mexico, and $20.95 all other countries, or $32.95

air mail. Single issue U.S. $3.50 ppd. *Perspectives on the Climate Crisis*, a compiled special edition of an article series "The Cooling," is available for $6 ppd.) Contact: *Solstice*, John David Mann, 1110 East Market, Suite #16E, Charlottesville, VA 22901.

SOIL REMINERALIZATION

For further updated
information write to:

Soil Remineralization
c/o Joanna Campe
152 South Street
Northampton, MA 01060

NOTE TO *APPENDIX F*

The authors and the publisher wish to make clear that the description of products and methods in this Appendix are offered for informational purposes. Neither the authors nor the publisher are making nor can make any assurance about the results of their use. Without the tender care or dedicated desire of the farmer to improve himself, the soil, and his crops, these products and methods are not alone the answer.

It should be particularly noted that the Bio-Dynamic Association does not endorse this book.

Bibliography

CHAPTER 1

Bio-Dynamics (quarterly magazine). Kimberton, PA: Bio-Dynamic Literature.
Index to Bio-Dynamics 1941–1975. Wyoming, RI: Bio-Dynamic Literature, 1976.
Koepf, H. H. *What is Bio-Dynamic Agriculture?* Wyoming, RI: Bio-Dynamic
     Literature, 1976.
Koepf, H. H., Petersson, B., and Schaumann, Wolfgang. *Bio-Dynamic Agricul-
     ture*. Kimberton, PA: Bio-Dynamic Association National Office, 1976. For
     further information write to Bio-Dynamic Farming and Gardening Asso-
     ciation, Inc., P.O. Box 550, Kimberton, PA, 19442 or call (215) 935–7797.
Storl, Wolf D. *Culture and Horticulture*. Wyoming, RI: Bio-Dynamic Literature,
     1979.

CHAPTER 2

Blaser, Peter, and Pfeiffer, E. E. *Bio-Dynamic Composting on the Farm—How
     Much Compost Should We Use*. Wyoming, RI: Bio-Dynamic Literature,
     1954.
Koepf, H. H. *Bio-Dynamic Sprays*. Wyoming, RI: Bio-Dynamic Literature, 1971.
———. *Compost—What It Is, How It Is Made, What It Does*. Wyoming, RI: Bio-
     Dynamic Literature, 1980.
Pfeiffer, E. E. *How Much Compost Should We Use?* Wyoming, RI: Bio-Dynamic
     Literature, 1951.

CHAPTER 3

Bockemühl, Jochen. *In Partnership with Nature*. Kimberton, PA: Bio-Dynamic
     Association, 1981.
Fyfe, Agnes. *Moon and Planet. (Die Signatur des Mondes im Pflanzenreich.)*
     Stuttgart: Verlag Freies Geistesleben, 1967.
Klocek, Dennis. *The Bio-Dynamic Book of Moons*. Wyoming, RI: Bio-Dynamic
     Literature, 1983.
Kolisko, E. L. *Agriculture of Tomorrow*. Bournemouth, England: Kolisko Ar-
     chives Publishers, 1978.
Koenig, Karl. *Earth and Man*. Wyoming, RI: Bio-Dynamic Literature, 1982.
Kranich, E. M. *Planetary Influences upon Plants*. Wyoming, RI: Bio-Dynamic
     Literature, 1988.
Lakhovsky, Georges. *L'origine de la vie*. Paris: Editions Niesson, 1925.
———. *The Secret of Life*. London: W. Heinemann Ltd., 1939.

Pfeiffer, E. E. *Bio-Dynamic Treatment of Fruit Trees, Berries, Shrubs.* Wyoming, RI: Bio-Dynamic Literature, 1976.

CHAPTER 4

Heckel, Alice. *The Pfeiffer Garden Book.* Stroudsburg, PA: Bio-Dynamic Farming and Gardening Association, 1967.
Pfeiffer, Ehrenfried. *Bio-Dynamics, Three Introductory Articles.* Wyoming, RI: Bio-Dynamic Literature, 1948.
———. *Bio-Dynamic Farming and Gardening.* Spring Valley, NY: Mercury Press, 1983. 3 vols.
———. *Chromatography Applied to Quality Testing.* Wyoming, RI: Bio-Dynamic Literature, 1984.
———. *Sensitive Crystallization Processes.* Hudson, NY: Anthroposophic Press, 1975.
Philbrick, Helen Louise. *Companion. Plants and How to Use Them.* New York: Devin-Adair Co., 1966.
Philbrick, John H. *Gardening for Health and Nutrition.* Blauvelt, NY: Rudolf Steiner Publications, 1971.

CHAPTER 5

Block, Bartley C. *Man, Microbes, and Matter.* New York: McGraw-Hill, 1975.
Darwin, Charles. *Darwin on Humus and the Earthworm.* London: Faber and Faber, 1966.
———. *The Formation of Vegetable Mould, Through the Action of Worms.* London: J. Murray, 1897.
Ford, Brian J. *Microbe Power: Tomorrow's Revolution.* New York: Stein & Day, 1978.
Gest, Howard. *The World of Microbes.* Garden City, NY: Doubleday, 1965.
Gross, Cynthia S. *The New Bacteriology.* Minneapolis, MN: Lerner Publications, 1988.
Lee, Kenneth Ernest. *Earthworms.* Orlando, FL: Academic Press, 1985.
Margulis, Lynn. *Microcosmos.* New York: Summit Books, 1986.
Minnich, Jerry. *The Earthworm Book.* Emmaus, PA: Rodale Press, 1977.
Selsam, Millicent Ellis. *Microbes at Work.* New York: Morrow, 1953.
Voisin, Andre. *L'herbe tendre.* Paris: Laffont, 1986.
Webb, William Larkin. *Brief Biography and Popular Account of the Unparalleled Discoveries of T. J. J. See.* Lynn, MA: T.P. Nichols & Son Co., 1913.

CHAPTER 6

Podolinsky, Alex. *Bio-Dynamic Agriculture Introductory Lectures.* Sydney, Australia: Gavemer Foundation Publishers, 1985.

CHAPTER 7

Berry, Wendell. *The Unsettling of America.* San Francisco, CA: Sierra Club Books, 1977.
Jackson, Wes, ed. *Meeting the Expectations of the Land.* San Francisco, CA: North Point Press, 1984.

CHAPTER 8

Bircher, Ralph. *Hunsa, das Volk, das keine Krankheit kannte.* Bad Homburg vdH Erlenbach-Zurich, Bircher-Benner, 1980.
Howard, Albert. *The Soil and Health.* New York: Schocken Books, 1972.
McCarrison, Howard. *Nutrition and National Health.* London: Faber & Faber Ltd., 1944.
Steffen, Robert. *Introduction to Organic Farming Methods and Organic Markets.* Emmaus, PA: Rodale Press, 1971.
———. *Organic Farming: Methods and Markets.* Emmaus, PA: Rodale Press, 1972.
Taylor, Renee. *Come Along to Hunza.* Minneapolis, MN: T. S. Denison, 1974.
Tobe, John H. *Guideposts to Health and Vigorous Long Life.* St. Catherines, Ontario: Modern Publications, 1965.
———. *How to Be Healthy and Live Longer.* Don Mills, Ontario: Greywood, 1973.
———. *I Found Shangri-La.* St. Catherines, Ontario: Provoker Press, 1970.

CHAPTER 9

Alexandersson, Olof. *Living Water.* Wellingborough, Northamptonshire: Turnstone Press, 1982.
Flanagan, C. Patrick. *Beyond Pyramid Power.* Santa Monica, CA: DeVorss, 1975.
———. *Pyramid Power.* Glendale, CA: Pyramid Publishers, 1973.
Flanagan, C. Patrick, and Gail Crystal. *Elixir of the Ages: You Are What You Drink.* Flagstaff, AZ: Vortex Press, 1986.
Ott, John. *Work as You Like It.* New York: Julian Messner, 1979.
Prigogine, Ilya. *Etude thermodynamique des phenomenes irreversibles.* Paris: Dunod, 1947.
———. *The Molecular Theory of Solutions.* Amsterdam: North Holland Publishing Co., 1957.
———. *Order Out of Chaos.* New York: Bantam Books, 1984.
———. *Superfluidite et equation de transport quantique.* Bruxelles, 1960.
Riddick, Thomas M. *Control of Colloid Stability Through Zeta Potential.* Wynnewood, PA: Published for Zeta-Meter, Inc., by Livingstone Publishing Co., 1968.
Schwenk, Theodor. *Sensitive Chaos.* New York: Schocken Books, 1976.
Sutphen, Richard. *Dick Sutphen Presents Sedona.* Malibu, CA: Valley of the Sun Publishing Co.: 1986.
Watson, Lyall. *The Romeo Error.* Garden City, NY: Anchor Press, 1975.

CHAPTER 10

Bingham, Hiram. *Lost City of the Incas.* Westport, CT: Greenwood Press, 1981.
———. *Machu Picchu, a Citadel of the Incas.* New York: Hacker Art Books, 1979.
Dokuchayev, Vasily. *Russkii Chernozem (The Black Soil of Russia),* 1936.
Fawcett, Percy Harrison. *Exploration Fawcett.* London: Heron, 1969.
Glinka, Konstantin D. *Treatise on Soil Science.* Washington, D.C.: National Science Foundation, 1963.
International Symposium. "Humus et Planta." Prague: Publishing House of the

Czechoslovak Academy of Sciences, 1962.

Joffe, Jacob Samuel. *The ABC of Soils*. New Brunswick, NJ: Pedology Publications, 1949.

Pochvennyi Institut Imeni V.V. Dokuchaeva. *Microorganisms and organic matter of soils*. Jerusalem: Israel Program for Scientific Translations, 1970. Available from the United States Department of Commerce Clearinghouse for Federal Scientific and Technical Information, Springfield, VA.

Schatz, Albert. *The Story of Microbes*. New York: Harper Brothers, 1952.

———. *Teaching Science with Soil*. Emmaus, PA: Rodale Press, 1972.

"Advancing Frontiers of Plant Sciences." Conference held in New Delhi, India, 1963.

"Chelation as a Biological Weathering Factor in Pedogenesis." Proceedings of the Pennsylvania Academy of Sciences, 1954.

CHAPTER 11

Catchpole, Clive. *Vocal Communication in Birds*. Baltimore, MD: University Park Press, 1979.

Jellis, Rosemary. *Bird Sounds and Their Meaning*. London: British Broadcasting Corporation, 1977.

Retallack, Dorothy L. *The Sound of Music and Plants*. Santa Monica, CA: DeVorss, 1973.

Saunders, Aretas Andrew. *A Guide to Bird Songs*. Garden City, NY: Doubleday, 1951.

———. *An Introduction to Bird Life for Bird Watchers*. (Formerly titled *The Lives of Wild Birds*.) New York: Dover Publications, 1964.

CHAPTER 12

Doyle, Jack. *Altered Harvest*. New York: Viking, 1985.

Jabs, Carolyn. *The Heirloom Gardener*. San Francisco, CA: Sierra Club Books, 1984.

Kapuler, Alan, and Olaf Brentnar. *Catalog and Research Journal*. Corvallis, OR: Peace Seeds, 1987.

Mooney, Patrick R. *Seeds of the Earth*. Ottawa: Published by Inter Pares for the Canadian Council for International Cooperation and the International Coalition for Development Action, 1979.

Nabhan, Gary Paul. *The Desert Smells Like Rain*. San Francisco, CA: North Point Press, 1982.

———. *Gathering the Desert*. Tucson, AZ: University of Arizona Press, 1985.

Tracy, W. W. *American Varieties of Vegetables*. Washington, D.C.: United States Department of Agriculture, 1903.

Whealy, Kent. *The Garden Seed Inventory*. Decorah, IA: Seed Saver Publications, 1985.

———. *Winter Year Book*. Decorah, IA: Seed Savers Exchange, 1986.

CHAPTER 13

Cocannouer, Joseph A. *Farming with Nature*. Norman, OK: University of Oklahoma Press, 1954.

———. *Organic Vegetable Gardening, the Better Way*. New York: Arco Publishing Inc., 1977.

——. *Weeds, Guardians of the Soil.* New York: Devin-Adair Co., 1950.
Gregg, Evelyn, and Motch. *The Herb Chart.* Wyoming, RI: Bio-Dynamic Literature, 1980.
Jackson, Wes. *Man and the Environment.* Dubuque, IA: W. C. Brown, 1978.
——. *Meeting the Expectations of the Land.* San Francisco, CA: Sierra Club Books, 1977.
——. *New Roots for Agriculture.* San Francisco, CA: Friends of the Earth, Salina, Kansas, Land Institute, 1980.
Kummer, Anna Pederson. *The Role of Weeds in Maintaining the Plains Grasslands.* Chicago, IL: University of Chicago Press, 1951.
Pfeiffer, E. E. *Weeds and What They Tell.* Wyoming, RI: Bio-Dynamic Literature, 1981.

CHAPTER 14
Bryson, R. S. *Environmental Conservation Education.* Danville, IL: Interstate, 1975. Vol. 2.
Croll, James. *Climate and Time in Their Geological Relations.* London: Daldy, Tsbister & Co., 1875.
Ephron, Larry. *The End.* Berkeley, CA: Celestial Arts, 1988.
Emiliani, Cesare. *Dictionary of Physical Sciences.* New York: Oxford University Press, Inc., 1987.
——. *The Scientific Companion.* New York: Wiley, 1988.
Fodor, R. V. *Frozen Earth.* Hillside, NJ: Enslow Publishers, 1981.
Gates, William Lawrence. *Climatic Change.* Santa Monica, CA: Rand Corp., 1975.
Hamaker, John D., and Donald A. Weaver. *The Survival of Civilization.* Burlingame, CA: Hamaker-Weaver Publishers, 1982.
Imbrie, John. *Ice Ages.* Hillside, NJ: Enslow Publishers, 1979.
Kovda, Viktor Ambramovich. *Microelements in the Soils of the USSR.* Springfield, VA: United States Department of Commerce Clearinghouse for Federal Scientific and Technical Information, 1966.
——. Personal communication. Moscow, USSR: Institute for Soil Science and Soil Melioration, USSR Academy of Sciences, Moscow, 1987.
Kukla, George J. *Variations of Arctic Cloud Cover During Summer, 1979,* part I, Technical Report 84-2. New York: Columbia University, Lamont-Doherty Geological Observatory, 1984.
Kukla, George J., and B. Choudhury. "Impact of CO_2 on Cooling and Water Surfaces," *Nature,* vol. 280, 1979.
Libby, Willard F. *Solar System Physics and Chemistry; and Papers for the Public.* Santa Monica, CA: UCLA Press, 1981.
Milankovich, Milutin. *Canon of Insolation and the Ice Age Problem.* Springfield, VA: United States Department of Commerce, 1969.
NATO *Advanced Workshop on Milankovitch and Climate.* Hingham, MA: Dordrecht Boston, D. Reidel Publishing Co. Sold and distributed in the USA and Canada by Kluwer Academic Publishers, 1984.
Potential Implications of Trends in World Population, Food Production and Climate. Washington, DC: Office of Research and Development, Central Intelligence Agency, 1979.
"The Present Interglacial . . . : When Will it End?" Conference held at Brown

University, Providence, RI, 1972. Reported in *Science*, vol. 78, 1972, p. 198.

A *Study of Climatological Research as it Pertains to Intelligence.* Washington, DC: Office of Research and Development, Central Intelligence Agency, 1974.

CHAPTER 15

Bretscher, Mark S. "How Animal Cells Move," *Scientific American*, December 1987.

Epstein, Emanuel. *Mineral Nutrition of Plants: Principles and Perspectives.* New York: Wiley, 1971.

Pfeiffer, Ehrenfried. *A Condensation of Bio-Dynamic Farming and Gardening.* Pauma Valley, CA, 1973.

Rateaver, Bargyla. *The Organic Method Primer.* San Diego, CA, 2nd edition.

CHAPTER 16

Tansley, David V. *Dimensions of Radionics.* Holsworthy, Health Science Press, 1977.

Westlake, Aubrey T. *Life Threatened: Menace and Way Out.* London: Stuart & Watkins, 1967.

CHAPTER 17

Abderhalden, Rudolf. *Medizinische Terminologie.* Basel, Switzerland: B. Schwabe, 1948.

Dikkers, Melchior T. *The Story of Trace Minerals.* Privately printed.

——. *Unintentional Suicide, It's Time to Stop Killing Ourselves.* Phoenix, AZ: Printed by Lebeau Print. Co., 1971.

CHAPTER 18

Baker, Richard St. Barbe. *My Life—My Trees.* London: Lutterworth Press, 1970.

——. *I Planted Trees.* London: Lutherhill Press, 1944.

——. *Men of the Trees.* New York: The Dial Press, 1931.

Duke, James A. *CRC Handbook of Agricultural Energy Potential of Developing Countries.* Boca Raton, FL: CRC Press, 1987.

——. *CRC Handbook of Medicinal Herbs.* Boca Raton, FL: CRC Press, 1985.

——. *Culinary Herbs.* New York: Trado-Medic Books, 1985.

——. *Medicinal Plants of the Bible.* New York: Trado-Medic Books, 1983.

CHAPTER 19

Based on visits with John T. Brown and Vasant V. Parange, Maryland; Mato, Modrić, Rovinj, Yugoslavia; Lech Stefansk, Warsaw, Poland; and others.

CHAPTER 20

Callahan, Philip S. *The Evolution of Insects.* New York: Holiday House, 1972.

——. *Insects and How They Function.* New York: Holiday House, 1971.

——. *Insect Molecular Bio-Electronics.* College Park, MD: Entomological Society of America, 1967.

——. *A Rucksack Naturalist in Ireland.* Greenwich, CT: Devin-Adair Co., 1988.

——. *The Soul of the Ghost Moth.* Old Greenwich, CT: Devin-Adair Co., 1981.

———. *Tuning in to Nature*. Old Greenwich, CT: Devin-Adair Co., 1975.
Laithwaite, Eric Roberts. *Force*. New York: Franklin Watts, 1986.

CHAPTER 21
Barrow, Lennox. *The Round Towers of Ireland*. Dublin: Academy Press, 1979.
Callahan, Philip S. *Ancient Mysteries, Modern Visions*. Kansas City, MO: Acres, USA, 1984.
Praeger, Robert Lloyd. *The Way That I Went*. Dublin: A. Figgis, 1980.

CHAPTER 22
Based on visits with Galen and Sarah Hieronymus, Lakemont, GA, 1987.

CHAPTER 23
Jahn, Robert G. *Physics of Electric Propulsion*. New York: McGraw-Hill, 1968.
Small, Machaelle Wright. *Behaving as If the God in All Life Mattered*. Jeffersonton, VA: Perelandra, 1983.
———. *The Perelandra Garden Book*. Jeffersonton, VA: Perelandra, 1983.

EPILOGUE
Based on visits to the *Otdel Teoreticheskikh Problem* (Department for Theoretical Problems), U.S.S.R. Academy of Sciences, Moscow.
Eddington, Arthur Stanley. *Report on the Relativity Theory of Gravitation*. London: Fleetway Press, Ltd., 1920.
———. *Stellar Movements and the Structure of the Universe*. London: Macmillan and Co. Ltd., 1914.
Jahn, Robert G., and Brenda J. Dunne. *Margins of Reality*. San Diego: Harcourt Brace Jovanovich, 1987.
Young, Arthur M. *The Geometry of Meaning*. New York: Delacorte Press, 1976.
———. *The Reflexive Universe*. New York: Delacorte Press, 1976.
———. *The Role of Consciousness in the Physical World*. Boulder, CO: Westview Press, 1981.
The Foundations of Science: The Missing Parameter and *Science and Astrology: The Relationship Between the Measure Formulae and the Zodiac*. Mill Valley, CA: Robert Briggs Associates.

APPENDIX A
Blavatsky, Helena Petrovna. *The Book of the Secret Wisdom of the Earth*.
———. *From the Caves and Jungles of Hindustan*. Delhi: Indological Book House, 1892.
———. *Isis Unveiled*. Pasadena, CA: Theosophical University Press, 1972.
———. *The Key to Theosophy*. Pasadena, CA: Theosophical University Press, 1972.
———. *Personal Memoirs*. Wheaton, IL: Theosophical Publishing House, 1967.
———. *The Secret Doctrine*. Pasadena, CA: Theosophical University Press, 1977.
Campbell, Bruce F. *Ancient Wisdom Revived*. Berkeley, CA: University of California Press, 1900.
Olcott, Henry Steel. *Inside the Occult*. Philadelphia, PA: Running Press, 1975.
———. *People from the Other World*. Rutland, VT: C. E. Tuttle, Co., 1972.

Appendix B

Carrington, Hereward. *Death, Its Causes and Phenomena*. New York: Arno Press, 1977.

———. *Your Psychic Powers and How to Develop Them*. Hollywood, CA: Newcastle Publishing Co., 1975.

Crookall, Robert. *The Jung Jaffe View of Out-of-Body Experiences*. London: Published for the Churches Fellowship for Psychical & Spiritual Studies by the World Fellowship Press, 1970.

———. *The Mechanisms of Astral Projection*. Moradabad, India: Darshana International, 1968.

———. *Out-of-the-Body Experiences*. New Hyde Park, NY: University Books, 1970.

———. *The Study and Practice of Astral Projection*. New Hyde Park, NY: University Books, 1966.

Monroe, Robert A. *Further Journeys*. New York: Doubleday, 1985.

———. *Journeys Out of the Body*. New York: Doubleday, 1971.

Muldoon, Sylvan Joseph. *The Phenomena of Astral Projection*. New York: Samuel Weiser Inc., 1969.

Appendix C

Besant, Annie Wood. *The Ancient Wisdom*. Wheaton, IL: Theosophical Publishing House, 1966.

———. *Occult Chemistry: Investigations by Clairvoyant Magnification into the Structure of Atoms of the Periodic Table and of Some Compounds*. Wheaton, IL: Theosophical Publishing House, 1951.

———. *Thought Power, Its Control and Culture*. Wheaton, IL: Theosophical Publishing House, 1966.

Leadbeater, Charles Webster. *The Chakras*. Wheaton, IL: Theosophical Publishing House, 1972.

———. *The Inner Life*. Wheaton, IL: Theosophical Publishing House, 1978.

———. *Man Visible and Invisible*. Wheaton, IL: Theosophical Publishing House, 1969.

Mauldin, John H. *Particles in Nature*. Blue Ridge Summit, PA: Tab Books, 1906.

Phillips, Stephen M. *Extral-sensory Perception of Quarks*. Wheaton, IL: Theosophical Publishing House, 1900.

Smith, Ernest Lester. *Occult Chemistry Re-Evaluated*. Wheaton, IL: Theosophical Publishing House, 1982.

Appendix D

Some five hundred volumes by or about Rudolf Steiner are listed in the Library of Congress. Most of his works are obtainable from Anthroposophic Press, Bell's Pond, Star Route, Hudson, NY, 12534. Tel: (518) 851-2054. Many printed and taped lectures, newly translated, are available from the Rudolf Steiner Research Foundation, Box 1760, Redondo Beach, CA 90278. Tel. (213) 437-5438.

Coroze, Paul Maxime Victor. *A Road to the Spirit*. London: Anthroposophical Publications, 1950.

Culpeper, Nicholas. *Culpeper's Complete English Physician Enlarged and Im-

proved, or, An Universal Medical, Herbal and Botanical and Astrological Practice of Physic. London: Alex. Hogg, 1802.

————. *Culpeper's English Physician, and Complete Herbal.* London: Lewis and Roden, 1802.

————. *The English Physician.* London: printed by P. Cole, 1652.

Easton, Stewart C. *Man and World in the Light of Anthroposophy.* Spring Valley, NY: Anthroposophic Press, 1975.

Lievegoed, Bernardus Cornelis Johannes. *The Developing Organization.* Millbrae, CA: Celestial Arts, 1980.

Steiner, Rudolf. *Agriculture.* London: The Biodynamic Agricultural Association, 1984.

————. *The Ahrimanic Deception.* Spring Valley, NY: Anthroposophic Press, 1985.

————. *Christianity as Mystical Fact and the Mysteries of Antiquity.* West Nyack, NY: Rudolf Steiner Publications, 1961.

————. *The Cycle of the Year as Breathing-process of the Earth.* Spring Valley, NY: Anthroposophic Press, 1984.

————. *The East in the Light of the West.* Blauvelt, NY: Spiritual Science Library, 1986.

————. *Egyptian Myths and Mysteries.* New York: Anthroposophic Press, 1971.

————. *From Jesus to Christ.* New York: Anthroposophic Press, 1930.

————. *Health and Illness.* Spring Valley, NY: Anthroposophic Press, 1981.

————. *The Influences of Lucifer and Ahriman.* London: Rudolf Steiner Publishing Co., 1954.

————. *The Inner Nature of Music and the Experience of Tone.* Spring Valley, NY: Anthroposophic Press, 1983.

————. *Life, Birth, Death, and Reincarnation.* Hudson, NY: Anthroposophic Press, 1987.

————. *The Mission of the Archangel Michael.* New York: Anthroposophic Press, 1961.

————. *Mystery Knowledge and Mystery Centres.* (Fourteen lectures given in Dornach, 23rd November to 23rd December 1923). London: Rudolf Steiner Press, 1973.

————. *Nine Lectures on Bees.* Blauvelt, NY: Rudolf Steiner Publications, 1964.

————. *Nutrition and Health.* Hudson, NY: Anthroposophic Press, 1987.

————. *The Occult Significance of Blood.* London: Rudolf Steiner Press, 1967.

————. *The Reappearance of Christ in the Etheric.* Spring Valley, NY: Anthroposophic Press, 1983.

————. *Rudolf Steiner and Holistic Medicine.* York Beach, ME: Nicholas-Hays, 1987.

————. *Secrets of the Threshold.* Hudson, NY: Anthroposophic Press, 1987.

————. *The Submerged Continents of Atlantis and Lemuria, Their History and Civilization.* Chicago, IL: The Rajput Press, 1911.

————. *Supersensible Knowledge.* Hudson, NY: Anthroposophic Press, 1987.

————. *The Temple Legend.* London: Rudolf Steiner Press, 1985.

————. *Theosophy of the Rosicrucians.* London: Rudolf Steiner Publishing Co., 1953.

Wachsmuth, Guenther. *Earth and Man. (Erde und Mensch, ihre Bildkraefte.)*

Kreuzlingen, Germany: Archimedes Verlag, 1945.

———. *The Etheric Formative Forces in the Cosmos*. London: Anthroposophic Press, 1932.

———. *The Life and Work of Rudolf Steiner from the Turn of the Century to His Death*. New York: Whittier Books, 1955.

———. *Le monde etherique*. Paris: Association de la Science Spirituelle, 1933.

Appendix E

Koenig, Karl. *Earth and Man*. Wyoming, RI: Bio-Dynamic Literature, 1982.

Index